Mathematical Modelling
in Biomedicine

Mathematical Modelling in Biomedicine

Editor

Vitaly Volpert

MDPI • Basel • Beijing • Wuhan • Barcelona • Belgrade • Manchester • Tokyo • Cluj • Tianjin

Editor
Vitaly Volpert
University Lyon 1
France

Editorial Office
MDPI
St. Alban-Anlage 66
4052 Basel, Switzerland

This is a reprint of articles from the Special Issue published online in the open access journal *Mathematics* (ISSN 2227-7390) (available at: https://www.mdpi.com/journal/mathematics/special_issues/Mathematical_Modelling_Biomedicine).

For citation purposes, cite each article independently as indicated on the article page online and as indicated below:

LastName, A.A.; LastName, B.B.; LastName, C.C. Article Title. *Journal Name* **Year**, *Article Number*, Page Range.

ISBN 978-3-03943-493-0 (Hbk)
ISBN 978-3-03943-494-7 (PDF)

© 2020 by the authors. Articles in this book are Open Access and distributed under the Creative Commons Attribution (CC BY) license, which allows users to download, copy and build upon published articles, as long as the author and publisher are properly credited, which ensures maximum dissemination and a wider impact of our publications.

The book as a whole is distributed by MDPI under the terms and conditions of the Creative Commons license CC BY-NC-ND.

Contents

About the Editor vii

Preface to "Mathematical Modelling in Biomedicine" ix

Timur Gamilov, Philipp Kopylov, Maria Serova, Roman Syunyaev, Andrey Pikunov, Sofya Belova, Fuyou Liang, Jordi Alastruey, Sergey Simakov
Computational Analysis of Coronary Blood Flow: The Role of Asynchronous Pacing and Arrhythmias
Reprinted from: *Mathematics* **2020**, *8*, 1205, doi:10.3390/math8081205 1

Sergey Simakov, Alexander Timofeev, Timur Gamilov, Philip Kopylov, Dmitry Telyshev and Yuri Vassilevski
Analysis of Operating Modes for Left Ventricle Assist Devices via Integrated Models of Blood Circulation
Reprinted from: *Mathematics* **2020**, *8*, 1331, doi:10.3390/math8081331 17

Anastasiia Mozokhina and Rostislav Savinkov
Mathematical Modelling of the Structure and Function of the Lymphatic System
Reprinted from: *Mathematics* **2020**, *8*, 1467, doi:10.3390/math8091467 35

Sergey Pravdin, Pavel Konovalov, Hans Dierckx, Olga Solovyova and Alexander Panfilov
Drift of Scroll Waves in a Mathematical Model of a Heterogeneous Human Heart Left Ventricle
Reprinted from: *Mathematics* **2020**, *8*, 776, doi:10.3390/math8050776 53

Arsenii Dokuchaev, Olga Solovyova and Alexander V. Panfilov
Myocardial Fibrosis in a 3D Model: Effect of Texture on Wave Propagation
Reprinted from: *Mathematics* **2020**, *8*, 1352, doi:10.3390/math8081352 67

Fyodor Syomin, Albina Khabibullina, Anna Osepyan and Andrey Tsaturyan
Hemodynamic Effects of Alpha-Tropomyosin Mutations Associated with Inherited Cardiomyopathies: Multiscale Simulation
Reprinted from: *Mathematics* **2020**, *8*, 1169, doi:10.3390/math8071169 83

Nikolai Bessonov, Anne Beuter, Sergei Trofimchuk and Vitaly Volpert
Dynamics of Periodic Waves in a Neural Field Model
Reprinted from: *Mathematics* **2020**, *8*, 1076, doi:10.3390/math8071076 103

Maxim Kuznetsov
Mathematical Modeling Shows That the Response of a Solid Tumor to Antiangiogenic Therapy Depends on the Type of Growth
Reprinted from: *Mathematics* **2020**, *8*, 760, doi:10.3390/math8050760 121

Maxim Kuznetsov and Andrey Kolobov
Optimization of Dose Fractionation for Radiotherapy of a Solid Tumor with Account of Oxygen Effect and Proliferative Heterogeneity
Reprinted from: *Mathematics* **2020**, *8*, 1204, doi:10.3390/math8081204 141

Alexander Churkin, Stephanie Lewkiewicz, Vladimir Reinharz, Harel Dahari and Danny Barash
Efficient Methods for Parameter Estimation of Ordinary and Partial Differential Equation Models of Viral Hepatitis Kinetics
Reprinted from: *Mathematics* **2020**, *8*, 1483, doi:10.3390/math8091483 161

Nikolai Bessonov, Gennady Bocharov, Andreas Meyerhans, Vladimir Popov and Vitaly Volpert
Nonlocal Reaction–Diffusion Model of Viral Evolution: Emergence of Virus Strains
Reprinted from: *Mathematics* **2020**, *8*, 117, doi:10.3390/math8010117 . . . . **193**

About the Editor

Vitaly Volpert is a senior researcher at the National Center for Scientific Research (CNRS, France). He Twas awarded his Ph.D. by the Institute of Chemical Physics of the Russian Academy of Sciences and Habilitation at the University Claude Bernard Lyon 1. His research interests in mathematics include partial different equations, elliptic and parabolic problems, and reaction-diffusion waves. He works in different fields of mathematical modelling, such as biomedical problems, population dynamics, morphogenesis and plant growth, combustion and explosion, frontal polymerization and miscible fluids. He is author of about 350 scientific publications including 4 monographs. He is a founder and Editor-in-Chief of the journal *Mathematical Modelling of Natural Phenomena* and a member of editorial board of a number of other journals.

Preface to "Mathematical Modelling in Biomedicine"

Mathematical modelling in biomedicine is a rapidly developing scientific field due to the fundamental importance of scientific research and applications to public health. Cardiovascular diseases, cancer, and infectious diseases are the main causes of mortality and morbidity in the world, and they represent a major challenge for society. Mathematical modelling of physiological processes in normal and pathological situations can help us understand the underlying processes and develop efficient treatments. In spite of the considerable progress in this area during the last decade, many questions remain open because of their complexity, and interpatient variability.

This Special Issue contains 11 papers devoted to various topics in biomedical modelling, such as blood circulation and the lymphatic system, heart and brain modelling, tumor growth under anti-angiogenic and radiotherapy, viral infection, and immune response. These works present the state of the art in these very different areas of biomedical modelling and present interesting perspectives for the future research in this field.

Vitaly Volpert
Editor

Article

Computational Analysis of Coronary Blood Flow: The Role of Asynchronous Pacing and Arrhythmias

Timur Gamilov [1,2], Philipp Kopylov [2], Maria Serova [2], Roman Syunyaev [1,2], Andrey Pikunov [1], Sofya Belova [3], Fuyou Liang [2,4], Jordi Alastruey [2,5] and Sergey Simakov [1,2,*]

[1] Moscow Institute of Physics and Technology, 141701 Dolgoprudny, Russia; gamilov@crec.mipt.ru (T.G.); roman.syunyaev@gmail.com (R.S.); pikunov@phystech.edu (A.P.)
[2] Institute of Personalized Medicine, Sechenov University, 119992 Moscow, Russia; fjk@inbox.ru (P.K.); yamarfa@yandex.ru (M.S.); fuyouliang@sjtu.edu.cn (F.L.); jordi.alastruey-arimon@kcl.ac.uk (J.A.)
[3] Institute of Psychology of Russian Academy of Sciences, 129366 Moscow, Russia; sbelova@gmail.com
[4] School of Naval Architecture, Ocean and Civil Engineering, Shanghai Jiao Tong University, Shanghai 200240, China
[5] King's College London, London SE1 7EH, UK
* Correspondence: simakov.ss@mipt.ru

Received: 15 June 2020; Accepted: 17 July 2020; Published: 22 July 2020

Abstract: In this work we present a one-dimensional (1D) mathematical model of the coronary circulation and use it to study the effects of arrhythmias on coronary blood flow (CBF). Hydrodynamical models are rarely used to study arrhythmias' effects on CBF. Our model accounts for action potential duration, which updates the length of systole depending on the heart rate. It also includes dependency of stroke volume on heart rate, which is based on clinical data. We apply the new methodology to the computational evaluation of CBF during interventricular asynchrony due to cardiac pacing and some types of arrhythmias including tachycardia, bradycardia, long QT syndrome and premature ventricular contraction (bigeminy, trigeminy, quadrigeminy). We find that CBF can be significantly affected by arrhythmias. CBF at rest (60 bpm) is 26% lower in LCA and 22% lower in RCA for long QT syndrome. During bigeminy, trigeminy and quadrigeminy, respectively, CBF decreases by 28%, 19% and 14% with respect to a healthy case.

Keywords: 1D haemodynamics; systole variations; coronary circulation; cardiac pacing; tachycardia; bradycardia; interventricular asynchrony; long QT syndrome; premature ventricular contraction

---

1. Introduction

Coronary blood flow (CBF) supplies the myocardium tissue with oxygen and other essential nutrients. The distal coronary vessels are immersed in the myocardium. Myocardium contractions produce external pressure to the immersed blood vessels and substantially elevate the terminal hydraulic resistance. Thus, the dependence of the blood flow in coronary arteries (CA) on the characteristics of the heart cycle is an essential feature of CBF.

The characteristics of the heart cycle are controlled by the electrical activity of the sinoatrial node (SAN). The SAN activity is modified by a variety of factors: signals from the sympathetic and the parasympathetic nervous system and humoral factors. Some pathological processes and artificial electric stimulation of the myocardium are also among the factors, which can modify the heart activity.

Pathological change of the myocardium contractions due to asynchronous cardiac pacing and arrhythmias produces violations of the CBF, which, in turn, results in decreased myocardium supply with nutrients and possible ischemic events. In this work, we use computational modelling to study changes in CBF, which are produced by several types of pathological heart rhythms including interventricular asynchrony due to inappropriate pacing, several types of arrhythmia including bradycardia, tachycardia, long QT syndrome and premature ventricular contraction (bigeminy, trigeminy, quadrigeminy).

Interventricular asynchrony refers to the decoordination between RV and LV contraction due to the failure in the electrical conducting system of the heart or asynchronous cardiac pacing. Cardiac pacing is performed by the pacemakers. The pacemakers stimulate the heart by electrical impulses and prompt heart beating at a regular rate. Asynchronous electrical activation of the ventricles during ventricular pacing produces irregular patterns of the mechanical stress [1–3] which, in turn, results in abnormal ventricle contractions and violation of CBF. Finally, it causes deficiencies in perfusion and glucose uptake even in the absence of CA disease [4,5]. In this work, we perform numerical simulations to study the effects of the asynchronous performance caused by pacemakers on the CBF. These results are also valid for interventricular asynchrony caused by other factors. We simulate asynchrony by modifying terminal hydraulic resistance in coronary vessels according to their spatial position relative to the pacemaker. We add shifts in time-dependent hydraulic resistance functions according to the asynchronous pulse generation.

Abnormal tachycardia is a stable, permanent increase of atria and/or ventricular contractions at rest above 85–100 beats per minute (bpm). It decreases the atria and ventricular filling and, thus, decreases the heart output. Various abnormalities in action potential initiation and propagation are the common reasons of tachycardia [6,7]. Increased physical load and emotional stress are the possible reasons of tachycardia in elderly people. Myocardial infarction often produces ventricular tachycardia. Tachycardia is a significant reason for morbidity and mortality in patients with ischemic heart disease. Tachycardia itself may be a reason for ischemic events and disease due to the changes in CBF. Bradycardia is a stable, permanent decrease of atria and/or ventricular contractions at rest below 55 bpm. It decreases the heart activity. Analysis of CBF during both tachycardia and bradycardia is rarely addressed in the literature.

Long-QT syndrome is a life-threatening cardiac arrhythmia syndrome that may cause sudden cardiac death [8,9]. The problems in myocardium repolarisation after a heart contraction produce increased QT intervals on the electrocardiogram (ECG) by more than 480 ms. Long-QT syndrome is associated with high duration of systole and decreased length of diastole. It may decrease CBF with an associated decrease of nutrient delivery.

Premature ventricular contractions (PVC) produce extra, abnormal heartbeats that disrupt regular heart rhythm, sometimes causing a skipped beat or palpitations [10]. They can originate from dysfunctional Purkinje fibres, ventricular or atrial tissue. In this work, we focus on PVCs with a full compensatory pause: the following SAN impulse occurs on time based on the sinus rate. The heart rhythms with one, two, or three regular heartbeats between each PVC are called bigeminy, trigeminy, or quadrigeminy [11]. PVC may increase the risk of developing arrhythmias and lead to chaotic heart rhythms and cardiomyopathy. Similar to other cases of heart rhythm failure, PVC affects myocardium contractions and, thus, the average CBF.

Mathematical models of haemodynamics utilise a variety of approaches to simulate blood flow in a network of vessels [12,13]. Mathematical modelling of the impact of tachycardia and bradycardia on blood flow [1] and electrophysiology [14] has been a subject of many studies. Coronary circulation and related haemodynamic indices in the presence of atherosclerosis are also popular topics of mathematical modelling [2,15–19]. To the best of our knowledge, the blood flow in CA during abnormal heart rhythms is rarely studied in clinics and by mathematical modelling. In this work, we present a modification of the previously developed one-dimensional (1D) mathematical model of the coronary circulation, which was applied to the simulations of blood flow in CA during cardiac pacing and tachycardia [20]. We apply the new model to the computational evaluation of changes in CBF during asynchronous cardiac pacing and some types of arrhythmia. Due to the lack of appropriate patient-specific data on coronary vessels' structure and properties, we use anatomically correct arterial networks [21] and typical values of cardiovascular characteristics (vessels' elasticity, cardiac output, etc.). A similar dataset can be collected or estimated by using non-invasive data such as CT scans, arterial pressure, lifestyle conditions (sex, age, body mass index, smoking, sport) of the patient and others [22–24]. In our previous work, we considered the stroke volume (SV) and the ratio between systolic and diastolic phases of the heart cycle as predefined constants based on known values from the literature. Our present model updates the length of systole depending on the heart rate (HR). This model is based on the simulated duration of an action potential. We also include a regression relating SV to HR, which is based on clinical data. We show that these novel features provide substantially different numerical results on the variations of CBF during tachycardia and bradycardia. We also apply the new model to simulations of CBF during interventricular asynchrony, some types of arrhythmias, including long QT syndrome and premature ventricular contraction (bigeminy, trigeminy, quadrigeminy).

2. Materials and Methods

2.1. 1D Mathematical Model of Blood Flow in the Coronary Vascular Network

The blood flow in the coronary vascular network and the aorta is simulated by a 1D reduced-order model of unsteady flow of viscous incompressible fluid through the network of elastic tubes. Reviews and details of the 1D models of haemodynamics can be found in [13,22,25]. The 1D models were adapted and applied to the coronary circulation in [20,26,27]. In this section, we briefly present this approach and describe some novel features of the model to account for interactions with myocardium contractions. The flow in every vessel is described by mass and momentum conservation in the form

$$\frac{\partial \mathbf{V}}{\partial t} + \frac{\partial \mathbf{F}(\mathbf{V})}{\partial x} = \mathbf{G}(\mathbf{V}), \quad (1)$$

$$\mathbf{V} = \begin{pmatrix} A \\ u \end{pmatrix}, \mathbf{F}(\mathbf{V}) = \begin{pmatrix} Au \\ u^2/2 + p(A)/\rho \end{pmatrix}, \mathbf{G}(\mathbf{V}) = \begin{pmatrix} 0 \\ \psi \end{pmatrix},$$

where t is the time, x is the distance along the vessel counted from the vessel's junction point, ρ is the blood density (constant), $A(t,x)$ is the vessel cross-sectional area, p is the blood pressure, $u(t,x)$ is the linear velocity averaged over the cross-section, ψ is the friction force,

$$\psi = -8\pi\mu\frac{u}{\rho A}, \quad (2)$$

μ is the dynamic viscosity of blood. Blood density $\rho = 1\,g/cm^3$ and blood viscosity $\mu = 4\,cP$. Properties and characteristics of the blood are considered to be constant throughout the computational domain. This assumption is considered to be accurate in arteries with diameters larger than 1 mm under physiological conditions [22,25]. The elasticity of the vessel's wall material is characterised by the $p(A)$ relationship

$$p(A) = \rho c^2 f(A), \quad (3)$$

where c is the velocity of small disturbances propagation in the material of the vessel wall, and $f(A)$ is the monotone S-like function (see [28] for a review)

$$f(A) = \begin{cases} \exp(\eta - 1) - 1, & \eta > 1 \\ \ln \eta, & \eta \leqslant 1 \end{cases}, \eta = \frac{A}{\tilde{A}}, \quad (4)$$

where $\tilde{A}$ is the cross-sectional area of the unstressed vessel.

The boundary conditions at the vessel's junctions include the mass conservation condition and the continuity of the total pressure,

$$\sum_{k=k_1,k_2,\ldots,k_M} \varepsilon_k A_k(t,\tilde{x}_k) u_k(t,\tilde{x}_k) = 0, \quad (5)$$

$$p_k\left(A_k\left(t,\tilde{x}_k\right)\right) + \frac{\rho u^2\left(t,\tilde{x}_k\right)}{2} = p_{k+1}\left(A_{k+1}\left(t,\tilde{x}_{k+1}\right)\right) + \frac{\rho u^2\left(t,\tilde{x}_{k+1}\right)}{2}, \quad k = k_1, k_2, \ldots, k_{M-1}, \quad (6)$$

where k is the index of the vessel, M is the number of the connected vessels, $\{k_1, \ldots, k_M\}$ is the range of the indices of the connected vessels, $\varepsilon = 1, \tilde{x}_k = L_k$ for incoming vessels, $\varepsilon = -1, \tilde{x}_k = 0$ for outgoing vessels.

The boundary conditions at the aortic root include the blood flow from the heart, which is set as a predefined time function $Q_H(t)$,

$$u(t,0)\,A(t,0) = Q_H(t). \quad (7)$$

The outflow boundary conditions assume that a terminal artery with index k is connected to the venous pressure reservoir with the pressure $p_{veins} = 8$ mmHg by the hydraulic resistance R_k. It is described by Poiseuille's pressure drop condition

$$p_k\left(A_k\left(t,\tilde{x}_k\right)\right) - p_{veins} = R_k A_k\left(t,\tilde{x}_k\right) u_k\left(t,\tilde{x}_k\right). \quad (8)$$

The hyperbolic system (1) is numerically solved within every vessel by the second-order grid-characteristic method [29]. The systems of nonlinear algebraic equations, which represent boundary conditions at the vessel's junctions (5) and (6), aortic root (7) and at the end points of terminal arteries (8) are numerically solved by Newton's method. All formulations of boundary conditions include a compatibility condition along the characteristic curve of the hyperbolic system (1), which extends outside the integration domain for every incoming and/or outgoing vessel.

The computational domain is the network of vessels including the aortic root, aorta, left and right coronary arteries and their branches (Figure A2). Parameters of the aorta, aortic root and coronary arteries correspond to the physiological values for an adult human (Table A1). The structure of the coronary vascular network (Figure A1) was derived from a general anatomical model [21].

2.2. Effects of the Heart Rhythm on the Coronary Circulation

The distal coronary arteries are immersed in the myocardium. Thus, myocardium contractions cause a substantial effect on the blood flow in these vessels. The variations of the heart rhythm change how the myocardium works. There are three essential features of the heart function, which cause a substantial effect on the CBF: dependency of the length of systole on the HR; dependency of the SV on the HR and dramatic increase in peripheral resistance due to compression of the terminal CAs by the myocardium during systole.

The dependency of systole duration on HR is complex. In this work, we use the action potential duration at 80% repolarisation (APD80) in a single cardiac cell as an estimation of systole duration. The action potential waveform and APD of human cardiac cells were simulated with the O'Hara-Rudy model [30]. We use the implementation of this model with a custom C++ code using the Rush-Larsen integration technique [31] with an adaptive step [32]. The minimal time step was set to 5×10^{-3} ms. The model was paced at every HR to steady-state (100 beats). The simulated restitution curve (i.e., APD80 dependence on HR) is shown in Figure 1. It gives the approximation

$$\tau = 287.09[\text{ms}] - \frac{30685.24[\text{ms}^2]}{T[\text{ms}]}, \quad (9)$$

where τ is the length of systole and T is the period of the cardiac cycle.

In this work, we use a simple approximation of the heart outflow $Q_H(t)$ in the time domain. We define it as a half-sine function during ventricular systole and set it to zero otherwise,

$$Q_H(t) = \begin{cases} SV\dfrac{\pi}{2\tau}\sin\left(\dfrac{\pi t}{\tau}\right), & 0 \leqslant t \leqslant \tau, \\ 0, & \tau < t \leqslant T, \end{cases} \quad (10)$$

where SV is the stroke volume of the left ventricle. Thus, we have

$$SV = \int_0^T Q_H(t)\, dt. \tag{11}$$

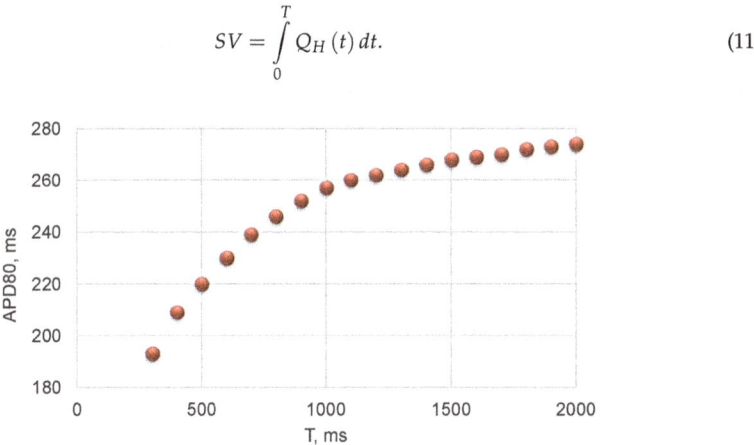

Figure 1. Relationship between the action potential duration at 80% repolarisation (APD80) and the period of the cardiac cycle (T).

The heart outflow is a function of two parameters: the SV and the length of the ventricular systole τ. When studying cardiovascular events with different or variable heart rhythms, it is especially important that both SV and τ depend on HR. In the wide range of HR values, cardiac output (Q_{CO}) remains constant under pacing conditions [33,34]; i.e.,

$$Q_{CO} = SV \cdot HR. \tag{12}$$

Thus, SV should be inversely proportional to HR. However, some experimental and clinical studies on the hearts of conscious dogs, which were paced by an implanted right atrial electrode [35], and the hearts of normal human foetus, which were auditory stimulated by a sound emitter placed on the mother's abdomen [36], report a linear relationship between SV and HR. We use the data from clinical studies on eighteen vascular surgery patients having general anaesthesia, mechanical ventilation with tidal volume 6 mL/kg and a transesophageal atrial pacemaker [37]. In this study, increasing HR from 80 to 110 bpm caused a reduction in SV from 72 to 57 mL. We use these values to obtain the linear relationship

$$SV = 112[\text{mL}] - \frac{HR[\text{bpm}]}{2[\text{bpm/mL}]}. \tag{13}$$

In this work, we consider the range of HR from 40 to 160 bpm where the linear regression (13) is close to the inverse relationship. Remarkable nonlinear behaviour is observed for the HR values above 200 bpm [38].

Compression of the terminal CAs during systole by the myocardium is an essential feature of coronary haemodynamics. To account for this compression, we set $R_k = R_k(t)$ for the boundary condition (8) in the terminal CAs. Similar to our previous works [20,27,39,40] we assume that the dimensionless time profile of $R_k(t)$ is the same as the dimensionless time profile of cardiac output (10).

$$R_k(t) = \begin{cases} R_k + (R_k^{max} - R_k)\sin\left(\dfrac{\pi t}{\tau}\right), & 0 \leqslant t \leqslant \tau \\ R_k, & \tau < t \leqslant T \end{cases} \tag{14}$$

The peak value of the peripheral resistance during systole is set to $R_k^{max} = 3R_k$, where R_k is the terminal resistance during diastole [41]. It is sufficient for the complete blockage of the flow in terminal CAs during systole. The values of R_k are set by the following algorithm. We assume that the total arterio-venous resistance R_{total} of the systemic circulation produces the pressure drop $\Delta P = 100$ mmHg [42], thus

$$R_{total} = \frac{\Delta P}{Q_{CO}} \qquad (15)$$

Then we perform three steps. First, we split R_{total} between the terminal resistance of the aorta $R_a = R_{total}/0.9$ and the total terminal resistance of the CAs $R_{cor} = R_{total}/0.1$. These values produce the ratio of CBF to CO of about 3–6%. Second, we split R_{cor} between the total effective resistances of major CAs and their branches: right coronary artery (RCA) and branches of left coronary artery (LCA)(circumflex artery and left anterior descending artery). R_{cor} is divided depending on the diameters of the major CAs according to Murray's law with a power of 2.27 [41]. Third, we divide the total effective resistances of major CAs between their terminal branches according to Murray's law with a power of 2.27. We note that the error in measuring diameters of the terminal CAs may be substantial due to the quality of CT data. The major CAs generally have good visualisation on CT. The error in measuring their diameters is relatively small. Thus, the second step allows us to decrease the total error of R_k assessment.

3. Results

3.1. Interventricular Dyssynchrony Due to Asynchronous Cardiac Pacing

Failures in the electrical conducting system of the heart or asynchronous cardiac pacing produce dyssynchrony between RV and LV contraction. We simulate interventricular dyssynchrony by introducing a time difference of 30 and 60 ms in resistance functions (14) for the terminal branches of RCA. The disagreements among regional pre-ejection periods greater than 50 ms are considered as significant [43]. We associate the early right ventricle contraction with a pacemaker location in the right ventricle and the late right ventricle contraction with a pacemaker location in the left ventricle.

We calculate the ratios of the average blood flow in the LCA and RCA in the asynchronous and normal (synchronous) pacing conditions at different HRs. The relative change in the average blood flow in LCA is less than 0.5% in all cases because resistance functions (14) for the left CAs are always synchronous to the heart outflow (7) and, therefore, blood flow in LCA is synchronised to the contractions of the left ventricle.

From Figure 2 we observe that interventricular asynchrony causes significant changes in the average blood flow in RCA at the normal values of HR. The 60 ms early contraction of the right ventricular produces 12% increase, while 60 ms late contraction produces approximately the same decrease. The elevation of HR leads to a decrease in the relative changes in blood flow in RCA, which tends to the value corresponding to the normal pacing conditions.

Since both RCA and LCA are the branches of the aorta, the left ventricle contractions supply both RCA and LCA with the blood. It explains the asymmetric behaviour of the relative change of the blood flow in RCA and LCA.

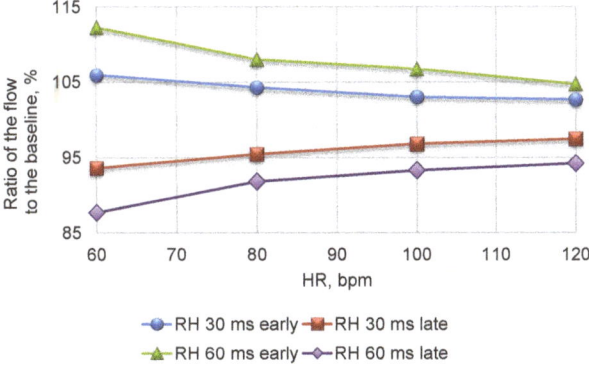

Figure 2. Relationship between the relative changes in the average blood flow in the right coronary artery (RCA) for early (right ventricular pacing) and late (left ventricular pacing) right ventricular contractions and heart rate. Baseline corresponds to the no pacing conditions.

3.2. Tachycardia and Bradycardia

In this section, we present results of CBF simulations at constant HR in the range from 40 to 160 bpm. The values of HR lower than 55 bpm are generally associated with bradycardia, and the values of HR higher than 85 bpm are related to tachycardia. We take into account the changes in systole duration (9) and stroke volume (13) with the variations in HR. Figure 3 shows the dependency between the ratio of average CBF and the average cardiac output (12) in the cases of variable (9) and constant (35%) ratio of the systole duration to the heart period from HR.

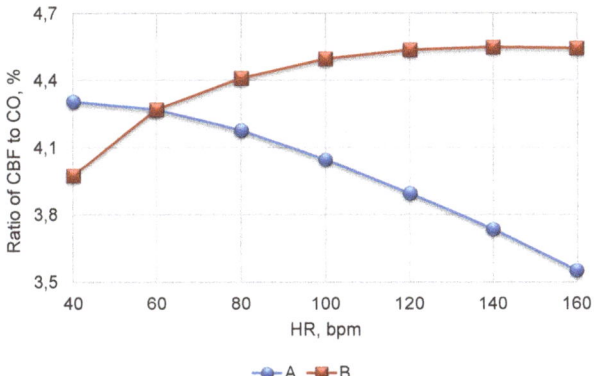

Figure 3. Dependence of fraction of average coronary blood flow (CBF) relative to the average cardiac output (CO) on the heart rate (HR) for two ways of systole representation: (**A**) systolic interval calculated using Equation (9); (**B**) systolic interval calculated as 35% of the cycle length ($\tau = 0.35 \cdot \frac{60}{HR}$)

From Figure 3 we observe a substantial difference in the fraction of CBF depending on the model assumptions. A constant fraction of systole duration leads to the elevation of the fraction of CBF with the increase in HR. It contradicts the clinical data [44]. The model with a variable fraction of systole results in the decrease in CBF fraction with the increase in HR. According to (9) the length of systole increases with the decreasing period of the cardiac cycle (and increasing HR). It leads to the increase in the relative length of systole and decrease in the relative length of diastole within a cardiac cycle. Thus, the relative time period with increased peripheral resistance (14) increases.

We also note that the absolute value of the average CO achieves its maximum at 100–120 bpm (see Figure 4). Figure 5 presents the corresponding values of the average CBF during the constant and variable fraction of systole. From Figure 5 we observe that average CBF achieves its maximum at 100 bpm for the variable and at 100–120 bpm for the constant fraction of systole. The early decrease in the average CBF, which is simulated by the model with a variable fraction of systole, accounts for the integrated effect of the decrease in both SV and diastolic phase of the cardiac cycle. The sensitivity of the average CBF to the fraction of systole is low for low values of HR. For HR values above 80 bpm, the assumption of a constant fraction of systole produces overestimated values of the average CBF. The difference of the absolute values of average CBF at 50, 100 and 160 bpm from the standard value at 60 bpm (see Figure 4) is less than 15% despite the permanent decrease in the fraction of average CBF to average CO (see Figure 3A), which produces 20% difference at 160 bpm.

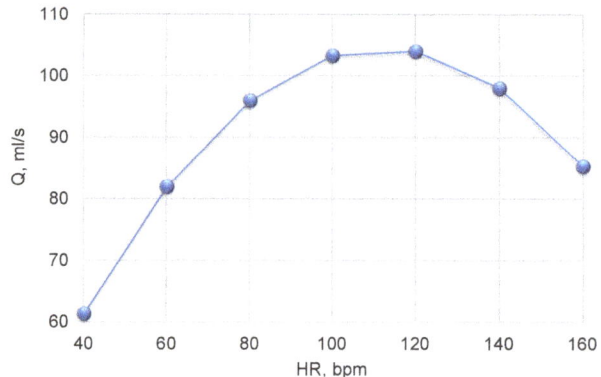

Figure 4. Average cardiac output.

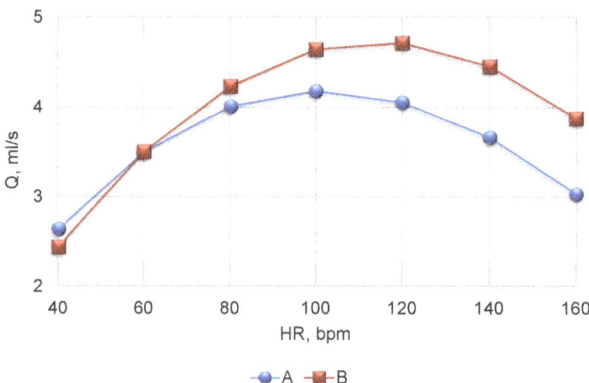

Figure 5. Absolute value of average CBF. Variable fraction (9) of systole (**A**), constant fraction (35%) of systole (**B**).

3.3. Long QT Syndrome

Long QT interval on ECG, more than 480 ms for 60 bpm, reflects the delay in myocardium repolarisation and myocardium relaxation after a heart contraction. It is unclear how the length of systole changes with HR in the case of a long QT syndrome. We assume that this relation is similar to (9): $\tau_{LongQT} = a - \frac{b}{T_{LongQT}}$, where T_{LongQT} is a period of the heart cycle during long QT syndrome, and a, b are constants. We estimate constants a and b from clinical data: for 60 bpm the length of systole

$\tau_{LongQT} = 480$ ms [8]; for 113 bpm the length of systole $\tau_{LongQT} = 353$ ms [9]. As a result, we derive $a = 624$ ms and $b = -143774$ ms^2.

Figure 6 shows the results of blood flow simulations in RCA and LCA during long QT syndrome for values of HR from 60 to 120 bpm. From Figure 6 we observe that long QT syndrome produces a significant decrease in CBF in both RCA and LCA (Figure 6), which is more than 25%. It accounts for the decrease in the diastolic period, which is accompanied by low values of peripheral resistance according to (14). Elevation of HR results in an increase in blood flow in both RCA and LCA, which tends to the baseline values. The baseline value is the blood flow simulated by the model with a variable fraction of systole according to (9).

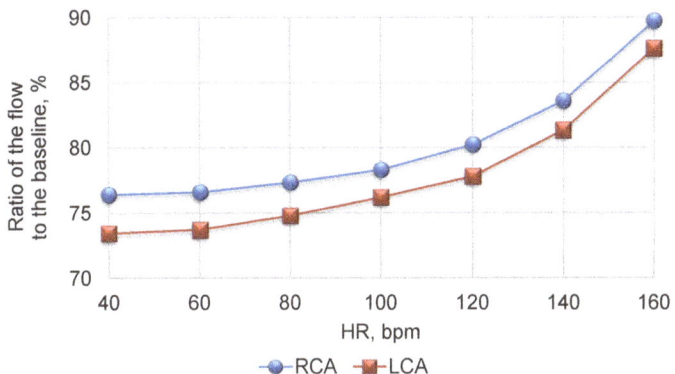

Figure 6. Relative average blood flow in RCA and LCA during long QT syndrome. Baseline corresponds to the absence of the long QT syndrome.

3.4. Premature Ventricular Contraction

PVC is an abnormal heartbeat, which occurs earlier than a scheduled normal heartbeat, which should be initiated by an action potential. The normal heartbeat is missed because the ventricles have been emptied by PVC. PVC also causes an effect on the subsequent normal pulses due to the increased ventricular filling after PVC. A regular PVC pattern includes PVC every second (bigeminy), third (trigeminy) or fourth (quadrigeminy) cardiac cycle.

We simulate PVC by modifying heart outflow function (7) in the following way. We suppose that systole starts 0.25T earlier than a periodic schedule; SV of the PVC is 71% less than SV of a normal beat due to insufficient ventricle filling [11]; the beat after PVC occurs on a schedule; the total length of PVC heartbeat and the following beat is 2T; SV of the heartbeat after PVC is 18% more than SV of a normal beat due to increased time of ventricle filling [11]. Figure 7 shows an example of modified heart outflow in the case of quadrigeminy and a HR value of 120 bpm.

We simulate average CBF in the cases of bigeminy, trigeminy and quadrigeminy for HR values from 40 to 120 bpm. Figure 8 presents the results, which include the fraction of the average CBF overall cardiac cycles of PVC pattern to the value of average CBF without PVC. From Figure 8, we observe a substantial decrease in the relative average CBF at low and normal HR. This value decreases with the increasing HR. The most pronounced effect (more, than 25% decrease at 40 bpm and 30% decrease at 120 bpm) is observed in the case of bigeminy as it produces the most frequent occurrence of PVC.

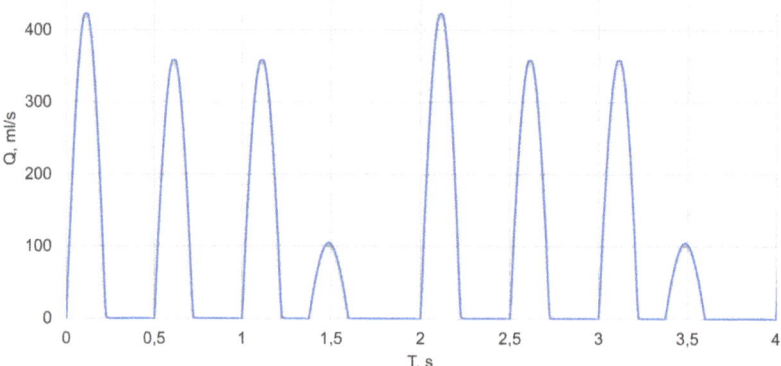

Figure 7. Modified heart outflow $Q_H(t)$ at $HR = 120$ bpm and quadrigeminy.

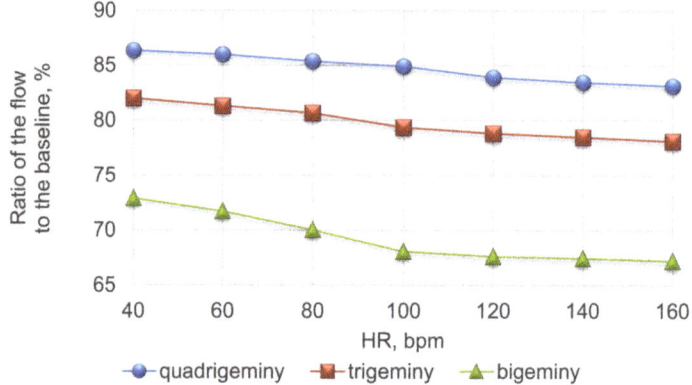

Figure 8. Relative average CBF with PVC. Baseline corresponds to the absence of PVC.

4. Discussion

In this work, we have presented a modified model of 1D haemodynamics in the network of coronary vessels, which includes dependencies of systole duration and stroke volume on the length of the cardiac cycle (or the heart rate). We have simulated interventricular dyssynchrony, which is caused by asynchronous cardiac pacing of the left and right ventricles. We have analysed coronary blood flow in the cases of several abnormal heart rhythms, including tachycardia, bradycardia, long QT syndrome and premature ventricular contraction.

The variable length of systole and SV produces results that are in agreement with clinical data [44]. In contrast to our previous works [20,27,39] we suggest that only terminal resistance rises due to myocardium contractions during systole, which is different from the models of general muscle contractions [26,45] and better corresponds to the anatomical features of major CAs laying outside the myocardium.

Based on experimental and clinical observations [35,36] we used linear regression for the dependency of SV on HR (13), which works well for the arrhythmia and pacing conditions. We note that this relationship becomes nonlinear for the values of HR above 160 bpm [38], and our model is not valid in this case. Some other factors (e.g., regulatory processes) may produce nonlinear behaviour even in the range from 40 to 160 bpm (e.g., physical exercise conditions).

We implicitly assumed that the length of the heart systole is approximately equal to the ventricular systole. We neglected atria systole as all applications in this work relate to the activity of the ventricles.

This assumption may produce some systematic error to the results. More detailed models are needed for the accurate simulations of the heart rhythm. We also use the single-cell model [30] for simulating action potential duration restitution curves in a cardiac cell. We assumed that the dependency of action potential duration at 80% repolarisation (APD80) on the duration of a cardiac cycle correlates with the similar dependence of the length of the ventricular systole (9). A more accurate approach should simulate the propagation of the action potential over myocardium and mechanical contractions. We imply that this process is instantaneous. It also may be a source of systematic error. We note that approximation (9) correlates with the results of QT interval measurements for different values of HR derived from ECG data. The length of the QT interval is associated with the length of the electrical and mechanical systole [46].

The cardiac muscle mechanical contraction is a result of Ca^{2+} concentration increase in the cytoplasm, which, in turn, is induced by sarcolemma depolarisation and consequent calcium-induced-calcium release. Thus, despite quantitative differences in duration, all stages of the mechanical contraction are closely interrelated. Moreover, previous experiments on non-failing human hearts demonstrated that the difference between calcium transients measured at 80% recovery (CaTD80) and APD80 are approximately the same at every pacing cycle length (approximately 80 ± 20 ms for subendocardial tissue) [47]. A similar correlation was observed between twitch force duration and calcium transients in rat papillary muscle [48]. On the other hand, cardiac contraction is a complex, spatially distributed process that could be affected by numerous factors. For example, the blood vessel pressure changes and consequent baroreflex affect the sympathetic system, which has unequal effects on action potential, calcium transients and mechanical contraction [49]. However, for the purpose of this study, which is to qualitatively investigate the principal effects of HR on the CBF, APD80 provides a reasonable estimate of the mechanical contraction duration.

The developed model of CBF under various heart rhythms allows simulating the effects of medical treatment. We mention anticoagulant therapy as the primary strategy to prevent atrial fibrillation (AF) complications, including thromboembolism and stroke. Oral anticoagulants (OACs) such as vitamin K antagonists, direct thrombin inhibitors, or factor Xa inhibitors are administered routinely in patients with AF [50,51]. Anticoagulation therapy changes rheological properties of the blood and modifies oxygen and nutrients delivery to the myocardium by CBF. Adherence to OACs is the extent to which a patient takes his medication as prescribed. It is estimated as appropriate in 76.6% of patients with hypertension and other cardiovascular diseases [52]. Persistence is the act of taking drugs for the prescribed treatment duration. Persistence with OACs therapy is about 50% in a two-year perspective [53]. Thus, non-adherent and non-persistent patients make a tangible contribution to negative treatment outcomes. Multiple patient-level factors, as well as social, economic, health system causes contribute to poor adherence and persistence to OACs making this phenomenon a challenge for population studies [54]. Mathematical modelling of the CBF during arrhythmia episodes controlling for OACs therapy adherence and persistence may clarify their effects and give predictions with respect to possible negative outcomes. Some approaches to mathematical modelling of medication therapy effects on the CBF have been proposed in [55]. Anticoagulant therapy is not considered yet. We anticipate further elaboration of our model with respect to anticoagulation therapy as a possible ongoing endeavour.

Author Contributions: Conceptualization, S.S., P.K., M.S., R.S.; methodology, S.S., T.G., R.S., F.L. and J.A.; software, T.G., R.S. and A.P.; validation, S.S., T.G., R.S. and F.L.; formal analysis, S.S., P.K., T.G., R.S. and S.B.; writing-original draft preparation, S.S., T.G., R.S., S.B.; writing-review and editing, S.S., T.G., F.L., J.A. and S.B.; visualization, T.G., A.P.; supervision, S.S. and P.K. All authors have read and agreed to the published version of the manuscript.

Funding: This research was funded by the Russian Foundation for Basic Research grant numbers 18-00-01661, 18-00-01524, 18-00-01659, 18-00-01326, and 18-31-20048. Electrophysiological simulation and corresponding discussion was done by RS and AP and supported by Russian Scientific Foundation grant 18-71-10058.

Conflicts of Interest: The authors declare no conflicts of interest.

Abbreviations

The following abbreviations are used in this manuscript:

APD80	Action potential duration at 80% repolarisation
AF	Atrial fibrillation
bpm	Beats per minute
CA	Coronary arteries
CBF	Coronary blood flow
CO	Cardiac output
ECG	Electrocardiogram
HR	Heart rate
LV	Left ventricle
LCA	Left coronary artery
OAC	Oral anticoagulant
PVC	Premature ventricular contraction
RCA	Right coronary artery
RV	Right ventricle
SAN	Sinoatrial node
SV	Stroke volume

Appendix A

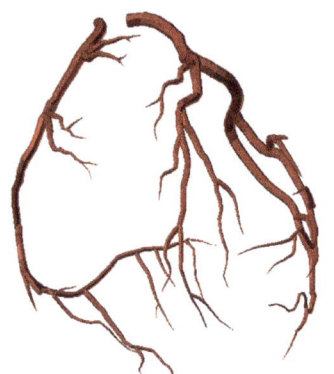

Figure A1. 3D anatomical model of the coronary arteries [21].

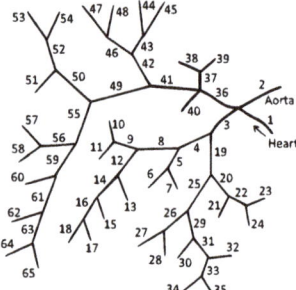

Figure A2. The 1D structure of the 3D anatomical model from Figure A1. Vessels 3–35 are the branches of LCA, vessels 36–65 are the branches of RCA. Parameters of the vessels are presented in Table A1.

Table A1. Parameters of the arterial network in Figure A2. k is the index of the vessels, l_k is length, d_k is diameter, c_k is pulse wave velocity index (see (3)).

k	l_k, cm	d_k, mm	$c_k, \frac{cm}{s}$	k	l_k, cm	d_k, mm	$c_k, \frac{cm}{s}$
1	5.2	21.7	700	34	1.23	0.9	1200
2	20	25	1094	35	0.71	1.87	1200
3	2.61	4.96	1200	36	1.74	3.46	1300
4	1.83	4.14	1200	37	2.35	1.84	1300
5	2.45	1.78	1200	38	0.38	0.9	1300
6	0.65	0.9	1200	39	0.27	0.88	1300
7	1.58	0.9	1200	40	2.05	1.95	1300
8	2.04	3.04	1200	41	2.42	3.26	1300
9	2.76	1.96	1200	42	0.81	2.53	1300
10	3.3	0.89	1200	43	1.86	1.56	1300
11	1.98	0.96	1200	44	0.75	0.9	1300
12	1.32	2.31	1200	45	0.62	0.88	1300
13	2.66	1.11	1200	46	2.95	1.59	1300
14	3.67	1.78	1200	47	0.47	0.92	1300
15	2.26	0.98	1200	48	0.76	0.93	1300
16	1.94	1.05	1200	49	4.53	2.57	1300
17	0.97	0.9	1200	50	1.84	1.97	1300
18	1.84	0.9	1200	51	1.34	1.07	1300
19	3.13	3.92	1200	52	2.34	1.52	1300
20	4.97	2.91	1200	53	3.17	0.72	1300
21	2.16	1.3	1200	54	1.05	0.54	1300
22	4.05	1.84	1200	55	4.6	1.85	1300
23	2.49	0.9	1200	56	3.37	1.41	1300
24	1.97	0.88	1200	57	2.34	0.6	1300
25	2.47	3.02	1200	58	1.88	0.67	1300
26	2.45	1.78	1200	59	2.42	1.5	1300
27	1.5	1.06	1200	60	3.14	0.88	1300
28	1.11	1.03	1200	61	0.66	1.34	1300
29	2.58	2.39	1200	62	1.47	0.9	1300
30	1.34	1.07	1200	63	0.87	1.15	1300
31	0.71	1.87	1200	64	2.75	0.6	1300
32	2.1	1.02	1200	65	1.23	0.42	1300
33	2.22	1.44	1200				

References

1. Abdi, M.; Karimi, A.; Navidbakhsh, M.; Pirzad, G.; Hassani, J.K. A lumped parameter mathematical model to analyze the effects of tachycardia and bradycardia on the cardiovascular system. *Int. J. Numer. Model. Electron. Netw. Devices Fields* **2015**, *3*, 346–357. [CrossRef]
2. Arthurs, C.J.; Lau, K.D.; Asrress, K.N.; Redwood, S.R.; Figueroa, C.A. A mathematical model of coronary blood flow control: Simulation of patient-specific three-dimensional hemodynamics during exercise. *Am. J. Physiol.-Heart Circ. Physiol.* **2016**, *310*, H1242–H1258. [CrossRef] [PubMed]
3. Bashore, T.M.; Stine, R.A.; Shaffer, P.B.; Bush, C.A.; Leier, C.V.; Schaal, S.F. The noninvasive localization of ventricular pacing sites by radionuclide phase imaging. *Circulation* **1984**, *70*, 681–694. [CrossRef]
4. Prinzen, F.W.; Augustijn, C.H.; Arts, T.; Allessie, M.A.; Reneman, R.S. Redistribution of myocardial fiber strain and blood flow by asynchronous activation. *Am. J. Physiol. Heart Circ. Physiol.* **1990**, *259*, H300–H308. [CrossRef] [PubMed]
5. Delhaas, T.; Arts, T.; Prinzen, F.W.; Reneman, R.S. Regional fibre stress–Fibre strain area as an estimate of regional blood flow and oxygen demand in the canine heart. *J. Physiol.* **1994**, *477*, 481–496. [CrossRef] [PubMed]
6. Tarumi, T.; Takebayashi, S.; Fujita, M.; Nakano, T.; Ito, M.; Yamakado, T. Pacing tachycardia exaggerates left ventricular diastolic dysfunction but not systolic function and regional asynergy or asynchrony in patients with hypertrophic cardiomyopathy. *EP Eur.* **2010**, *12*, 1308–1315. [CrossRef] [PubMed]

7. Numan, M.T.; Maposa, D.; Kantharia, B. Supraventricular tachycardia significantly reduces stroke volume and causes minimal reduction of cardiac output: Study of pediatric patients. *Heart Rhythm* **2011**, *8*, 1826. [CrossRef]
8. Khan, I.A. Long QT syndrome: Diagnosis and management. *Am. Heart J.* **2002**, *143*, 7–14. [CrossRef]
9. Patel, P.J.; Borovskiy, Y.; Killian, A.; Verdino, R.J.; Epstein, A.E.; Callans, D.J.; Marchlinski, F.E.; Deo, R. Optimal QT interval correction formula in sinus tachycardia for identifying cardiovascular and mortality risk: Findings from the Penn Atrial Fibrillation Free study. *Heart Rhythm* **2016**, *13*, 527–535. [CrossRef]
10. Gerstenfeld, E.P.; De Marco, T. Premature Ventricular Contractions. *Circulation* **2019**, *140*, 624–626. [CrossRef]
11. Cohn, K.; Kryda, W. The influence of ectopic beats and tachyarrhythmias on stroke volume and cardiac output. *J. Electrocardiol.* **1981**, *14*, 207–218. [CrossRef]
12. Van de Vosse, F.N. Mathematical modelling of the cardiovascular system. *J. Eng. Math.* **2003**, *47*, 175–183. [CrossRef]
13. Bessonov, N.; Sequeira, A.; Simakov, S.; Vassilevski, Y.; Volpert, V. Methods of blood flow modelling. *Math. Model. Nat. Phenom.* **2016**, *11*, 1–25. [CrossRef]
14. Jackowska-Zduniak, B.; Fory's, U. Mathematical model of the atrioventricular nodal double response tachycardia and double-fire pathology. *Math. Biosci. Eng.* **2016**, *13*, 1143–1158. [CrossRef] [PubMed]
15. Ge, X.; Liu, Y.; Yin, Z.; Tu, S.; Fan, Y.; Vassilevski, Y.; Simakov, S.; Liang, F. Comparison of instantaneous wave-free ratio (iFR) and fractional flow reserve (FFR) with respect to their sensitivities to cardiovascular factors: A computational model-based study. *J. Interv. Cardiol.* **2020**, *2020*, 4094121. [CrossRef] [PubMed]
16. Ge, X.; Liu, Y.; Tu, S.; Simakov, S.; Vassilevski, Y.; Liang, F. Model-based analysis of the sensitivities and diagnostic implications of FFR and CFR under various pathological conditions. *Int. J. Numer. Methods Biomed. Eng.* **2019**. [CrossRef]
17. Boileau, E.; Pant, S.; Roobottom, C.; Sazonov, I.; Deng, J.; Xie, X.; Nithiarasu, P. Estimating the accuracy of a reduced-order model for the calculation of fractional flow reserve (FFR). *Int. J. Numer. Methods Biomed. Eng.* **2017**, *34*, e2908. [CrossRef]
18. Bezerra, C.; Lemos, P.A.; Pinton, F.A.; Müller, L.; Bulant, C.; Talou, G.M.; Feijóo, R.A.; Esteves, A.E.F.; Blanco, P. Tct-619 comparison of one-dimensional (1d) and three-dimensional (3d) models for the estimation of coronary fractional flow reserve through cardiovascular imaging. *J. Am. Coll. Cardiol.* **2018**, *72* (Suppl. 13), B248. [CrossRef]
19. Sinclair, M.D.; Lee, J.; Cookson, A.N.; Rivolo, S.; Hyde, E.R.; Smith, N.P. Measurement and modeling of coronary blood flow. *Wiley Interdiscip. Rev. Syst. Biol. Med.* **2015**, *7*, 335–356. [CrossRef]
20. Gamilov, T.M.; Liang, F.Y.; Simakov, S.S. Mathematical modeling of the coronary circulation during cardiac pacing and tachycardia. *Lobachevskii J. Math.* **2019**, *40*, 448–458. [CrossRef]
21. Plasticboy Pictures CC. 2009. Available online: http://www.plasticboy.co.uk/store/ (accessed on 20 June 2020).
22. Vassilevski, Y.; Olshanskii, M.; Simakov, S.; Kolobov, A.; Danilov, A. *Personalized Computational Hemodynamics: Models, Methods, and Applications for Vascular Surgery and Antitumor Therapy*; Academic Press: Cambridge, MA, USA, 2020.
23. Danilov A.; Ivanov, Y.; Pryamonosov, R.; Vassilevski, Y. Methods of graph network reconstruction in personalized medicine. *Int. J. Numer. Methods Biomed. Eng.* **2016**, *32*, e02754. [CrossRef] [PubMed]
24. Gamilov, T.; Alastruey, J.; Simakov, S. Linear optimization algorithm for 1D hemodynamics parameter estimation. In *Proceedings of the 6th European Conference on Computational Mechanics: Solids, Structures and Coupled Problems, ECCM 2018 and 7th European Conference on Computational Fluid Dynamics, ECFD 2018*; Owen R., de Borst R., Reese J., Pearce C. Eds.; CIMNE: Barcelona, Spain, 2020; pp. 1845–1850.
25. Van de Vosse, F.N.; Stergiopulos, N. Pulse wave propagation in the arterial tree. *Annu. Rev. Fluid Mech.* **2011**, *43*, 467–499. [CrossRef]
26. Gamilov, T.; Simakov, S. Blood flow under mechanical stimulations. *Adv. Intell. Syst. Comput.* **2020**, *1028*, 143–150.
27. Gamilov, T.; Kopylov, P.; Simakov, S. Computational simulations of fractional flow reserve variability. *Lect. Notes Comput. Sci. Eng.* **2016**, *112*, 499–507.
28. Vassilevski, Y.V.; Salamatova, V.Y.; Simakov, S.S. On the elasticity of blood vessels in one-dimensional problems of hemodynamics. *Comput. Math. Math. Phys.* **2015**, *55*, 1567–1578. [CrossRef]
29. Magomedov, K.M.; Kholodov, A.S. *Grid-Characteristic Numerical Methods*; Urite: Moscow, Russia, 2018.

30. O'Hara, T.; Virág, L.; Varró, A.; Rudy, Y. Simulation of the undiseased human cardiac ventricular action potential: Model formulation and experimental validation. *PLoS Comput. Biol.* **2011**, *7*, e1002061. [CrossRef]
31. Rush, S.; Larsen, H. A practical algorithm for solving dynamic membrane equations. *IEEE Trans. Biomed. Eng.* **1978**, *25*, 389–392. [CrossRef]
32. Smirnov, D.; Pikunov, A.; Syunyaev, R.; Deviatiiarov, R.; Gusev, O.; Aras, K.; Gams, A.; Koppel, A.; Efimov, I.R. Genetic algorithm-based personalized models of human cardiac action potential. *PLoS ONE* **2020**, *15*, e0231695. [CrossRef]
33. Linhart, J.W. Myocardial function in coronary artery disease determined by atrial pacing. *Circulation* **1972**, *44*, 203–212. [CrossRef]
34. Geddes, L.A.; Wessale, J.L. Cardiac output, stroke Volume, and pacing rate: A review of the literature and a proposed technique for selection of the optimum pacing rate for an Exercise responsive pacemaker. *J. Cardiovasc. Electrophysiol.* **1991**, *2*, 408–415. [CrossRef]
35. Noble, M.I.M.; Trenchord, D.; Guz, A. Effect of changing heart rate on cardiovascular function in the conscious dog. *Circ. Res.* **1966**, *19*, 206–213. [CrossRef]
36. Kenny, J.; Plappert, T.; Doubilet, P.; Salzman, D.; Sutton, M.G. Effects of heart rate on ventricular size, stroke volume, and output in the normal human fetus: A prospective Doppler echocardiographic study. *Circulation* **1987**, *76*, 52–58. [CrossRef]
37. Roeth, N.A.; Ball, T.R.; Culp, W.C.; Bohannon, W.T.; Atkins, M.D.; Johnston, W.E. Effect of increasing heart rate and tidal volume on stroke volume variability in vascular surgery patients. *J. Cardiothorac. Vasc. Anesth.* **2014**, *28*, 1516–1520. [CrossRef]
38. Kumada, M.; Azuma, T.; Matsuda, K. The cardiac output-heart rate relationship under different conditions. *Jpn. J. Physiol.* **1967**, *17*, 538–555. [CrossRef]
39. Gamilov, T.M.; Kopylov, P.Y.; Pryamonosov, R.A.; Simakov, S.S. Virtual fractional flow reserve assessment in patient-specific coronary networks by 1d hemodynamic model. *Russ. J. Numer. Anal. Math. Model.* **2015**, *30*, 269–276. [CrossRef]
40. Vassilevski, Y.V.; Danilov, A.A.; Gamilov, T.M.; Simakov, S.S.; Ivanov, Y.A.; Pryamonosov, R.A. Patient-specific anatomical models in human physiology. *Russian J. Numer. Anal. Math. Model.* **2015**, *30*, 185–201. [CrossRef]
41. Carson, J.M.; Pant, S.; Roobottom, C.; Alcock, R.; Blanco, P.J.; Carlos Bulant, C.A.; Vassilevski, Y.; Simakov, S.; Gamilov, T.; Pryamonosov, R.; et al. Non-invasive coronary CT angiography-derived fractional flow reserve: A benchmark study comparing the diagnostic performance of four different computational methodologies. *Int. J. Numer. Methods Biomed. Eng.* **2019**, *35*, e3235 [CrossRef]
42. Barret, K.; Brooks, H.; Boitano, S.; Barman, S. *Ganong's Review of Medical Physiology*, 23th ed.; The McGraw-Hill: New York, NY, USA, 2010.
43. Ghio, S.; Constantin, C.; Klersy, C.; Serio, A.; Fontana, A.; Campana, C.; Tavazzi, L. Interventricular and intraventricular dyssynchrony are common in heart failure patients, regardless of QRS duration. *Eur. Heart J.* **2004**, *25*, 571–578. [CrossRef]
44. Heusch, G. Heart rate in the pathophysiology of coronary blood flow and myocardial ischaemia: Benefit from selective bradycardic agents. *Br. J. Pharmacol.* **2008**, *153*, 1589–1601. [CrossRef]
45. Simakov, S.; Gamilov, T.; Soe, Y.N. Computational study of blood flow in lower extremities under intense physical load. *Russ. J. Numer. Anal. Math. Model.* **2013**, *28*, 485–503. [CrossRef]
46. Kovács, S.J. The duration of the QT interval as a function of heart rate: A derivation based on physical principles and a comparison to measured values. *Am. Heart J.* **1985**, *110*, 872–878. [CrossRef]
47. Lou, Q.; Fedorov, V.V.; Glukhov, A.V.; Moazami, N.; Fast, V.G.; Efimov, I.R. Transmural heterogeneity and remodeling of ventricular excitation-contraction coupling in human heart failure. *Circulation* **2011**, *123*, 1881–1890. [CrossRef] [PubMed]
48. Kassiri, Z.; Myers, R.; Kaprielian, R.; Banijamali, H.S.; Backx, P.H. Rate-dependent changes of twitch force duration in rat cardiac trabeculae: A property of the contractile system. *J. Physiol.* **2000**, *524 Pt 1*, 221–231. [CrossRef]
49. Lang, D.; Holzem, K.; Kang, C.; Xiao, M.; Hwang, H.J.; Ewald, G.A.; Yamada, K.A.; Efimov, I.R. Arrhythmogenic remodeling of $\beta 2$ versus $\beta 1$ adrenergic signaling in the human failing heart. *Circ. Arrhythmia Electrophysiol.* **2015**, *8*, 409–419. [CrossRef]
50. Manav, S. Contemporary management of stroke prevention in atrial fibrillation following the European Society of Cardiology guidelines. *Eur. Cardiol.* **2017**, *12*, 38–39.

51. January, C.T.; Wann, L.S.; Alpert, J.S.; Calkins, H.; Cigarroa, J.E.; Clevel, J.C., Jr.; Conti, J.B.; Ellinor, P.T.; Ezekowitz, M.D.; Field, M.E.; et al. 2014 AHA/ACC/HRS guideline for the management of patients with atrial fibrillation: A report of the American College of Cardiology/American Heart Association task force on practice guidelines and the heart rhythm society. *J. Am. Coll. Cardiol.* **2014**, *64*, e1–e76. [CrossRef] [PubMed]
52. DiMatteo, M.R. Variations in Patients' Adherence to Medical Recommendations. *Med. Care* **2004**, *42*, 200–209. [CrossRef] [PubMed]
53. Lowres, N.; Giskes, K.; Hespe, C.; Freedman, B. Reducing stroke risk in atrial fibrillation: Adherence to guidelines has improved, but patient persistence with anticoagulant therapy remains suboptimal. *Korean Circ. J.* **2019**, *49*, 883–907. [CrossRef] [PubMed]
54. Banerjee, A.; Benedetto, V.; Gichuru, P.; Burnell, J.; Antoniou, S.; Schilling, R.J.; Strain, W.D.; Ryan, R.; Watkins, C.; Marshall, T.; et.al. Adherence and persistence to direct oral anticoagulants in atrial fibrillation: A population–Based study. *Heart* **2020**, *106*, 119–126. [CrossRef]
55. Guala, A.; Leone, D.; Milan, A.; Ridolfi, L. In silico analysis of the anti-hypertensive drugs impact on myocardial oxygen balance. *Biomech. Model. Mechanobiol.* **2017**, *16*, 1035–1047. [CrossRef]

 © 2020 by the authors. Licensee MDPI, Basel, Switzerland. This article is an open access article distributed under the terms and conditions of the Creative Commons Attribution (CC BY) license (http://creativecommons.org/licenses/by/4.0/).

Article

Analysis of Operating Modes for Left Ventricle Assist Devices via Integrated Models of Blood Circulation

Sergey Simakov [1,2,*], Alexander Timofeev [3], Timur Gamilov [1,2], Philip Kopylov [2], Dmitry Telyshev [2,4] and Yuri Vassilevski [1,2,5]

1. Moscow Institute of Physics and Technology, 141701 Dolgoprudny, Russia; gamilov@crec.mipt.ru (T.G.); yuri.vassilevski@gmail.com (Y.V.)
2. Institute of Personalized Medicine, Sechenov University, 119991 Moscow, Russia; fjk@inbox.ru (P.K.); telyshev@bms.zone (D.T.)
3. Faculty of Computational Mathematics and Cybernetics, Lomonosov Moscow State University, 119991 Moscow, Russia; richardstallman42@gmail.com
4. Institute of Biomedical Systems, National Research University of Electronic Technology, 124498 Moscow, Russia
5. Marchuk Institute of Numerical Mathematics of the Russian Academy of Sciences, 119991 Moscow, Russia
* Correspondence: simakov.ss@mipt.ru

Received: 1 July 2020; Accepted: 3 August 2020; Published: 10 August 2020

Abstract: Left ventricular assist devices provide circulatory support to patients with end-stage heart failure. The standard operating conditions of the pump imply limitations on the rotation speed of the rotor. In this work we validate a model for three pumps (Sputnik 1, Sputnik 2, Sputnik D) using a mock circulation facility and known data for the pump HeartMate II. We combine this model with a 1D model of haemodynamics in the aorta and a lumped model of the left heart with valves dynamics. The model without pump is validated with known data in normal conditions. Simulations of left ventricular dilated cardiomyopathy show that none of the pumps are capable of reproducing the normal stroke volume in their operating ranges while complying with all criteria of physiologically feasible operation. We also observe that the paediatric pump Sputnik D can operate in the conditions of adult circulation with the same efficiency as the adult LVADs.

Keywords: rotary blood pump; 1D haemodynamics; lumped heart model

1. Introduction

Left ventricular assist devices (LVADs) provide a therapeutic option to treat patients with end-stage heart failure (HF). LVAD connects the left ventricle (LV) and the aortic arch (AA), provides pulsatile or continuous blood flow and maintains circulatory support. The total aortic flow is the sum of the LV and LVAD outflows. Thus, LVAD decreases the work of the LV on ejecting blood to the aorta. The outflow from the LVAD to the aorta depends on the pressure drop over the LVAD. The pressure drop over LVAD is a complex interplay of many factors including LV contraction and ejection, aortic valve function, aorta extensibility and outflow to the distal parts of the systemic arteries. Modern LVADs are the rotary blood pumps (RBPs) which produce continuous flow to maintain temporary and permanent circulatory support [1,2]. The area of the pressure–volume (P-V) loop of the LV represents its stroke work. Dynamic head pressure-bypass flow (H-Q) curves characterise the RBP function during the cardiac cycle. These two curves correlate with each other [3]. They help to analyze dynamic interaction between the LV and RBP [4]. The other useful parameters of such analysis are the stroke work, the hydraulic pump work, and the cardiac mechanical efficiency [5,6].

Clinical efficacy of LVADs has been recently proven [7], although their impact on the cardiovascular system is not always clear. Algorithms of autonomous optimal control for LVADs are still widely discussed. A lot of information on LVADs operation in different regimes is available from mock circulation facilities [8–10]. Blood flow near aortic valve after implantation of LVAD was simulated in [11]. However, systematic data of the impact of pump operation in patient's physiological conditions are limited due to complexity of measurements and a relatively small number of observable cases. In this work we develop an in silico model of the left heart and aorta with LVAD which allows us to simulate the impact of a pump in realistic physiological conditions. Our primary goal is to study behaviour of RBPs Sputnik 1, Sputnik 2 and Sputnik D [8–10,12,13] in patient's physiological conditions. We also compare models of the Sputnik devices with a model of the HeartMate II [14], the well known and widely used pump.

In Section 2 of this work, we present an integrated model of the left heart function with aortic and mitral valves dynamic opening and closing (see Section 2.3). The model includes two segments of the aorta which are simulated by a 1D haemodynamic model (see Section 2.2). We identify the LVAD model for pumps Sputnik 1, Sputnik 2, Sputnik D by fitting parameters with data from mock circulation facilities (see Section 2.1). The parameters of the LVAD HeartMate II model are set according to the literature [14]. The LVAD model is included in the integrated model as a nonlinear lumped compartment which connects the left ventricle and the aorta. In Section 3.1, the heart and aorta model is validated in healthy conditions using data from the literature. Essential conditions of the physiologically feasible pump function include the temporary opening of the aortic valve, the positive direction of the flow through the LVAD as well as absence of ventricular suction and recovery of the typical stroke volume [13]. In Section 3.2, we study the haemodynamic effect of every pump in case of HF accompanied with left ventricular dilated cardiomyopathy in a range of rotational speed which covers ranges defined by manufacturers. We observe that in the considered conditions none of the pumps are capable of restoring the normal stroke volume in the ranges recommended by the manufacturers and at the same time complying with all criteria of physiologically feasible operation. We show that although Sputnik D was initially designed for paediatric patients, it can operate in the conditions of adult circulation at a higher pump speed, with about the same efficiency as adult LVADs. In Section 4 we discuss the results, limitations, conclusions and our future work.

2. Materials and Methods

2.1. Identification of Pump Models

Head pressure–flow rate (H-Q) relationship is a mechanical characteristic of a pump which can be determined from the laboratory tests. It provides a convenient interface for incorporating the pump model to a model of the cardiovascular system as a nonlinear compartment. The two options for deriving the H-Q relationship are the usage of (semi-)empirical formulas and the derivation from the physical principles. In rather general form $H(Q, \omega)$ is a quadratic form which is sometimes extended with the terms of the flow and pump rotation accelerations $H\left(Q, \omega, \frac{dQ}{dt}, \frac{d\omega}{dt}\right)$. Here H is the head pressure, Q is the flow through the pump, ω is the rotation speed of the pump rotor. The Euler head equation with added quadratic term Q^2 due to experimental evidence and the flow inertia term [15] gives Model 1:

$$H = aQ^2 + bQ + c\omega^2 + d\frac{dQ}{dt}. \tag{1}$$

A similar model [16] with the rotational acceleration of the pump defines Model 2:

$$H = aQ^2 + bQ + c\omega^2 + d\frac{dQ}{dt} + e\frac{d\omega}{dt}. \tag{2}$$

The steady-flow model based on the conservation laws of mass, momentum and energy [17] yields Model 3:

$$H = aQ^2 + bQ\omega + c\omega^2. \tag{3}$$

An addition to Model 3 of unsteady-flow effects and periphery parts [14] results in Model 4:

$$P_{lv} - P_p = aQ^2 + bQ\omega + c\omega^2 + d\frac{dQ}{dt} + P_{rec} - H_{per}, \qquad (4)$$

where P_p is the pressure in the junction of the pump outlet and aorta, P_{lv} is the pressure in the LV,

$$P_{rec} = \begin{cases} 0, & Q > e\omega \\ R_{rec}(Q - e\omega)^2, & Q \leqslant e\omega \end{cases}, \qquad (5)$$

$$H_{per} = -L_{per}\frac{dQ}{dt} + R_{per}Q_p|Q_p|. \qquad (6)$$

We note that Model 4 is not actually an H-Q relationship. It also includes parameters of the external (periphery) part which connects the pump to the LV and the aorta. The physical background of (4) is discussed in [14]. The theoretical Euler head equation gives the terms proportional to ω^2 and $Q\omega$. The fluid friction losses produce quadratic growth (Q^2) with the flow elevation. The flow detachment at the leading and trailing edges of the blade produces eddy and separation losses proportional to ω^2, $Q\omega$ and Q^2. Part-load recirculation in the blade channels occurs below the design flow rate. It partly blocks the channels, decreases their effective diameter and increases the head pressure introducing $(Q - e\omega)^2$ term. The flow inertia term is proportional to $\frac{dQ}{dt}$. Fluid friction and inertia frequency-dependent losses in the peripheral part are included via H_{per} in (6). See [14] and references herein for more details.

All the pump models (1)–(4) have physical interpretation. They were successfully validated with experimental data from different pumps in [14–17]. We take all these models as possible candidates for the H-Q mathematical relationship of non-pulsatile axial flow LVADs Sputnik D, Sputnik 1 and Sputnik 2. We use data from laboratory experiments with physical models of the paediatric mock circulation with Sputnik D [8–10] and the adult mock circulation with Sputnik 1 and Sputnik 2 [8,12] for validation. Sensors are placed as close to the pumps as possible, thus, we exclude peripheral term H_{per} from (4) at the model fitting stage. Experimental setup imitates physiological conditions, including the Frank-Starling autoregulation mechanism of the heart which regulates the cardiac output depending on the ventricle preload. The 32% aqueous glycerol solution was used as the model fluid. Head pressure–flow rate (H-Q) curves for Sputnik D, Sputnik 1 and Sputnik 2 were measured at various constant pump speeds. For Sputnik D, the data from the range 6×10^3–12×10^3 rpm with the step 10^3 rpm were used as the training dataset, and the data from the range 13×10^3–15×10^3 rpm were used as the test dataset. For Sputnik 1 and Sputnik 2 the data from the range 5×10^3–10^4 rpm with the step 200 rpm and a contractility factor of the artificial LV 0.25 were used as the training dataset and the data from the range 5×10^3–10^4 rpm and the contractility factor of the artificial LV 0.5 were used as the test dataset. The contractility factor [8] is a coefficient which decreases the end-systolic elasticity.

We set head pressure H as a target variable. The parameters of the models were identified by the damped least-squares method (Levenberg–Marquardt algorithm) [18,19]. We smooth up the raw data by Savitzky–Golay filter [20] for computing time derivatives of the flow and the rotational speed of the pump. The coefficient of determination R^2 was used as the best-fit criterion. According to results presented in Table 1, Model 4 provides the best fit with experimental data for all Sputnik pumps. Table 2 comprises identified parameters of Sputnik pumps for Model 4, as well as Model 4 parameters of the LVAD HeartMate II from [14]. Due to the lack of experimental data for Sputnik pumps periphery, we use mean values of the corresponding parameters for HeartMate II from [14].

In the following sections, we incorporate Model 4 into a lumped model of the heart coupled with a 1D model of the aorta.

Table 1. Coefficients of determination (R^2) for models (1)–(4).

	Sputnik D	Sputnik 1	Sputnik 2
Model 1	−0.2	0.92	0.93
Model 2	−0.19	0.92	0.94
Model 3	0.87	0.37	0.39
Model 4	0.96	0.93	0.97

Table 2. Parameters of model (4).

Parameter	Unit	Sputnik D	Sputnik 1	Sputnik 2	HeartMate II [14]
a	mmHg/(L/min)^2	0.48	3.58×10^{-2}	−0.46	−0.86
b	mmHg/(rpm · L/min)	-1.52×10^{-3}	-9.86×10^{-4}	-5.64×10^{-4}	3.21×10^{-4}
c	mmHg/rpm^2	8.49×10^{-7}	1.74×10^{-6}	1.73×10^{-6}	9.54×10^{-7}
d	mmHg·s^2/L	−60.06	−83.50	−85.91	−22.97
e	(L/min)/rpm	4.92×10^{-5}	-2.18×10^{-4}	-3.70×10^{-4}	3.59×10^{-4}
R_{rec}	mmHg/(L/min)^2	5.63	4.15	5.59	3.07
L_{per}	mmHg·s^2/L	19.33	19.33	19.33	20
R_{per}	mmHg/(L/min)^2	0.35	0.35	0.35	0.38

2.2. 1D Mathematical Model of the Blood Flow in Aorta Segments

The blood flow in the aorta is simulated by a 1D reduced-order model of unsteady flow of viscous incompressible fluid in elastic tubes. The aorta is divided into two segments. The 1D model of the aorta is connected to the LV at the inlet, to the Windkessel compartment at the outlet and to the pump compartment between its segments I and II (see Figure 1).

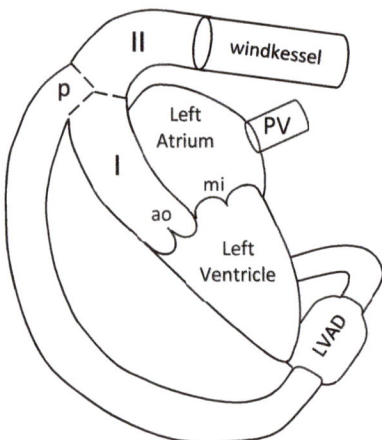

Figure 1. Scheme of the integrated model and notations used throughout the paper: left ventricle (lv), left atrium (la), pulmonary veins (pv), mitral valve (mi), aortic valve (ao), Windkessel compartment (WK), left ventricle assist device (LVAD), pump compartment (p).

Reviews and details of 1D haemodynamic models can be found in [21–25]. Algorithms of patient-specific parameter identification of such models were suggested in [26,27]. In this section, we briefly present this approach. We consider two 1D segments of the aorta which correspond to two parts of the ascending aorta (I and II in Figure 1). We assume that the pump is connected to the aorta at the aortic arch before the carotid arteries. The flow in every vessel is described by mass and momentum conservation equations

$$\frac{\partial \mathbf{V}}{\partial t} + \frac{\partial \mathbf{F}(\mathbf{V})}{\partial x} = \mathbf{G}(\mathbf{V}), \tag{7}$$

$$\mathbf{V} = \begin{pmatrix} A \\ u \end{pmatrix}, \mathbf{F}(\mathbf{V}) = \begin{pmatrix} Au \\ u^2/2 + p(A)/\rho \end{pmatrix}, \mathbf{G}(\mathbf{V}) = \begin{pmatrix} 0 \\ \psi \end{pmatrix},$$

where t is the time, x is the distance along the vessel counted from the vessel junction point, ρ is the blood density (constant), $A(t,x)$ is the vessel cross-section area, p is the blood pressure, $u(t,x)$ is the linear velocity averaged over the cross-section, ψ is the friction force

$$\psi = -8\pi\mu \frac{u}{\rho A}, \tag{8}$$

μ is the dynamic viscosity of the blood. The elasticity of the vessel wall material is characterised by the $p(A)$ relationship

$$p(A) = \rho c_0^2 f(A), \tag{9}$$

where c_0 is the velocity of small disturbances propagation in the vessel wall, $f(A)$ is the monotone S-like function (see [28] for the review of other options)

$$f(A) = \begin{cases} \exp(\eta - 1) - 1, & \eta > 1 \\ \ln \eta, & \eta \leqslant 1 \end{cases}, \eta = \frac{A}{A_0}, \tag{10}$$

A_0 is the cross-sectional area of the unstressed vessel.

The mass conservation condition at the aortic root includes the blood flow through the aortic root Q_{ao} which is also a variable of the heart model from Section 2.3:

$$u_I(t,0) A_I(t,0) = Q_{ao}(t). \tag{11}$$

Boundary conditions at the connection of aorta and the pump include mass conservation condition

$$u_I(t, L_I) A_I(t, L_I) + Q_{pump} = u_{II}(t, 0) A_{II}(t, 0) \tag{12}$$

and the continuity of the total pressure

$$p_I(A_I(t, L_I)) + \frac{\rho u_I^2(t, L_I)}{2} = p_{II}(A_{II}(t, 0)) + \frac{\rho u_{II}^2(t, 0)}{2} = p_p + \frac{\rho}{2}\left(\frac{Q_p}{S_p}\right)^2, \tag{13}$$

where p_p is the static pressure at the output of the pump, Q_p is the flow through the pump contributing to (4), S_p is the cross-section area of the tube which connects the output of the pump and the aorta.

The outflow boundary conditions assume that the terminal part of the aorta is connected to the Windkessel compartment which describes the rest of the systemic circulation

$$\frac{dQ}{dt} = \frac{1}{R_1}\left(\frac{dp_{II}(A_{II}(t, L_{II}))}{dt} - \frac{dp_{WK}}{dt}\right), \tag{14}$$

$$\frac{dp_{WK}}{dt} = \frac{Q}{C}\left(1 + \frac{R_1}{R_2}\right) - \frac{dp_{II}(A_{II}(t, L_{II})) - p_\infty}{R_2 C}, \tag{15}$$

$$Q = u_{II}(t, L_{II}) A_{II}(t, L_{II}), \tag{16}$$

where R_1, R_2, C, p_∞ are parameters presented in Table 3, p_{WK} is pressure in the Windkessel compartment.

The formulations of boundary conditions at the aortic root (11), at the connection of the aorta and the pump (12), (13) and at the terminal part of the aorta (14)–(16) include a numerical discretisation of compatibility condition along the characteristic curve of the hyperbolic system (7) which leaves the integration domain for every incoming and/or outgoing segment of the aorta (see [21–23] for details). The systems of nonlinear algebraic equations, which represent boundary conditions with the time discretisation of the differential part are numerically solved by the Newton's method.

The hyperbolic system (7) inside every segment is numerically solved by the second order grid-characteristic method (see [29] for the details of the method and [21,22] for the features of its implementation to the 1D model of the blood flow). The analysis of the characteristic curves of (7) and similar formulations of 1D blood flow model allows implementing discontinuous Galerkin method [30,31]. The deep analysis of the quasilinear effects in a hyperbolic model blood flow through compliant axi-symmetric vessels can be found in [32]. The generalised approach to the numerical implementation of the models describing various nonlinear wave process on graphs is described in [33].

The parameters of the 1D model are given in Table 3. The cross-sectional area and the length of the aortic segments I and II are set according to [34]. The parameters of the Windkessel compartment are set manually. These values allow us to achieve the well known systolic and diastolic aortal pressures in the normal conditions (rf. Section 3.1). For ρ and μ we use the well known values [35].

Table 3. Parameters of the 1D model of haemodynamics in the segments of the aorta.

Parameter	Unit	Value	Parameter	Unit	Value
$A_{0,I}$	cm^2	7.1	$A_{0,II}$	cm^2	5.7
$c_{0,I}$	cm/s	700	$c_{0,I}$	cm/s	700
L_I	cm	4.4	L_{II}	cm	3
R_1	Ba·s/mL	60	R_2	Ba·s/mL	1500
C	mL/Ba	10^{-3}	p_∞	Ba	7000
ρ	g/cm^3	1.04	μ	cP	4

2.3. Integrated Mathematical Model of the Heart Function, Pump and Aortic Flow

The two chamber model of the heart comprises the LV and the left atrium (LA), the mitral and aortic valves. It connects the pulmonary veins (PV) with the aorta. The nonlinear LVAD compartment connects the LV with the aorta (see Figure 1). The variable elasticity concept of the heart contractions [36,37] allows describing the heart chambers dynamics by the following lumped compartment model

$$I_k \frac{d^2 V_k}{dt^2} + R_k P_k \frac{dV_k}{dt} + E_k(t)\left(V_k - V_k^0\right) + P_k^0 = P_k, \qquad (17)$$

where $k \in \{lv, la\}$, indices lv and la refer to the LV and the LA, respectively, $V_k(t)$ is the volume of the chamber, V_k^0 is the reference volume of the chamber, $P_k(t)$ is the pressure in the chamber, P_k^0 is the reference pressure in the chamber, I_k is the inertia coefficient of the chamber, R_k is the hydraulic resistance coefficient of the chamber, $E_k(t)$ is variable elasticity which is approximated by

$$E_k(t) = E_{k,d} + \frac{E_{k,s} - E_{k,d}}{2} e_k(t), \qquad (18)$$

$E_{k,d}$ and $E_{k,s}$ are elasticity constants related to the end diastolic and end systolic states of chamber k (rf. Table 4). For the LV we set

$$e_{lv}(t) = \begin{cases} 1 - \cos\left(\dfrac{t}{T_{s1}}\pi\right), & 0 \leqslant t \leqslant T_{s1}, \\ 1 + \cos\left(\dfrac{t - T_{s1}}{T_{s2} - T_{s1}}\pi\right), & T_{s1} < t < T_{s2}, \\ 0, & T_{s2} \leqslant t \leqslant T, \end{cases} \qquad (19)$$

whereas for the LA

$$e_{la}(t) = \begin{cases} 0, & 0 \leqslant t \leqslant T_{pb}, \\ 1 - \cos\left(\dfrac{t - T_{pb}}{T_{pw}}2\pi\right), & T_{pb} < t < T. \end{cases} \qquad (20)$$

Here we modify the model [38] by adding to (17) the term proportional to $P_k \frac{dV_k}{dt}$, which accounts for viscoelasticity of the myocardium [39–41]. The values of constants $T_{s1}, T_{s2}, T_{pb}, T_{pw}$ are presented in Table 4.

The mass conservation law for the LV and LA states

$$\begin{aligned} \frac{dV_{lv}}{dt} &= Q_{mi} - Q_{ao} - Q_p, \\ \frac{dV_{la}}{dt} &= Q_{pv} - Q_{mi}, \end{aligned} \qquad (21)$$

where Q_{mi} is the flow through the mitral valve, Q_{ao} is the flow through the aortic valve, Q_p is the flow through the pump, Q_{pv} is the flow from the PV.

We set the pressure drop $\Delta P = P_{pv} - P_{la}$ for PV – LA connection, $\Delta P = P_{la} - P_{lv}$ for LA – LV connection, and $\Delta P = P_{lv} - p(A_I(t,0))$ for LV – AA connection. For unsteady flow in a channel with a variable cross-section, the pressure drop satisfies the relation [39,42]

$$\Delta P = L(g)\frac{dQ}{dt} + \alpha(g)Q + \beta(g)Q|Q|, \qquad (22)$$

where $g(\theta) = \{\theta^{min} \leqslant \theta \leqslant \theta^{max}, 0 \leqslant g(\theta) \leqslant 1\}$ is a smooth monotone function of the angle of a valve opening θ [43]:

$$g(\theta) = \begin{cases} \dfrac{(1 - \cos\theta^{min})^2}{(1 - \cos\theta^{max})^2}, & \theta < \theta^{min}, \\ \dfrac{(1 - \cos\theta)^2}{(1 - \cos\theta^{max})^2}, & \theta^{min} \leqslant \theta \leqslant \theta^{max}, \\ 1, & \theta > \theta^{max}. \end{cases} \qquad (23)$$

The value $g(\theta^{min})$ corresponds to the closed valve, while the value $g(\theta^{max}) = 1$ corresponds to the opened valve. For $L = 0$, $\beta = 0$ we have linear Poiseuille pressure drop condition which also accounts for the viscous friction losses. By analogy with [44,45] we neglect this term and set $\alpha = 0$ for all cases. For $L = 0$, $\alpha = 0$ we have the orifice pressure drop condition. The first term in (22) accounts for the inertia of non-stationary flow. The coefficient β is defined as [42,46,47]

$$\beta(A_*) = \frac{\rho}{2B_*}\left(\frac{1}{\tilde{A}_*} - \frac{1}{A_*}\right)^2, \qquad (24)$$

where parameters $\tilde{A}_{mi}, B_{ao}$ and B_{mi} are defined in Table 4 whereas $\tilde{A}_{ao} = A_I(t,0)$. For the PV – LA connection $\beta = const$. For both mitral and aortic valves, their cross-section A_* depends on the angle of the valve opening, $A_*(\theta) = A_*^{max}g(\theta)$.

The dynamics of the aortic and mitral valves is governed by the second Newton law. Pressure gradient across the valve, vorticity generation and shear forces acting on the valve leaflets [48] have to be accounted in the model, rf. [43,49]. In this work we set valve dynamics equations as

$$\frac{d^2\theta_{ao}}{dt^2} = -K_{ao}^f \frac{d\theta_{ao}}{dt} + (P_{lv} - P_{ao}) K_{ao}^p \cos\theta_{ao} - F_{ao}^r(\theta_{ao}),$$
$$\frac{d^2\theta_{mi}}{dt^2} = -K_{mi}^f \frac{d\theta_{mi}}{dt} + (P_{la} - P_{lv}) K_{mi}^p \cos\theta_{mi} - F_{mi}^r(\theta_{mi}). \tag{25}$$

where $\theta_{ao}(t)$ is the angle of the aortic valve opening, $\theta_{mi}(t)$ is the angle of the mitral valve opening, K_{ao}^f, K_{ao}^p, K_{mi}^f, K_{mi}^p are parameters presented in Table 4, the first term at the right-hand side corresponds to the friction force, the second term corresponds to the pressure force driving the valve motion, F^r is the force which helps to avoid physiologically abnormal valve positions ($\theta < \theta^{min}$ and $\theta > \theta^{max}$)

$$F^r(\theta) = \begin{cases} 0, & \theta^{min} \leqslant \theta \leqslant \theta^{max}, \\ e^{10^3(\theta - \theta^{max})} - 1, & \theta > \theta^{max}, \\ 1 - e^{10^3(\theta^{min} - \theta)}, & \theta < \theta^{min}. \end{cases} \tag{26}$$

The other forces are neglected.

Parameters of the lumped model of the left heart are summarized in Table 4. We take some values from [38,41,43] and set other values manually basing on the values from [38,43] and keeping them in the physiological range.

Table 4. Parameters of the lumped model of the left heart. * Parameter is set manually.

Parameter	Unit	Value	Reference	Parameter	Unit	Value	Reference
$E_{lv,s}$	$\frac{\text{mm Hg}}{\text{mL}}$	4.0	*	θ_{ao}^{min}		0°	[38]
$E_{lv,d}$	$\frac{\text{mm Hg}}{\text{mL}}$	0.09	*	θ_{ao}^{max}		75°	[43]
I_{lv}	$\frac{\text{mm Hg}\cdot\text{s}^2}{\text{mL}}$	10^{-7}	*	θ_{mi}^{min}		0°	[38]
R_{lv}	$\frac{\text{s}}{\text{mL}}$	1.5×10^{-3}	[41]	θ_{mi}^{max}		75°	[43]
$E_{la,s}$	$\frac{\text{mm Hg}}{\text{mL}}$	1.2	*	V_{lv}^0	mL	5	*
$E_{la,d}$	$\frac{\text{mm Hg}}{\text{mL}}$	0.3	*	V_{la}^0	mL	4	*
I_{la}	$\frac{\text{mm Hg}\cdot\text{s}^2}{\text{mL}}$	10^{-7}	*	T_{s1}	s	0.3	[43]
R_{la}	$\frac{\text{s}}{\text{mL}}$	1.5×10^{-3}	[41]	T_{s2}	s	0.35	[43]
T_{pw}	s	0.1	[43]	T_{pb}	s	0.9	[43]
K^p	$\frac{\text{rad}}{\text{s}^2 \cdot \text{mm Hg}}$	10^4	*	K^f	$\frac{\text{rad}}{\text{s}}$	50	[43]
P_{pv}	mm Hg	13	*	S_p	cm^2	1.1	*
L_{pv}	$\frac{\text{mm Hg}\cdot\text{s}^2}{\text{mL}}$	10^{-2}	*	β_{pv}	$\frac{\text{mm Hg}\cdot\text{s}^2}{\text{mL}^2}$	4×10^{-4}	*
L_{mi}	$\frac{\text{mm Hg}\cdot\text{s}^2}{\text{mL}}$	5×10^{-10}	*	B_{mi}		300	*
L_{ao}	$\frac{\text{mm Hg}\cdot\text{s}^2}{\text{mL}}$	5×10^{-5}	*	B_{ao}		500	*
$\tilde{A}_{mi}$	cm^2	5	*	A_{ao}^{max}	cm^2	4	*
T	s	1	*	A_{mi}^{max}	cm^2	4	*

3. Results

Validation of the pump model is addressed in Section 2.1. In Section 3.1, we validate the integrated model by comparing simulations of the normal heart function with known physiological data from the literature. In Section 3.2, we analyze the effect of four LVADs in patients with the end-stage HF associated with the LV dilated cardiomyopathy (DCM).

3.1. Validation of the Model

The parameters of the integrated model for healthy conditions are shown in Tables 2–4. Simulations with these parameters without LVAD produce values of stroke volume of LV, systolic and diastolic pressures in the aortic root which are in a good agreement with the well-known physiological data [35,50] (rf. Table 5). We observe remarkable difference in the end systolic and end diastolic volumes of the LV. We note, that these parameters are highly individual. Even in healthy cases they depend on many factors including age, sex, sports lifestyle etc. [50,51]. Our values fall in the physiological range: they are typical for men of the age 70–80, women of the age about 40–70 [51] and other individual cases.

The PV diagram of the LV and time curves of the LV volume, the aortic flow and the aortic pressure in the terminal point of segment II (rf. Figure 1) in healthy conditions without a pump are shown in Figure 2. The loop in the lower right part of the PV diagram (rf. Figure 2a) accounts for the backflow from LV to LA through the mitral valve in early systole. This backflow is a result of non-instant closing of the mitral valve. Such loop is observed in critically ill patients. It is typical both for the right ventricle and the atria, but also it may be monitored in the left ventricle [52]. We also observe aortic regurgitation at the end of systole which accounts for the non-instant closing of the aortic valve [35,50] (rf. Figure 2c for $t \approx 0.3$ s).

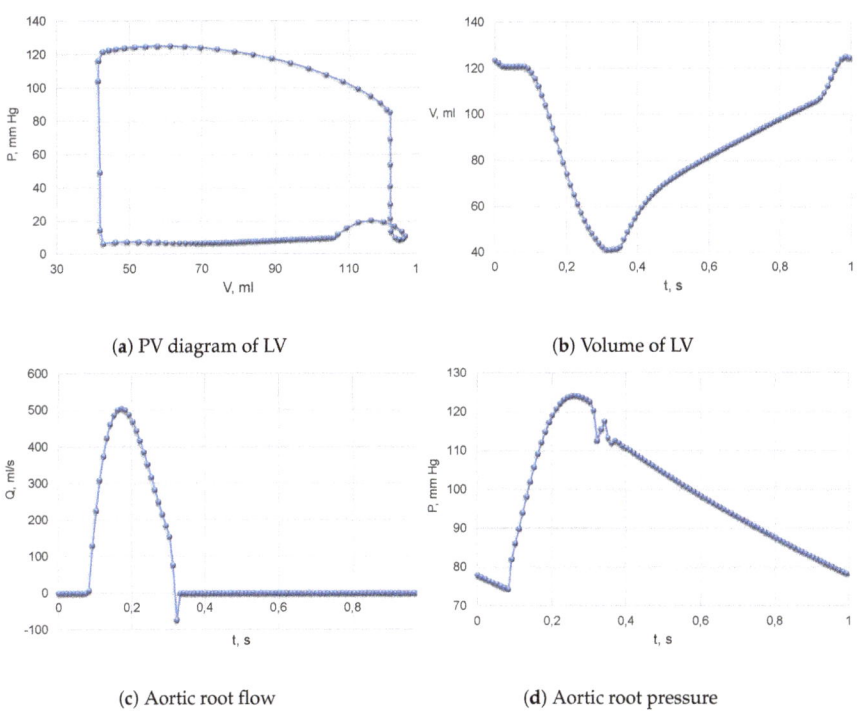

(a) PV diagram of LV

(b) Volume of LV

(c) Aortic root flow

(d) Aortic root pressure

Figure 2. Validation of the integrated model for the heart and the aorta.

Table 5. Validation of the integrated model for the heart and the aorta in healthy conditions without LVAD.

Parameter	Unit	[35,50]	Model
End systolic LV volume	mL	50–70	42
End diastolic LV volume	mL	130	121
Stroke volume of LV	mL	60–80	79
Systolic pressure in the aortic root	mm Hg	130	124
Diastolic pressure in the aortic root	mm Hg	78	76

3.2. Haemodynamic Simulations in the Aorta for LV DCM with LVAD

We simulate the haemodynamic characteristics in the left heart and the aorta in the presence of one of four LVADs (Sputnik 1, Sputnik 2, Sputnik D and HeartMate II) operating at various rotation speeds under HF conditions. RBPs are applied as long term circulatory support systems in patients with end-stage HF both as a bridge to heart transplantation and as an alternative to heart transplantation [53,54]. Some types of HF are accompanied with LV DCM which is the common indication for the long term LVAD installation. LV DCM is characterized by decreased LV contractility, thinning of the LV wall and increased cavity volume of the LV. These changes produce substantial decrease in the cardiac output and related cardiovascular dysfunction. LVAD unloads the LV and decreases its volume by pumping a portion of blood to the aorta. Thus, it supports the heart and sometimes may produce conditions for LV wall recovery. We update some parameters of the heart with LV DCM as shown in Table 6. The other parameters from Table 4 remain intact. The simulations with these parameters produce values which correlate with physiological data from the literature [53,54] (see Table 7 for comparison).

Table 6. Parameters of the heart model with LV DCM.

Parameter	Unit	Value
P_{pv}	mm Hg	10
$E_{lv,d}$	mm Hg/mL	0.04
$E_{lv,s}$	mm Hg/mL	0.44
$V_{0,lv}$	mL	20
$E_{la,s}$	mm Hg/mL	1.1
R_{lv}	s/mL	5×10^{-4}
R_{la}	s/mL	5×10^{-4}

Table 7. Comparison of LV DCM simulations without LVAD with the data from the literature [53,54].

Parameter	Unit	[13,53,54]	Model
End systolic LV volume	mL	215	227
End diastolic LV volume	mL	259	275
Stroke volume of LV	mL	44	48
Systolic pressure in the aortic root	mm Hg	83	81
Diastolic pressure in the aortic root	mm Hg	55	47

We compare the function of the four pumps by setting the same parameters of the heart function and the aorta for all cases. The normal operating conditions of the pump are defined according to [13]. In the context of our model we formulate them as follows:

1. The aortic valve should be opened within a part of cardiac cycle, i.e., the LV should eject some portion of blood to the aorta.
2. The flow through the pump should be positive, i.e., it always should be directed from the LV to the aorta.
3. The ventricular suction is not admitted.

4. The total ejected volume per cardiac cycle in the aorta should be possibly close to the physiological value in normal conditions (see Section 3.1).

Figure 3 shows the duration of the opening of the aortic and mitral valves. Duration of the opening of the aortic and mitral valves at zero speed of the pumps are in agreement with the data on valve function in the normal conditions [43,49]. Decrease of time of the aortic valve opening starts at 5×10^3 rpm for all four pumps. The permanent closure of the aortic valve is observed at 7×10^3 rpm for Sputnik 1 and Sputnik 2, at 12×10^3 rpm for Sputnik D and at 8×10^3 rpm for HeartMate II which produce the upper bounds for the settings of these devices in the considered conditions. The noticeable increase of the mitral valve opening time starts at 9.5×10^3 rpm for all four pumps. The permanent opening of the mitral valve is observed beyond the values 12×10^3 rpm for Sputnik 1, 11.5×10^3 rpm for Sputnik 2, 13.5×10^3 rpm for HeartMate II and is not observed for Sputnik D in the whole range.

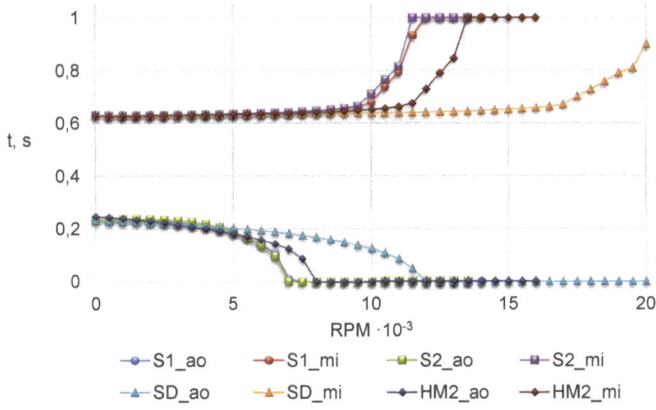

Figure 3. Duration of the opening of the aortic and mitral valves. S1—Sputnik 1, S2—Sputnik 2, SD — Sputnik D, HM2 — HeartMate II. The lower index ao stands for aortic valve, mi stands for mitral valve.

Figure 4 shows the minimum flow through the pump over the cardiac cycle. We observe that a negative flow through the pump disappears at 5.75×10^3 rpm for Sputnik 1 and Sputnik 2, at 8.5×10^3 rpm for Sputnik D and at 7×10^3 rpm for HeartMate II which sets the lower bound for the settings of these devices in the considered conditions.

Figure 5 shows the volume of the blood which is ejected through the aortic valve, through the pump and the total volume ejected to the ascending aorta per cardiac cycle. These volumes are calculated as the time integral of the corresponding flow. Thus, the negative value of the volume ejected through the pump means that the pump takes the blood from aorta back to the LV (see Figure 4). The total volume ejected to the ascending aorta per cardiac cycle is an analog of the stroke volume (SV). In the rest of the paper we refer to it as SV. We observe zero value of the volume ejected through the aortic valve for the values of the LVAD rotation speed which correspond to the permanent closure of the aortic valve (see Figure 3). The normal physiological value of the SV is observed at 8.5×10^3 rpm for Sputnik 1 and Sputnik 2, at 15×10^3 rpm for Sputnik D and at 10^4 rpm for HeartMate II. This means that none of the four pumps can produce the normal SV in the considered conditions jointly with the opening of the aortic valve (see Figure 3).

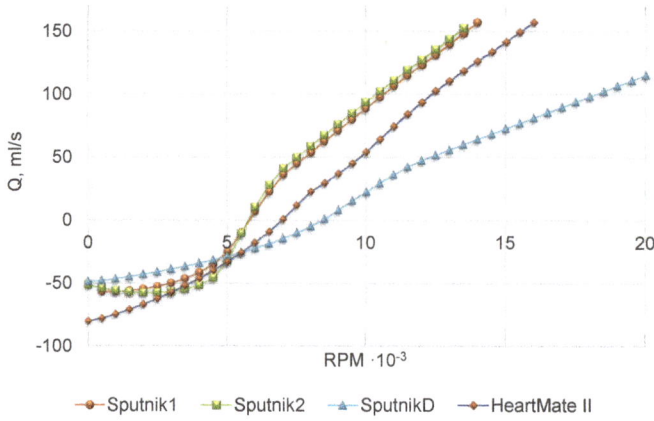

Figure 4. The minimum flow through the pump.

It is interesting to observe (see Figure 5) that the volume of blood passing through the aortic valve is close to the normal value at zero rotation speed of the pump rotor. This effect is due to the negative flow through the pump from the aorta back to the LV (rf. Figure 4). From Figure 5 we notice that the pumps with speed below 5×10^3 rpm produce the same SV as LV DCM model without a pump. The effect of the pumps on the SV becomes significant only after decrease of the aortic valve opening time (rf. Figure 3).

Figure 6 shows the systolic and diastolic pressures in the aortic root. We observe that in a particular range (up to 7×10^3 rpm for Sputnik 1 and Sputnik 2, up to 12×10^3 rpm for Sputnik D and up to 8×10^3 rpm for HeartMate II) the pump increases the diastolic pressure. The systolic pressure remains the same in this range. At the upper bound of this range, the systolic and diastolic pressures are almost equal, and the pulse pressure tends to zero. The elevation of the systolic pressure starts at the rotation speed corresponding to the permanent closure of the aortic valve (rf. Figure 3). The pulse pressure remains almost zero. The flow becomes non-pulsatile.

Figure 7 shows the work of the LV. It decreases with the increase of the pump speed and, thus, with the rise of the pump work. We observe three specific values of the pump speed. For values below 5×10^3 rpm, the LV work is almost constant due to the decreasing time of the aortic valve opening (rf. Figure 3). For values above 5×10^3 rpm, the LV work decreases. The kink of the work curve at 7×10^3 rpm for Sputnik 1 and Sputnik 2, 12×10^3 rpm for Sputnik D and 8×10^3 rpm for HeartMate II occurs due to the permanent closure of the aortic valve (rf. Figure 3). The LV performs almost zero work at 12×10^3 rpm for Sputnik 1 and Sputnik 2, 14×10^3 rpm for HeartMate II. Zero LV work is not observed for Sputnik D due to the permanent opening of the mitral valve (rf. Figure 3).

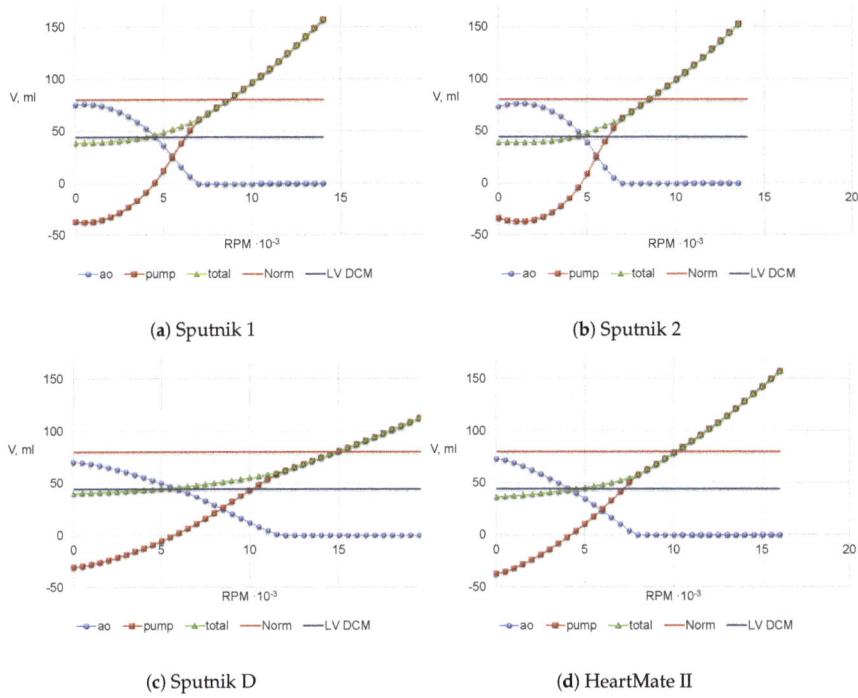

Figure 5. Ejected volume per cardiac cycle through the aortic valve, through the pump and the total value for the segment II of the aorta (see Figure 1). Norm—normal (healthy) value, LV DCM—left ventricular dilated cardiomyopathy value without pump.

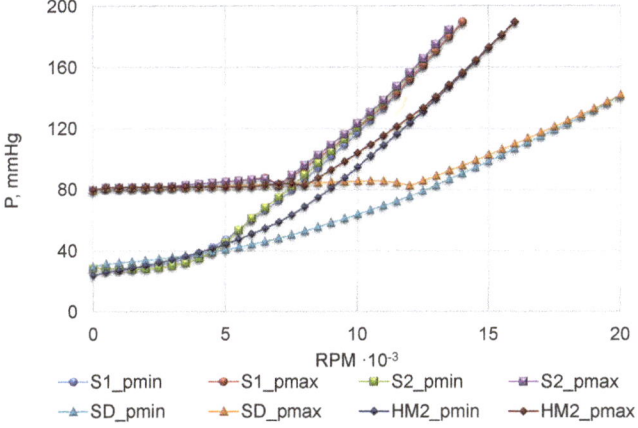

Figure 6. The systolic and diastolic pressures in the aortic root.

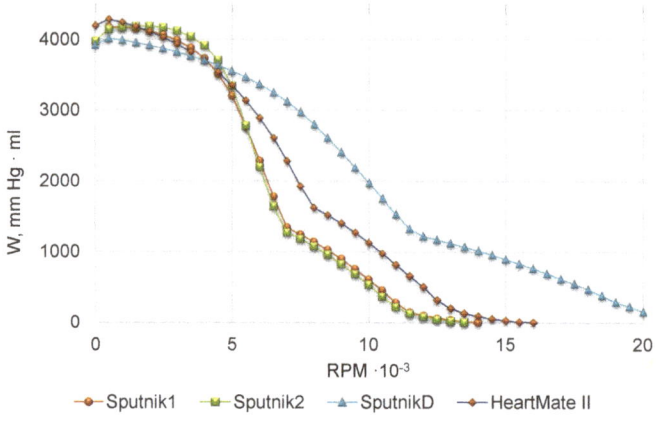

Figure 7. The work of the LV.

4. Discussion

In this work we study the impact of LVADs Sputnik 1, Sputnik 2, Sputnik D and HeartMate II on haemodynamics in the left heart and the aorta at various pump speeds. For every pump we validate the model of pressure – flow relationship using data from physical experiments. The pump model is combined with the 1D model of the aorta and the lumped model of the left heart with valves dynamics. The model without the pump reproduces successfully the known physiological characteristics of the heart in the healthy conditions: PV diagram of the LV, the LV volume, the aortic flow and the aortic pressure. We fit the model parameters to the LV DCM conditions and perform haemodynamic simulations with the pumps.

We observe different regimes of the heart function depending on the value of the pump speed. At low rotation speeds, the work of the pump is insufficient, and we observe reverse blood flow from the aorta to the LV through the pump. Formally, the volume of the blood ejected through the aortic valve is close to the standard value, but the pump backflow decreases the total amount of blood ejected to segment II of the aorta and makes it close to the value of LV DCM case without the pump. Increase of the rotation speed reduces the time of the aortic valve opening until valve's permanent closure. Three of the four operating conditions of the normal pump functioning (see Section 3.2) hold within this range. Unfortunately, none of the pumps are capable of recovering the standard SV within this range. Further increase of the rotation speed produces non-pulsatile flow and permanent opening of the mitral valve. The technical characteristics and clinical restrictions limit the in vivo change of the rotation speed. For Sputnik 1 and Sputnik 2 the range is 5×10^3–10^4 rpm, for Sputnik D the range is 6×10^3–2×10^4 rpm, for HeartMate II the range is 6×10^3–15×10^3 rpm.

In general, we observe the slight difference between Sputnik 1 and Sputnik 2 in all parameters (see Figures 3–7). These pumps produce similar impacts on the haemodynamics, although they have different technical characteristics. Therefore, Sputnik 2 should be preferred as it provides such benefits as lesser weight, size, etc. [8–10]. The HeartMate II has a wider range of the rotation speed with acceptable operating conditions and provides flexibility in tuning the device settings. However, this LVAD recovers the normal SV at a higher rotation speed (see Figures 5a,b,d, which may require more energy. Sputnik D was initially designed for paediatric patients. It has lesser weight and size, but it also has less power. In our simulations, we test this LVAD in the adult LV DCM conditions. We show that it is possible to achieve heart operating conditions which are similar to the conditions with the adult LVADs. Sputnik D produces these conditions at a higher pump speed.

In this work we validate the model using the general characteristics of the heart function in the healthy and LV DCM conditions and data from the test facility. Personalized parameters may produce a more accurate assessment of LVAD impact.

The following parameters cause a substantial impact on the haemodynamics: elasticity of the aorta and the heart chambers, the central venous pressure, insufficiency of the heart valves, parameters of the peripheral circulation (Windkessel model). The dependency of LVAD function on these parameters should be studied in a future work.

In Section 3.2 we observe that the normal SV is achieved at zero pulse pressure and the closed aortic valve. Healthy blood flow in systemic arteries is a nonlinear wave phenomenon. The absence of pulsations may decrease the blood velocity and promote conditions for blood coagulation and thrombus formation in distal arteries. Thus, the loss of pressure and flow pulsatility in the systemic circulation in the presence of LVAD should be analyzed.

The variation of the pump speed changes the haemodynamic parameters in the heart and the aorta. These changes may activate regulatory mechanisms of the heart function. For instance, the change of the heart rate is associated with a modified duration of the systole and the value of the LV output through the aortic valve. These effects are beyond the scope of this work.

The changes in the SV and the aortic pressure modify the central venous pressure which, in turn, changes the LA filling conditions and adjusts the other haemodynamic parameters of the heart function. A closed model of the cardiovascular system is needed to study such feedback system.

Author Contributions: Conceptualization, P.K., D.T. and Y.V.; methodology, S.S., P.K.; software, A.T. and T.G.; validation, S.S., P.K., A.T. and T.G.; formal analysis, P.K., D.T., Y.V. and S.S.; writing–original draft preparation, S.S., A.T., Y.V.; writing–review and editing, S.S., A.T., T.G., D.T. and Y.V.; visualization, S.S. and A.T.; supervision, Y.V. All authors have read and agreed to the published version of the manuscript.

Funding: This research was funded by Russian Foundation for Basic Research grant numbers 18-00-01661, 18-00-01524, 18-00-01659, 18-31-20048, and 19-51-45001 and the world-class research center "Moscow Center for Fundamental and Applied Mathematics" (agreement with the Ministry of Education and Science of the Russian Federation No. 075-15-2019-1624).

Conflicts of Interest: The authors declare no conflict of interest.

Abbreviations

The following abbreviations are used in this manuscript:

AA	Aortic arch
DCM	Dilated cardiomyopathy
HF	Heart failure
LA	Left atrium
LV	Left ventricle
LVAD	Left ventricular assist device
RBP	Rotary blood pump
SV	Stroke volume
PV	Pulmonary veins

References

1. Giridharan, G.A.; Lee, T.J.; Ising, M.; Sobieski, M.A.; Koenig, S.C.; Gray, L.A.; Slaughter, M.S. Miniaturization of mechanical circulatory support systems. *Artif. Organs* **2012**, *36*, 731–739. [CrossRef] [PubMed]
2. Misgeld, B.J.E.; Rüschen, D.; Schwandtner, S.; Heinke, S.; Walter, M.; Leonhardt, S. Robust decentralised control of a hydrodynamic human circulatory system simulator. *Biomed. Signal Process. Control.* **2015**, *20*, 35–44. [CrossRef]
3. Jahren, S.E.; Ochsner, G.; Shu, F.; Amacher, R.; Antaki, J.F.; Vandenberghe, S. Analysis of pressure head-flow loops of pulsatile rotodynamic blood pumps. *Artif. Organs* **2014**, *38*, 316–326. [CrossRef] [PubMed]

4. Yokoyama, Y.; Kawaguchi, O.; Kitao, T.; Kimura, T.; Steinseifer, U.; Takatani S. Prediction of the external work of the native heart from the dynamic H-Q curves of the rotary blood pumps during left heart bypass. *Artif. Organs* **2010**, *34*, 766–777. [CrossRef]
5. Amacher, R.; Weber, A.; Brinks, H.; Axiak, S.; Ferreira, A.; Guzzella, L.; Carrel, T.; Antaki, J.; Vandenberghe, S. Control of ventricular unloading using an electrocardiogram-synchronized Thoratec paracorporeal ventricular assist device. *J. Thorac. Cardiovasc. Surg.* **2013**, *146*, 710–717. [CrossRef]
6. Moscato, F.; Vollkron, M.; Bergmeister, H.; Wieselthaler, G.; Leonard, E.; Schima, H. Left ventricular pressure? Volume loop analysis during continuous cardiac assist in acute animal trials. *Artif. Organs* **2007**, *31*, 369–376. [CrossRef]
7. Teuteberg, J.J.; Cleveland, J.C., Jr.; Cowger, J.; Higgins, R.S.; Goldstein, D.J.; Keebler, M.; Kirklin, J.K.; Myers, S.L.; Salerno, C.T.; Stehlik, J.; et al. The Society of Thoracic Surgeons Intermacs 2019 Annual Report: The Changing Landscape of Devices and Indications. *Ann. Thorac. Surg.* **2020**, *109*, 649–660. [CrossRef]
8. Pugovkin, A.A.; Markov, A.; Selishchev, S.V.; Korn, L.; Walter, M.; Leonhardt, S.; Bockeria, L.A.; Bockeria, O.L.; Telyshev, D.V. Advances in hemodynamic analysis in cardiovascular diseases investigation of energetic characteristics of adult and pediatric Sputnik left ventricular assist devices during mock circulation support. *Cardiol. Res. Pract.* **2019**, *2019*, 4593174. [CrossRef]
9. Telyshev, D.; Denisov, M.; Pugovkin, A.; Selishchev, S.; Nesterenko, I. The progress in the novel pediatric rotary blood pump sputnik development. *Artif. Organs* **2018**, *42*, 432–443. [CrossRef]
10. Telyshev, D.V.; Pugovkin, A.A.; Selishchev, S.V. A mock circulatory system for testing pediatric rotary blood pumps. *Biomed. Eng.* **2017**, *51*, 83–87. [CrossRef]
11. Quaini, A.; Čanić, S.; Paniagua, D. Numerical charcterisation of hemodynamics conditions near aortic valve after implantation of left ventricular assist device. *Math. Biosci. Eng.* **2011**, *8*, 785–806. [PubMed]
12. Selishchev, S.V.; Telyshev, D.V. Optimisation of the Sputnik VAD design. *Int. J. Artif. Organs* **2016**, *39*, 407–414. [CrossRef] [PubMed]
13. Telyshev, D.; Petukhov, D.; Selishchev, S. Numerical modeling of continuous-flow left ventricular assist device performance. *Int. J. Artif. Organs* **2019**, *42*, 611–620. [CrossRef] [PubMed]
14. Boës, S.; Thamsen, B.; Haas, M.; Daners, M.S.; Meboldt, M.; Granegger, M. Hydraulic characterization of implantable rotary blood pumps. *IEEE Trans. Bio-Med. Eng.* **2018**, *66*, 1618–1627. [CrossRef] [PubMed]
15. Moscato, F.; Danieli, G.A.; Schima, H. Dynamic modeling and identification of an axial flow ventricular assist device. *Int. J. Artif. Organs* **2009**, *32*, 336–343. [CrossRef]
16. Pirbodaghi, T.; Mathematical modeling of rotary blood pumps in a pulsatile in vitro flow environment. *Artif. Organs* **2017**, *41*, 710–716. [CrossRef]
17. Shi, Y.; Korakianitis, T. Impeller-pump model derived from conservation laws applied to the simulation of the cardiovascular system coupled to heart-assist pumps. *Comput. Biol. Med.* **2018**, *93*, 127–138. [CrossRef]
18. Levenberg, K. A method for the solution of certain non-linear problems in least squares. *Q. Appl. Math.* **1944**, *2*, 164–168. [CrossRef]
19. Marquardt, D. An algorithm for least-squares estimation of nonlinear parameters. *SIAM J. Appl. Math.* **1963**, *11*, 431–441. [CrossRef]
20. Savitzky, A.; Golay, M.J.E. Smoothing and differentiation of data by simplified least squares procedures. *Anal. Chem.* **2016**, *36*, 1627–1639. [CrossRef]
21. Vassilevski, Y.; Olshanskii, M.; Simakov, S.; Kolobov, A.; Danilov, A. *Personalized Computational Hemodynamics: Models, Methods, and Applications for Vascular Surgery and Antitumor Therapy*; Academic Press: Cambridge, MA, USA, 2020.
22. Bessonov, N.; Sequeira, A.; Simakov, S.; Vassilevski, Yu, ; Volpert, V. Methods of blood flow modelling. *Math. Model. Nat. Phenom.* **2016**, *11*, 1–25. [CrossRef]
23. Gamilov, T.; Simakov, S. Blood flow under mechanical stimulations. *Adv. Intell. Syst. Comput.* **2020**, *1028*, 143–150. [CrossRef]
24. Simakov, S.S. Modern methods of mathematical modeling of blood flow using reduced order methods. *Comput. Res. Model.* **2018**, *10*, 581–604. [CrossRef]
25. Van de Vosse, F.N.; Stergiopulos, N. Pulse wave propagation in the arterial tree. *Annu. Rev. Fluid Mech.* **2011**, *43*, 467–499. [CrossRef]
26. Danilov, A.; Ivanov, Y.; Pryamonosov, R.; Vassilevski, Y. Methods of graph network reconstruction in personalized medicine. *Int. J. Numer. Methods Biomed. Eng.* **2016**, *32*, e02754. [CrossRef] [PubMed]

27. Gamilov, T.; Alastruey, J.; Simakov, S. Linear optimization algorithm for 1D hemodynamics parameter estimation. In *Proceedings of the 6th European Conference on Computational Mechanics: Solids, Structures and Coupled Problems, ECCM 2018 and 7th European Conference on Computational Fluid Dynamics, ECFD*; Owen, R., De Borst, R., Reese, J., Pearce, C., Eds.; CIMNE: Barcelona, Spain, 2018.
28. Vassilevski, Y.V.; Salamatova, V.Y.; Simakov, S.S. On the elasticity of blood vessels in one-Dimensional problems of hemodynamics. *Comput. Math. Math. Phys.* **2015**, *55*, 1567–1578. [CrossRef]
29. Magomedov, K.M.; Kholodov, A.S. *Grid-Characteristic Numerical Methods*; Urite: Moscow, Russia, 2018.
30. Sherwin, S.J.; Franke, V.; Peiró, J.; Parker, K. One-dimensional modelling of a vascular network in space-time variables. *J. Eng. Math.* **2003**, *47*, 217–250. [CrossRef]
31. Sherwin, S.J.; Formaggia, L.; Peiró, J.; Franke, V. Computational modelling of 1D blood flow with variable mechanical properties and its application to the simulation of wave propagation in the human arterial system. *Numer. Methods Fluids* **2003**, *43*, 673–700. [CrossRef]
32. Canić, S.; Kim, E.H. Mathematical analysis of the quasilinear effects in a hyperbolic model blood flow through compliant axi-symmetric vessels. *Math. Methods Appl. Sci.* **2003**, *26*, 1161–1186. [CrossRef]
33. Kholodov, Y.A. Development of network computational models for the study of nonlinear wave processes on graphs. *Comput. Res. Model.* **2019**, *11*, 777–814. [CrossRef]
34. Boileau, E.; Nithiarasu, P.; Blanco, P.J.; Müller, L.O.; Fossan, F.E.; Hellevik, L.R.; Donders, W.P.; Huberts, W.; Willemet, M.; Alastruey, J. A benchmark study of numerical schemes for one-dimensional arterial blood flow modelling. *Int. J. Numer. Methods Biomed. Eng.* **2015**, *31*, e02732. [CrossRef] [PubMed]
35. Schmidt, R.F.; Thews, G. *Human Physiology*, 2nd ed.; Springer-Verlag: Berlin/Heidelberg, Germany, 1989; Volume 2.
36. Suga, H. Cardiac energetics: From E_{MAX} to pressure-volume area. *Clin. Exp. Pharmacol. Physiol.* **2003**, *30*, 580–585. [CrossRef] [PubMed]
37. Walley, K.R. Left ventricular function: Time-varying elastance and left ventricular aortic coupling. *Crit. Care* **2016**, *20*, 1–11. [CrossRef] [PubMed]
38. Simakov, S.S. Lumped parameter heart model with valve dynamics. *Russ. J. Numer. Anal. Math. Model.* **2019**, *34*, 289–300. [CrossRef]
39. Liang, F.; Takagi, S.; Himeno, R.; Liu, H. Multi-scale modeling of the human cardiovascular system with applications to aortic valvular and arterial stenoses. *Med. Biol. Eng. Comput.* **2009**, *47*, 743–755. [CrossRef]
40. Sun, Y.; Beshara, M.; Lucariello, R.J.; Chiaramida, S.A. A comprehensive model for right-left heart interaction under the influence of pericardium and baroreflex. *Am. J. Physiol.* **1997**, *272*, H1499–H1515 [CrossRef]
41. Shroff, S.G.; Janicki, J.S.; Weber, K.T. Evidence and quantitation of left ventricular systolic resistance. *Am. J. Physiol.-Heart Circ. Physiol.* **1985**, *249*, H358–H370. [CrossRef]
42. Young, D.F.; Tsai, F.Y. Flow characteristics in models of arterial stenoses—II. Unsteady flow. *J. Biomech.* **1973**, *6*, 547–559. [CrossRef]
43. Korakianitis, T.; Shi, Y. Numerical simulation of cardiovascular dynamics with healthy and diseased heart valves. *J. Biomech.* **2006**, *39*, 1964–1982. [CrossRef]
44. Mynard, J.P.; Davidson, M.R.; Penny, D.J.; Smolich, J.J. A simple, versatile valve model for use in lumped parameter and one-Dimensional cardiovascular models. *Int. J. Numer. Methods Biomed. Eng.* **2012**, *28*, 626–641. [CrossRef]
45. Sun, Y.; Sjöberg, B.J.; Ask, P.; Loyd, D.; Wranne, B. Mathematical model that characterizes transmitral and pulmonary venous flow velocity patterns. *Am. J. Physiol.* **1995**, *268*, H476–H489. [CrossRef] [PubMed]
46. Seeley, B.D.; Young, D.F. Effect of geometry on pressure losses across models of arterial stenoses. *J. Biomech.* **1976**, *9*, 439–448 [CrossRef]
47. Young, D.F.; Tsai, F.Y. Flow characteristics in models of arterial stenoses—I. Steady flow. *J. Biomech.* **1973**, *6*, 395–410. [CrossRef]
48. Shi, Y.; Lawford, P.; Hose, R. Review of zero-D and 1-D models of blood flow in the cardiovascular system. *Biomed. Eng. Online* **2011**, *10*, 33. [CrossRef] [PubMed]
49. Korakianitis, T.; Shi, Y. A concentrated parameter model for the human cardiovascular system including heart valve dynamics and atrioventricular interaction. *Med Eng. Phys.* **2006**, *28*, 613–628. [CrossRef]
50. Barret, K.; Brooks, H.; Boitano, S.; Barman, S. *Ganong's Review of Medical Physiology*, 23rd, ed.; The McGraw-Hill: New York, NY, USA, 2010.

51. Maceira, A.M.; Prasad, S.K.; Khan, M.; Pennell, D.J. Normalized left ventricular systolic and diastolic function by steady state free precession cardiovascular magnetic resonance. *J. Cardiovasc. Magn. Reson.* **2006**, *8*, 417–426. [CrossRef]
52. García M.I.M., Santos A., Understanding ventriculo-arterial coupling. *Ann. Transl. Med.* **2020**, *8*, 795. [CrossRef]
53. Martina, J.R.; Bovendeerd, P.H.M.; de Jonge, N.; de Mol, B.A.J.M.; Lahpor, J.R.; Rutten, M.C.M. Simulation of changes in myocardial tissue properties during left ventricular assistance with a rotary blood pump. *Artif. Organs* **2013**, *37*, 531–540. [CrossRef]
54. Cox, L.G.E.; Loerakker, S.; Rutten, M.C.M.; de Mol, B.A.J.M.; van de Vosse, F.N. A mathematical model to evaluate control strategies for mechanical circulatory support. *Artif. Organs* **2009**, *33*, 593–603. [CrossRef]

© 2020 by the authors. Licensee MDPI, Basel, Switzerland. This article is an open access article distributed under the terms and conditions of the Creative Commons Attribution (CC BY) license (http://creativecommons.org/licenses/by/4.0/).

Article

Mathematical Modelling of the Structure and Function of the Lymphatic System

Anastasia Mozokhina [1,*] and Rostislav Savinkov [1,2]

1. Peoples' Friendship University of Russia (RUDN University), S.M. Nikolsky Mathematical Institute, 6 Miklukho-Maklaya St, 117198 Moscow, Russia
2. Marchuk Institute of Numerical Mathematics of the Russian Academy of Sciences, 119991 Moscow, Russia; r.savinkov@inm.ras.ru
* Correspondence: asm@cs.msu.ru

Received: 20 July 20; Accepted: 28 August 2020; Published: 1 September 2020

Abstract: This paper presents current knowledge about the structure and function of the lymphatic system. Mathematical models of lymph flow in the single lymphangion, the series of lymphangions, the lymph nodes, and the whole lymphatic system are considered. The main results and further perspectives are discussed.

Keywords: lymph flow; mathematical modelling; lymphatic vessels; lymph nodes

1. Introduction

The lymphatic system (LS) in the human body complements the cardiovascular system and provides drainage of interstitial fluid through a complex system of lymphatic vessels and lymph nodes. Various molecules, pathogens, and immune cells are transported in the organism along the LS. It performs drainage, transport, immune, and deposit functions, namely, lymph accumulation in the lymphatic capillary network, lymph nodes, and spleen.

For a long time, the role of the LS was underestimated and it was not sufficiently well studied. Only in the 20th century, its importance for normal functioning in humans and its role in the pathogenesis of various diseases were realized [1]. One of the most common disorders related to the LS is lymphedema, which can develop as a result of surgical treatment and because of other reasons. It is also reported that solid tumors can create metastases due to malignant cells travelling along the LS [2]. Recent observations correlate plastic bronchitis with dysfunction of the LS [3]. Other examples (lymphangioleiomyomatosis, Kaposi's sarcoma, lymphoceles, lymphatic malformations, and chylothorax) can be found in [1,4].

The presence of lymphatic structures—meningeal lymphatics or dural lymphatics—in the brain was recently discovered [5,6], opposing the previous belief that there are no lymphatic vessels in the brain. This discovery can be important for understanding the pathogenesis of neural diseases, including Alzheimer's disease, and for the development of efficient treatments of neurological disorders [7].

Despite progress in investigations of the LS, the parameters of lymph flow in the human body remain basically unknown because of their difficulty to be investigated: lymphatic vessels are small, so it is hard to place sensors; lymphs are low scattering and its signals are near the noise threshold, so the Doppler-type technique has failed until recently; the LS starts from the interstitial space, so contrast cannot be injected globally; and the LS contains lymph nodes filtrating lymphs, so regional injections allow contrasting lymphs only to the first node. Moreover, injections of contrast alter the interstitial pressure and, therefore, the measured parameters. Modern methods of lymph investigation in vivo [8] are more suitable for lymphangiography than for determination of the flow parameters. Data about lymph flow velocity are usually calculated by dividing the length of the imaged vessel by the time of observation [9–14], which allows getting mean lymph velocity but not the the characterization of lymph pulsations. Moreover, these techniques are invasive and can alter the flow [15]. Recently, a Doppler optical coherence tomography platform with Doppler algorithm operating on low signal-to-noise measurements has been developed [16] to get lymph flow characteristics. It allows getting parameters of lymph flow in vivo including visualization of its pulsating nature. This approach has certain limitations which do not allow using it widely. However, it can open new perspectives in the determination of lymph flow parameters.

Some data about lymph flow are available from in vitro experiments. In the common setup, a part of the cattle mesenteric lymphatic vessel is fixed in the bath with a special solution and the pressure at both ends, the diameter, and the output flux are measured [17–20]. These data can differ from those under physiological conditions [15], but they are used in mathematical models as model parameters and for checking the model output. Mathematical models are used for investigation of lymph flow in the series of contracting parts of lymphatic vessels called lymphangions. The underlying assumptions in these models are essential, and depending on them, the models give different, sometimes opposite results (e.g., [21,22], see Section 3.2). In what follows, we consider lymph transport in different parts of the LS, the influence of valves and contractions on the flow, their modeling, and results. We will illustrate the main approaches to lymph flow modeling on the example of the models described in [23,24] including models developed in the last years and will consider some possible applications.

2. Transport Function of the LS

The lymphatic system (LS) includes lymphatic vessels and lymph nodes. It complements the cardiovascular system and performs drainage, transport, immune, and deposit functions. About 10% of blood goes to the LS in the process of capillary filtration [25,26]. The total lymph flux returning to the venous system is 2–4 L/day [25–28]. The LS absorbs excess fluid, waste products, and large molecules from the interstitial space and transports it to the veins of the cardiovascular system—to the left and right venous angles between neck veins—so the LS starts from the interstitial space and opens into the upper vena cava (Figure 1A). The initial part of the LS is called initial lymphatics. They absorb lymph (tissue fluids, cells, and large extracellular molecules) from the interstitial space. Initial lymphatics have one layer of endothelial cells and no or almost no basal membrane. They connect with anchoring filaments to the surrounding extracellular matrices, and this positioning prevents them from collapsing even under increased pressure in the interstitial fluid. Endothelial cells of initial lymphatics form junctions, where big cells can go into, and prevent lymphs from going out from the initial lymphatics into the interstitial space (Figure 1B). These overlapping cells are called primary (flap) valves. Here and below, anatomical descriptions are provided according to [1,27,29–31].

From the initial lymphatics, the lymph goes to the collectors, which have three layers of cells similar to blood vessels. There are secondary valves in these collectors which prevent the backward flow of lymphs. These valves consist of two or three leaflets located along with the flow (Figure 2A). Valves divide lymphatic vessels into functional parts, and the parts between adjacent valves are called lymphangions (Figure 2B). These lymphangions can produce active contractions. Contractions in the vessels with valves provide the unidirectional flow of lymph (Figure 2B), which is a characteristic property of lymph flow in the LS.

The network of lymphatic vessels is highly unstructured and contains many lymph nodes, and its hierarchical structure is shown in Figure 3. Numerous lymph nodes contained in the LS are lymphoid organs actively participating in the transport function of the LS [32].

It is believed now that lymph flows in the LS are because of the pressure gradient between the interstitial space and the upper vena cava and because of passive (extrinsic) contractions which are caused by the movements of muscle tissues, the diaphragm, and big blood vessels nearby and of active (intrinsic) contractions which are spontaneous contractions of lymphangions. The precise mechanism of active contractions is not yet well known [33]. One of the open questions is whether active contractions can produce enough force to drive the lymphs throughout the whole LS. The lymph flow cycle includes the entrance of lymphs into the initial lymphatics, the flow into lymphangions, and the flow into lymph nodes. The biomechanics of lymph flow in all these steps is not yet elucidated.

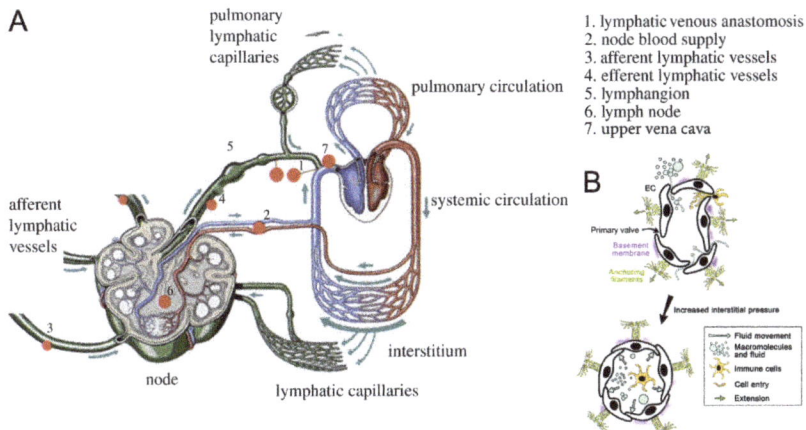

Figure 1. Structure of the lymphatic system (LS): (**A**) adapted from [23,34] with permission. Lymphs enter the LS through the initial lymphatics (lymphatic capillaries), goes to larger lymphatic vessels, and is filtrated in lymph nodes. Then, it is collected in the trunks and ducts, which are the largest lymphatic vessels with respect to their diameter. Ducts open to the neck veins, which in turn open to the upper vena cava, and thus, lymphs return to the blood circulatory system. (**B**) Initial lymphatics (lymphatic capillaries): adapted from [35] with permission. Lymphs enter the lymphatic capillary through junctions between endothelial cells (ECs). These junctions are called primary (flap) valves because they prevent lymphs from leaving the lymphatic capillary. Capillaries are surrounded by the intermittent basement membrane. ECs connect to the surrounding extracellular matrix with anchoring filaments. Anchoring filaments prevent lymph capillaries from collapsing even in increased interstitial pressure (black arrow).

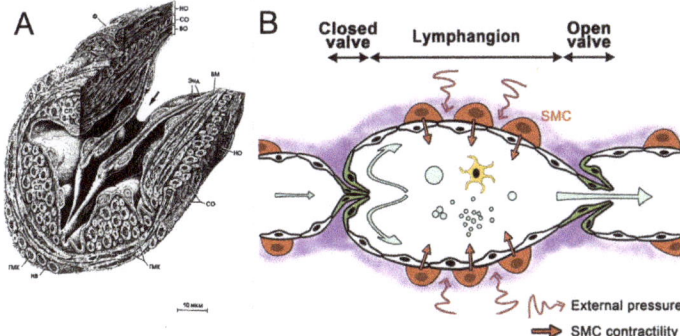

Figure 2. Structure of the lymphatic vessels: (**A**) microscopic view of an open secondary valve in the lymphatic vessel. Reprinted from [36] with permission. The arrow indicates the direction of lymph flow. (**B**) A scheme of lymph flow in the lymphatic vessel with secondary valves: republished from [35] with permission. Secondary valves allow lymphs to flow in one direction and restricts flow in the backward direction. The part of the lymphatic vessel between adjacent valves is called a lymphangion. There are smooth muscle cells (SMCs) in the vessel wall which allow wall contractions. Lymphs flow in the vessel under active (intrinsic) contractions of the lymphatic vessel wall and under passive (extrinsic) contractions having external nature.

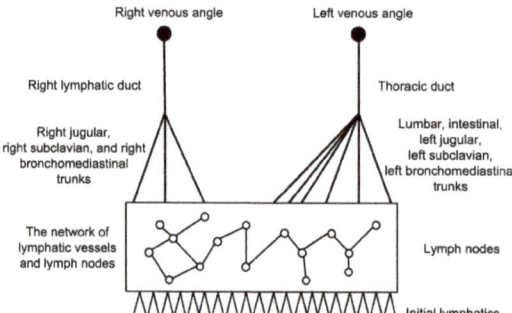

Figure 3. A schematic representation of the LS. From the bottom to the top: the LS starts from the initial lymphatics and then the network of lymphatic vessels and lymph nodes, followed by trunks and ducts. The right lymphatic duct collects lymphs from the right part of the head and body and opens to the right venous angle. The thoracic duct collects lymphs from the left part of the head and body and both lower extremities. It opens to the left venous angle between neck veins. About 1/4 of lymphs goes to the right lymphatic duct, and 3/4 go to the thoracic duct [27,28].

The majority of models of lymph flow in the lymphangion or series of lymphangions are based on the zero-dimensional (including lumped ones) [17,22,37,38] or one-dimensional (including quasi-one-dimensional ones) [18,39–41] approaches. "Lumped" models here are models based on electrical circuit theory, where the network of vessels is substituted with the circuit and where the voltage, current, and charge are equivalent to pressure, lymph flux, and mass, respectively [38,42,43]. Zero-dimensional models are models with no space distribution of velocity (flux) and pressure functions. Quasi-one-dimensional models are one-dimensional models in which the cross section may change. A three-dimensional formulation for lymph flow in the single lymphangion is considered in [44]. In the works [40,45], models of lymph flow in the whole LS are developed on the basis of zero-dimensional and quasi-one-dimensional approaches, respectively. The influence of valves and contractions on the flow distinguishes lymph flows from other physiological flows.

3. Flow in Lymphangions

We will describe lymph flow in this section using the quasi-one-dimensional approach [40] and other models when they are available. This approach is widely used to model systemic blood circulation [46,47] and, more recently, to model lymph flow in the lymphatic vessels [18,40,41]. In this approach, fluid flow is described by a system of hemodynamic equations which are derived from mass and momentum conservation (or from the Navier–Stocks equations [48]):

$$\frac{\partial s}{\partial t} + \frac{\partial us}{\partial x} = 0, \tag{1}$$

$$\frac{\partial u}{\partial t} + u\frac{\partial u}{\partial x} + \frac{1}{\rho}\frac{\partial p}{\partial x} = f, \tag{2}$$

where $u(x,t)$ is fluid velocity, $p(x,t)$ is pressure, $s(x,t)$ is cross sections area, $\rho = const$ is density, x is spatial coordinate, and t is time. The external force $f(x,t)$ corresponds to viscous friction. Systems (1) and (2) are completed by the relation between the cross-sectional area s and pressure p. While this system is suitable to describe blood flow in arteries and veins, it cannot be used to describe lymph flow in lymphatic vessels without modifications because of the presence of valves and vessel wall contractions.

3.1. Influence of Valves

Secondary valves in lymphatic vessels prevent backward flow of lymphs and provide its unidirectional flow. Some investigations report that valves are included in the contraction of lymphangion [31]. Leaflets of valves can influence the flow. The models of secondary valves are usually limited by the absence of backward flow and sometimes by the resistance to the incoming flow.

Secondary valves can prolapse in a high adverse pressure gradient, and the presence of backward flow is one of the diagnostic criteria for lower limb lymphedema [49]. However, it is not known whether it is the reason for or a consequence of the disease [17]. Some models of valves include the possibility of prolapse.

There are several ways to represent the influence of valves on the lymph flow. One of them is to consider adjacent lymphangions as two segments connected with the bifurcation point. In this representation, the flow in each vessel is described by System (1, 2) and a closing relation. At the point of bifurcation, conditions on the pressure and flux are stated. Valves are taken into account in the flux condition at the bifurcation point, and the simplest model for their behavior is as follows [17,37]:

$$q_v \geq 0, \tag{3}$$

where q_v is the lymph flux through the valve. Condition (3) restricts the backward lymph flow through the valve. Valve resistance to the incoming flow is considered in [37] on the basis of the data from [50].

A valve model in blood vessels for lumped and one-dimensional approximations is proposed in [51]. It is sufficiently simple and detailed to describe normal functioning of valves and their pathologies. This model is used in [39] to describe the valve influence on lymph flow. The equation for time-varying flux through valve q_v has the following form:

$$\frac{d}{dt} q_v = \frac{1}{L(\xi)} \left(\Delta p(t) - R(\xi) q_v - B(\xi) q_v |q_v| \right), \tag{4}$$

where Δp is pressure drop on the valve and where $L(\xi)$, $B(\xi)$, $R(\xi)$ are functions of lymphatic inertia (proportional to ls^{-1}), Bernoulli resistance (proportional to s^{-2}), and viscose resistance to flow (proportional to ls^{-2}), respectively.

In [22], the valve resistance R_v $\left(R = \frac{p}{q} \right)$ changes in time, depending on pressure gradient in the lymphangion:

$$R_v = R_{V_{min}} + R_{V_{max}} \left(\frac{1}{1 + \exp\left(s_{open} \left(\Delta p - p_{open} \right)\right)} + \frac{1}{1 + \exp\left(-s_{fail} \left(\Delta p - p_{fail} \right)\right)} - 1 \right). \tag{5}$$

The transition from the maximal resistance $R_{V_{max}}$ to the minimal $R_{V_{min}}$ occurs when pressure difference Δp before and after the valve reaches some threshold value p_{open} is sufficient to open a valve. The second sigmoid describes the prolapse effect at the large adverse pressure gradient p_{fail}. In this formulation, resistance to backward flow has a large but limited value.

Another way to model the valve influence on lymph flow is to take it into account in the right-hand part of Equation (2). The valve provides some force influencing the flow in one direction but not in the opposite direction. This approach is used in [52] to model valve functioning in blood vessels. In [40], the valves in trunks and ducts are modelled by Equation (3). The valve influence in the smaller vessels is modelled by the following force in the right-hand side of the momentum equation:

$$f = -8\pi\nu\left(u\right)\frac{u}{s}, \tag{6}$$

where $v(u)$ is a viscosity coefficient, considered as a function of velocity u having a sigmoidal form. This function characterizes friction force acting on flow in the backward direction. The sigmoidal form of this coefficient describes valve resistance to the upcoming flow. A similar approach was used in [41].

The choice of the valve model depends on the modeling task. If valve resistance to the flow is not important in the investigation, the simple boundary Condition (3) is sufficient. Formulas (4) and (5) allow for consideration of changing the valve resistance. Formula (6) is suitable for modeling multiple valves in big networks of lymphatic vessels [40]. It also allows for application of analytical methods to study lymph flow [53].

3.2. Influence of Contractions

Lymph flow in lymphatic vessels is influenced by active and passive contractions. Active contractions include a tonic part induced by intravascular pressure, and phase contractions are spontaneous and periodic (Figure 4). Active contractions are determined by lymphatic vessel wall properties and lymph chemical composition, while passive contractions occur due to some external reasons (e.g., muscle contractions or diaphragm movement). Biophysics of active contractions is not yet known. Many investigations consider the influence of different humoral factors on the amplitude, phase, and period of contractions, but a consistent theory does not exist. One more important question is about the coordination of contractions of adjacent lymphangions [17,22], since these data cannot be obtained from physiological experiments.

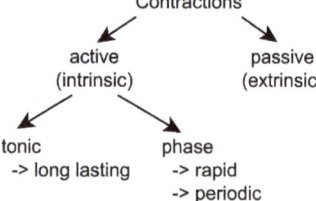

Figure 4. Classification of lymphatic contractions: Active (intrinsic) contractions are caused by the vessel wall properties and the lymph formulae. Passive (extrinsic) contractions occur because of external factors. Active contractions can be divided into two parts: tonic response and phase contractions. The tonic response is a long-lasting variation of elastic properties of the vessel wall, and phase contractions are rapid and periodic. All these types of contractions can be modelled together [17] or separately [18,22,37–39,41,54,55].

In early works, contractions of lymphangions were modeled using the same approach as in the models of heart pumping with time-varying elasticity [17]:

$$E(t) = \frac{P_t(t)}{V(t) - V_0}, \tag{7}$$

where V_0 is dead volume (theoretical volume at zero pressure), $V(t)$ is measured volume, and P_t is transmural pressure. Therefore, passive and active contractions are taken into account in this model. This model was extended in [21], where contractions of two and more (up to four) lymphangions and their coordination were investigated. In this series of works, it was demonstrated that coordination of lymphangions minimally affects lymph flow. However, later investigations showed opposite results [22]. Both works are supported by physiological experiments, and, as discussed in [23], the reason for such a difference can be related to their design, in particular, the presence of valves can be essential.

Lymphangion contractions depend on the pressure gradient. Lymphangions contract if the pressure gradient is negative, and they act as passive conduits if the pressure gradient is positive [56]. Active contractions decrease flow under a positive pressure gradient [17,18]. Therefore, under these conditions, more effective drainage occurs without contractions. This result can bring a new understanding of lymphedema pathogenesis and treatment [17]. In particular, the reduction of pumping activity in some edemas can facilitate drainage of excessive fluid [17], while it is conventionally considered a pathogenesis of the disease.

Both components of active contractions are considered together in (7) [17,21]. In other models, they can be singled out in the pressure–area relationship closing Systems (1) and (2). They are considered below in Sections 3.2.1 and 3.2.2, and passive contractions are considered in Section 3.2.3.

3.2.1. Modelling of Tonic Response

Tonic response determines how the vessel wall reacts to changing pressure inside the lumen. It is reported in [31] that elastic properties of lymphatic vessels are close to elastic properties of veins. More precise characteristic of lymph vessels is not yet determined.

In the quasi-one-dimensional approach, elastic properties of the vessel wall are usually taken into account in the third equation (also called state equation, tube law, or pressure–area relationship), which closes Systems (1) and (2) and describes dependence of the cross-sectional area on the intraluminal pressure:

$$s = s(p). \tag{8}$$

Despite the existence of models concerning the cell structure of blood vessels, in the majority of investigations, empirical curves for the s–p relationship are used [57]. Similar representations are also used for all existing models of lymph flow on the basis of physiological experiments [18,22,37,39]. Despite the similarity of s–p curves for different pressure–area dependencies, mathematical solutions (in particular, the velocity of small perturbations) can differ significantly [57]. Let us consider implementations of tonic reactions in different models of lymph flow in lymphatic vessels.

The simplest pressure–area relationship with a linear dependence is used in [40]:

$$s = s(p) = s_0 + \theta(p - p_0), \tag{9}$$

where s_0 is the minimum cross-sectional area at pressure p_0 and where θ is the coefficient characterising vessel elasticity. A nonlinear pressure–area relationship

$$p = p_{ext} + \frac{h}{r}\left(\frac{4E}{3}\frac{(r - r_0)}{r_0} + \sigma_{act}\right), \tag{10}$$

is considered in [37]. Here, h is the lymphangion wall thickness, r is the time-dependent radius, r_0 is the resting radius, and E is the elasticity coefficient (Young's module). Let us note that E in (10) has an opposite meaning compared to θ in (9): for larger E, the wall is more rigid. The tonic part is represented by the first term in brackets, while σ_{act} characterises phase contractions and p_{ext} is external passive pressure considered as a given function. In this formulation of the pressure–area relationship, the assumption of the thin vessel wall is imposed, i.e., wall thickness is supposed to be much smaller than lymphangion radius. A thick wall model is considered in [18]:

$$\Delta p = E\Delta r_{out}\frac{r_{out}^2 - r_{in}^2}{2(1 - \sigma^2)r_{in}^2 r_{out}} - 2\pi rT\frac{\partial^2 r}{\partial x^2} + 2\pi r\gamma\frac{\partial r}{\partial t}, \tag{11}$$

where $\Delta p = p - p_e$ is the transmural pressure (gradient between pressure inside of the vessel p and outside of the vessel p_e), r_{in} and r_{out} are the internal and external vessel radius respectively, Δr_{out} is the radius variation caused by the pressure difference, and σ is the Poisson ratio. This formulation takes into account tube tension T resulting from longitudinal bending of the vessel and the damping term with coefficient γ. The damping term γ serves as a natural regularization for the explicit numerical scheme [18], helping to avoid the numerical instability observed in [45].

Another dependence was used in [22,55]:

$$\Delta p = p_d \left(\exp(d/d_d) - \left(\frac{d_d}{d}\right)^3 \right), \tag{12}$$

where d is diameter, and p_d and d_d are some parameters. The exponent describes wall-stiffening with strain at positive transmural pressure, and the cubic term describes a progressively diminishing vessel compliance at negative transmural pressure. Two more refined forms of this relation are used by the same authors in [58]. A similar approach with more parameters is used in [39]. These parameters are determined by fitting the data from physiological experiments.

These models describe the elastic properties of lymphatic vessels. The parameters of Equation (9) can be directly measured from the physiological experiments (for s linearly depending on p). In (10) and (11), the additional model for Young's module E is required [59]. Formula (12) closely represents the considered physiological experiment. However, as discussed in [57], the same s–p dependence can give different solutions depending on the exact s–p function. Since the level of uncertainty in parameter determination for lymph flow is very high, a more complex model does not necessarily work better. Therefore, the final choice of the model depends on the modelling task.

3.2.2. Phase Contractions

Phase contractions represent another type of active contraction. They provide lymph propagation in the low or inverse pressure gradient between the interstitial space and the upper vena cava. Lymphedema is associated with failure of these contractions. The precise mechanism of phase contractions is not known. It is generally recognized that phase contractions are periodic and have relaxation and contraction times and that the amplitude of contractions seems to depend on the lymph volume inside the lymphangion just before contraction (diastolic volume). During the contraction phase, the lymphangion diameter can decrease up to 40% compared to its relaxed state [60]. The lymphangion length also decreases in the contraction, but existing mathematical models do not take it into account because of its small contribution compared to the diameter decrease. Furthermore, in some investigations, valves are reported to actively participate in contraction activity [31]. In the existing mathematical models, the valves are considered as a passive mechanism restricting backward flow.

It is not known whether the contractions can produce sufficient force to provide lymph flow against the gravity force in low or inverse pressure gradients between the interstitial space and upper vena cava. The models of lymph flow in the series of contracting lymphangions and in the whole LS are developed to answer this question [17,22,37,38,61]. In [61], and this question is answered positively with the developed mathematical model (Section 4). However, the evaluated parameters exceed the physiological values reported in [31,62].

A 3D model for different types of phase contractions shows that lymph flow velocity can be approximated by Poiseuille profile [44].

Since the precise mechanism of phase contractions is not known, they are modelled by some periodic, usually trigonometric, function. For example, active contractions in [40] have the following form:

$$p_{ph} = A \sin\left(\frac{2\pi}{\lambda}(x - at)\right), \tag{13}$$

where p_{ph} is the phase part of active contractions to be added in Equation (9) and where A, λ, and a are the parameters of a contraction wave: amplitude, wave length, and wave velocity, respectively.

An analytical estimate of pumping efficiency in the vessel with valves described by (6) is presented in [53]. It is shown that increased frequency of contractions leads to an increase in output flux. The same is true for increased amplitude. These results are extended in [63], where lymph flow is considered in a detailed graph of the LS (Figure 5c).

The trigonometric function is also used in [37], taking into account the current volume of lymphs in lymphangions and relax time: if these thresholds are not reached, contraction does not happen; otherwise, contractions are modelled by sinus function with an amplitude depending on lymphangion volume. A threshold value for the pressure to activate a fixed impulse is also used in [38], and this is a rare example of the work where lymphangion contraction duality is taken into account. Namely, contractions under negative pressure gradient are triggered if internal pressure in the lymphangion rises above a pressure threshold value, and it acts as a simple conduit under positive pressure gradient. This model reproduces a pulsating flow, and it can be used for the construction of big vessel networks.

Both sinus waves, simple like in (13) and with time delay like in [37], are used in [18], showing good correlation with the results of physiological experiments. It is supposed that contraction waves in lymphatic vessel networks propagate in the backward direction, from downstream to upstream, and that this behavior should be more efficient. Backward propagation (against the valves) of contraction waves was recorded in these experiments [18]. Cosine waves are used in [22] for every segment in the chain of lymphangions, and their propagation is investigated. All models reproduce the pulsating nature of lymph flow.

Phase contractions with switching from the minimal to the maximal values

$$p_{ph}\left(\frac{s}{s_0},\kappa\right) = \kappa\left(K_{max} - K_{min}\right)\psi\left(s,s_0\right), \tag{14}$$

in tube law $p = p(s) = p_p(s) + p_{ph}(s) + p_e$ are modelled in [39]. Tonic p_p, phase p_{ph}, and passive p_e components are taken into account. Parameter $\kappa \in [0,1]$ is a state of contractions (designated by s in original work), which is a solution of the Electro-Fluid-Mechanical contraction (EFMC) model based on the FitzHugh–Nagumo model for action potentials [64], and ψ is a function of a cross-sectional area similar to (12).

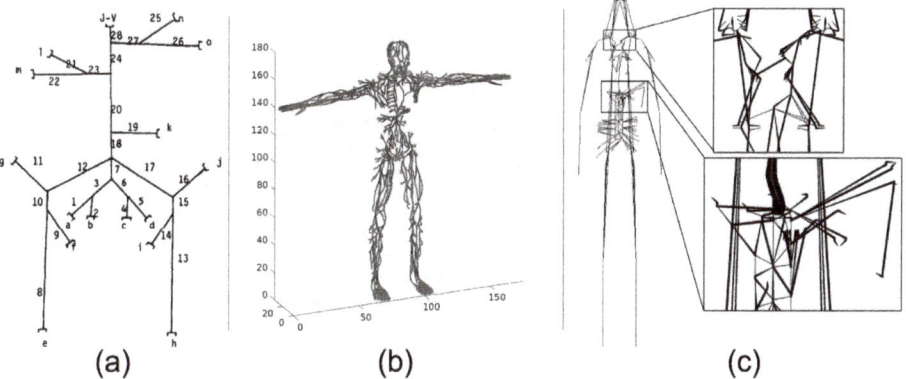

Figure 5. Examples of graphs of the LS: (**a**) The graph from [45], reprinted with permission. The graph contains 29 vessels including 297 lymphangions. The thoracic duct and its feeders are presented, and other parts have less detailed descriptions. (**b**) The graph from [65], reprinted with permission. The graph contains the majority of vessels presented in the anatomical model. The algorithm resolves mismatches from the anatomical model. Both ducts are presented. (**c**) The graph from [40]. The graph is spatially oriented, and thus calculations with respect to gravity force are possible [61]. The vessel network includes lymph nodes. Both ducts are presented. Also presented are zoom-ins on the region with the right and the left venous angles (top) and the region of the cisterna chyli (bottom).

In this model, lymphatic vessel contraction activation is similar to heart pacemakers. According to this assumption, the FitzHugh–Nagumo model is modified and used to describe calcium-dependent contractions. Modeling results are in agreement with the experimental trends. To our knowledge, this is the first model of lymph flow dynamics where humoral factors influencing contractions are taken into account. The EFMC model opens new perspectives in the study of humoral influence on the lymph flow contractions in models of lymph flow dynamics.

3.2.3. Passive Contractions

Passive contractions are determined by external factors, such as muscle contractions, massage, diaphragm movement, etc. They are independent of the characteristics of lymphatic vessels and of the lymph chemical composition. Passive contractions are usually modeled as some function of pressure outside the vessel wall, and they are taken into account in p_e ($\Delta p = p - p_e$, see (11)). Consideration of passive contractions can be essential in models of systemic lymph flow, e.g., in [40], because big muscles, especially of the lower limbs, and diaphragm excursions can have a significant effect on lymph flow. In models of contractions in one lymphangion or a series of lymphangions, the function of external pressure corresponds to the pressure inside the bath containing lymph vessel segment in in vitro experiments.

4. Systemic Lymph Flow and Graphs of the LS

The LS is highly unstructured, and it has almost no tree structure (Figure 3). As a consequence, the construction of graphs of the LS is quite difficult. However, there are several works where this attempt was implemented.

A graph of the LS was constructed in [45] (Figure 5a), and to our knowledge, it is the first reported attempt to consider systemic lymph flow. This graph starts from the interstitial space in several organs and contains feeders of the thoracic duct and the thoracic duct itself, while the right lymphatic duct is not included. It has 296 lymphangions, some of which represent lymph nodes. Lymph flow in the lymphangions is described with the zero-dimensional model from [37], and the pressure equilibrium and flux conservation are imposed in the bifurcation points. This model gives a description of fluid exchange in the interstitial space with equations similar to those describing flow in lymphangions but including flux from the interstitial space instead of flux from the previous lymphangion. Flux in the interstitial space is given by a model of Starling law for fluid exchange, and it takes into account hydraulic and osmotic pressure. The model for tissue fluid exchange was further developed in [62] with the one-dimensional approach, including the influence of anchoring filaments and primary valves in the initial lymphatics. This systemic model shows the pulsating nature of lymph flow from the thoracic duct that is consistent with physiological observations. The parameters of the flow were in agreement with available physiological data.

The algorithm of automatic construction of a graph of the LS is suggested in [65]. The data from the Plastic Boy project [66] is used as a source of anatomical structure. These data are not suitable for calculations as-is, and the constructed algorithm turns this image into a consistent graph of the LS (Figure 5b). Structural analysis of the LS is provided, including the average degree of lymph nodes (how many vessels connected to the node). This algorithm can be applied to other anatomical models. The presence of a graph of the lymphatic system makes it possible to analyze the redistribution of lymph flow in the human lymphatic system in case of damage and/or difficulty in the outflow of lymphs from individual lymph nodes, which will allow for prediction of the effects of various injuries and changes in the organs of the human lymphatic system as well as for prediction of the best methods for correcting these problems.

One more graph of the LS is constructed in [40] (Figure 5c). It is anatomically adequate, spatially oriented, and topology consistent with a similar graph of the cardiovascular system [47]. The graph contains 271 collecting lymphatic vessels, thousands of lymphangions, and 161 lymph nodes. Groups of lymph nodes form the basis of the graph, and it contains 46 main groups [54]. Flow in the vessels is described by the quasi-one-dimensional equations including valve and active contraction influence, and the conditions of pressure equilibrium and mass conservation are imposed in the bifurcation points. The model of active pumping in this graph and its influence on the total flow are investigated without the influence of gravity force [63] and with influence of gravity force [61]. The increase in amplitude and frequency of phase contractions is shown to increase the output flux. Also, the elasticity and contractions of lymph nodes have a considerable effect on the output flux [63]. Therefore, the lymph nodes affect the transport function of the LS, and it is consistent with physiological observations [32]. The frequency of contractions, required to get output flux in physiology adequate intervals, is higher than reported in the physiological experiments [31,62]; however, in vivo parameters can also differ from them [15]. The consideration of passive contractions, which are not included in the model now, may allow for a reduction in the frequency. The results show that active contractions can produce enough force to make lymphs flow though the whole LS against the pressure gradient [61].

The approach used to construct a spatially oriented graph in [65] requires source images to contain information about both lymph nodes and lymphatic vessels. Another approach is used to create a graph in [40]. The basis of this graph is the regional groups of lymph nodes which connect according to the communication matrix filled in with data from [29]. This approach can be used to construct personalized graphs of the LS by extracting information about lymph nodes from MRI images and by connecting them with vessels using the communication matrix according to some anatomical source (i.e., [29]). The described concept could be the first step towards the creation of personalized graphs of the LS.

5. Lymph Drainage and Flow in Initial Lymphatics

The LS starts from the initial lymphatics (lymphatic capillaries, terminal lymphatics), where lymphs arrive from the interstitial space. Characteristic features of lymph drainage from the interstitial space are the influence of primary valves, which allow lymphs to enter the LS and which stop it from flowing out of the lymphatic capillaries, and the presence of anchoring filaments, which prevent initial lymphatics from collapse even in excess pressure in the interstitial fluid (Figure 1B).

In contrast to models of secondary valves, models of the primary valves often take into account the geometry of the valve—the curvature of the cell [67,68].

Flow into the initial lymphatics can be described by a system of ordinary differential equations, including flux and pressure balance and taking into account osmotic pressure according to Starling's hypothesis [45]. Another model takes into account the primary valves influence given by Condition (3) and anchoring filaments with additional force in the pressure–area relationship similar to (10) [62]. This model describes pumping of income lymph flow in the LS due to fluctuations in the interstitial fluid pressure and to the suction effect of adjacent pumping lymphatic vessels.

Primary valve dynamics can also be described by the Stockes equation [67]:

$$\frac{dp}{dx} = -\frac{12Q\mu}{w(x)^3 T}, \tag{15}$$

where T is a unit tissue thickness, $p(x)$ is pressure inside the fluid gap between the cells, Q is flux through the gap, and μ is dynamic viscosity. This equation must be solved simultaneously with the equation for cell deflection $w(x)$:

$$\frac{d^4 w(x)}{dx^4} = \frac{1}{EI}\Delta p(x), \tag{16}$$

where $\Delta p = p(x) - p_L$ is transendothelial pressure drop, p_L is pressure inside the lumen, and E is Young's module. Flux through the gap is given by the following expression:

$$Q = \int_0^{w(x)} \mathbf{u}(x)\, dy, \tag{17}$$

where $\mathbf{u}(x)$ is flow velocity. Simulation results demonstrate primary valve dynamics for this model: the flux nonlinearly depends on pressure p with positive pressure gradient Δp, and it equals 0 for negative pressure gradient.

This model was extended in [69], where a more precise description of flow around the overlapping endothelial cells is considered in the same geometry. The computational results correspond to the available physiological data.

The geometrical structure of the interstitial space with initial lymphatics and blood capillaries is considered in [68], and a mechanical model for primary valve dynamics concerning their curvatures is used to describe flow through the junctions between endothelial cells. As a result of this two-dimensional analysis, the formula for lymph flux per unit length q in terms of the pressure difference between blood $\hat{p}$ and lymphatic capillaries p_0 was obtained:

$$q = \begin{cases} 0, & \hat{p} < p_0 + p_{crit} \\ \dfrac{\zeta \alpha \pi r_b}{\alpha \pi r_b + 2\zeta} \dfrac{k}{\mu} (\hat{p} - (p_0 + p_{crit})), & \hat{p} > p_0 + p_{crit} \end{cases}, \qquad (18)$$

where p_{crit} is the critical pressure needed to open a valve, ζ is a dimensionless parameter that describes the effect of geometry on the fluid flux, k is permeability of intestitium, μ is viscosity of interstitial fluid, and r_b is radius of the blood capillary. This model describes fluid exchange between blood and lymphatic capillaries through the interstitial space.

All considered models describe periodical entering of lymphs into the initial lymphatics. They provide more precise boundary conditions for models of flow in the lymphangions, and they can be used to describe lymph drainage in the investigations of lymphedema pathology.

6. The Structure and Function of the Lymph Node

Lymph nodes (LNs), which represent an important part of the lymphatic system, in addition to merging lymph flows, perform a number of functions related to the human immune system. This is due to the fact that they contain localized populations of cells that perform protective functions in the body. Thus, populations of T-lymphocytes are located in the so-called T-cell zone and B-lymphocytes are grouped in B-cell follicles, which are involved in the production of antibodies specific for the antigens entering the body. Dendritic and fibroblastic reticular cells are also present in the T-cell zone and are involved in all aspects of the functioning of lymph nodes. Figure 6 shows the key elements of the structure of the lymph node. These data are provided according to [70–72].

In the lymph node, T and B lymphocytes are in an inactive, naive state. The populations of these cells are regulated by fibroblastic reticular cells (FRC) secreting interleukin-7, which is a survival factor for naive lymphocytes [73]. The FRC network is also involved in maintaining the structural integrity of the entire organ through threads formed by polymerized collagen (Figure 6) secreted by fibroblastic reticular cells. The process of immune response in LNs begins with the helper T lymphocyte that has detected the antigen or with another activated antigen-presenting cell [74]. In this case, a cascade process occurs: T helper lymphocytes start activation of resting antigen-presenting cells, T effector lymphocytes and B-lymphocytes, and interleukin-2 is produced to increase the proliferative activity of T and B-lymphocytes [75]. Activated antigen-presenting cells (APC) fix onto the FRC network for a period of about 6 hours with the possibility of subsequent reactivation. After that, APCs begin to express interferon-α. Under the influence of interferon-α, naive T effector lymphocytes specialize according to the components of the antigen, presented by the antigen-presenting cells' major histocompatibility complex class I, II (MHC class I, II) [76]. At the same time, the FRC network is actively involved in stimulating immune response, producing chemokines CCL19 and CCL21. According to existing data, the lymphocytes interacting with these chemokines exhibit increased motor activity. This effect does not improve immune response alone, but in synergy with the chemotaxis effect used by cells to search for foci of infection, chemokines shortened the time from activation to interaction with the target. Activated B-lymphocytes can also activate resting cells, but their main function is the production of antibodies specific for the target antigen [74]. At the same time, activated T effector lymphocytes search for and destroy cells expressing signals characteristic of the desired antigen. T effector lymphocytes mainly destroy body cells infected with intracellular parasites (viruses or bacteria). Macrophages are also involved in the process of the immune response, destroying free viruses and bacteria. After activation in the lymph nodes, cells of the immune system move into tissues in which antigens are present to fulfill functions of the immune response. In connection with such a heterogeneous structure of the lymph nodes, it becomes extremely important to understand how the lymph flows into the node are distributed ([77,78]). The size of particles that can enter different zones of the lymph node is limited [77]. Thus, it turns out that huge viruses and bacteria cannot penetrate into the T-cell zone through the walls of the FRC network due to too large sizes. Therefore, they get there either through blood vessels or inside the cells that came from outside the lymph node. It should be noted that the detection of antigens is difficult not only because of the limited methods of their entry into the T-cell zone. About 93% of lymphs generally passes through the subcapsular sinus of the LN. This is shown, inter alia, in [78]. It should also be taken into account that part of the lymph from the lymph node enters the blood vessels system inside LNs.

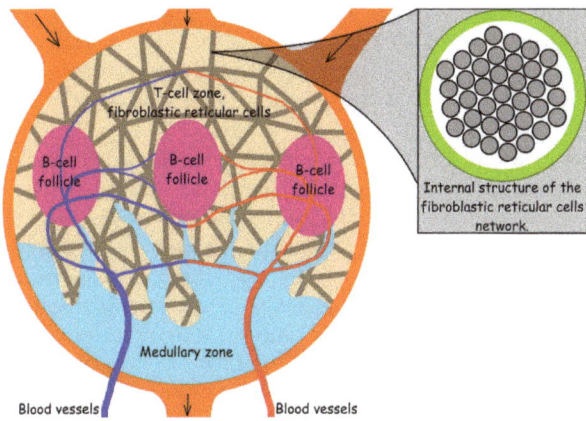

Figure 6. Lymph node scheme illustrating its key elements: the scheme shows subcapsular sinus (orange), medullary zone (azure), B-cell follicles (purple), T-cell zone, a network of fibroblastic cells (gray lines against the background of the T-cell zone), afferent (introducing lymph) and efferent (excreting lymph), as well as blood vessels.

Due to the complexity of the processes occurring in the lymph nodes and the heterogeneity of the lymph nodes, it becomes necessary to be able to form various configurations of the lymph nodes to simulate the processes of cell interactions, fluid flow, and collagen coagulation. Since it is expensive and time-consuming to restore structures using layer-by-layer scanning, an important area of research is the development of algorithmic methods for constructing lymph nodes. Algorithms were developed, and 3D models of the components of the lymph node were built in [79,80]. Three-dimensional models of key elements of LNs in a voxel format were created with further conversion to a solid-state 3D model: the shell of the lymph node, trabecular and medullary sinuses, B-cell follicles, and a network of fibroblastic reticular cells.

Separately, algorithms were developed to build a 3D model of a network of fibroblastic reticular cells [80]. It appears that a highly damaged network (up to 50%) can still support lymph flow. This is important for the case of HIV infection where the structure of the fibroblastic reticular cells as well as the ducts formed by them are gradually destroyed. Damage of the network significantly affects not only the ability of the T-cell zone of the lymph node to pass lymphs but also the homeostasis of cell populations and the immune response to HIV in general [80,81].

7. Conclusions and Perspectives

The main goal of investigations of lymph flow is to understand the underlying mechanisms in order to propose effective therapeutic techniques. In the 90s, the specific lymphatic factor (VEGF–C) was discovered, and it led to the explosive investigation of LS [82]. The mechanisms of lymph flow are more clear now, but there are still many open questions.

Mathematical modeling is used with in vitro experiments to get more data, to check the hypothesis about lymph flow, and to construct new experiments. In [17,21], the authors established conditions for pumping and conduit acting. In negative pressure difference, the lymphangion acts as an active pump, and in positive pressure, it acts as a simple conduit. Moreover, in large positive pressure gradients, active contractions are shown to inhibit lymph flow. This result is supported by a physiological experiment [18,56]. The meaning of this result in the application to lymphedema pathogenesis and treatment is discussed in [17]. For some edemas, the inhibition of lymphangion pumping may have a positive effect.

Pathogenesis and treatment of lymphedema are one of the most important applications of models of lymph flow in lymphatic vessels. Direct models of lymph flow in lymphatic vessels are not the only way to address edema pathology: in [83], lymph filtration through the extremity was considered with diffusion equations instead of considering flow in separate vessels, and some clinical recommendations (for using of bandages) were proposed.

Another important goal of lymph flow modelling is the investigation of drug distribution in the LS and various tissues and organs. Insufficient lymphatic circulation is one of possible reasons for ineffective drug therapy, and such a model for systemic circulation can find its applications not only for lymphedema treatment but also for other drug-delivery tasks. Lymph drainage from the interstitial fluid and sorption effect of the secondary lymphatic circulation (vessels with contractions and secondary valves) are essential for understanding the mechanisms of more efficient drainage in the cases of different pathologies. The proposed concept for construction of personalized graphs of the LS could be the first step towards achieving a personalized model of human LS. It opens new perspectives in these investigations.

Lymph flow in the lymph node may influence the immune response occurring in them. Modeling the processes of cellular interactions in the lymph nodes as well as the processes of transfer between nodes through the lymphatic system are critical for understanding the processes that occur in HIV-infected people. It gives an opportunity to develop the best strategies for drug therapy, to analyze the effectiveness of existing methods, and to find key relationships in complex immune response processes. The human immune system depends on many populations of cells represented in the T-cell zone of the lymph nodes: T-lymphocytes (helpers and effectors); B-lymphocytes, which produce antibodies when pathogens enter the body; macrophages; and antigen-presenting cells. Certain cell populations, such as fibroblast reticular cells, are involved in maintaining the homeostasis of naive lymphocyte populations, and the whole lymphatic system provides a reserve of bandwidth for the fluid circulation in the body and acts as another filter for intercellular fluid collected by lymphatic capillaries from tissues.

There are still many open questions in lymph flow mechanics, but close cooperation between physiologists and mathematicians will help understand it better.

Author Contributions: Writing—original draft preparation, A.M. and R.S. All authors have read and agreed to the published version of the manuscript.

Funding: This work is supported by the Ministry of Science and Higher Education of the Russian Federation: agreement no. 075-03-2020-223/3 (FSSF-2020-0018).

Conflicts of Interest: The authors declare no conflict of interest.

References

1. Choi, I.; Lee, S.; Hong, Y.K. The New Era of the Lymphatic System: No Longer Secondary to the Blood Vascular System. *Cold Spring Harb. Perspect. Med.* **2012**, *2*, 23. [CrossRef] [PubMed]
2. Filchenkov, A.A. Lymphangiogenesis and metastasis of tumors. *Creat. Surg. Oncol.* **2010**, *No 3*, 80–90. (In Russian)
3. Itkin, M.; McCormack, F.; Dori, Y. Diagnosis and Treatment of Lymphatic Plastic Bronchitis in Adults Using Advanced Lymphatic Imaging and Percutaneous Embolization. *Ann. Am. Thorac. Soc.* **2016**, *13*, 1689–1696. [CrossRef] [PubMed]
4. Pamarthi, V.; Pabon-Ramos, W.M.; Marnell, V.; Hurwitz, L.M. MRI of the Central Lymphatic System: Indications, Imaging Technique, and Pre-Procedural Planning. *Top. Magn. Reson. Imaging* **2017**, *26*, 175–180. [CrossRef] [PubMed]
5. Louveau, A.; Smirnov, I.; Keyes, T.J.; Eccles, J.D.; Rouhani, S.J.; Peske, J.D.; Derecki, N.C.; Castle, D.; Mandell, J.W.; Lee, K.S.; et al. Structural and functional features of central nervous system lymphatic vessels. *Nature* **2015**, *523*, 337–341. [CrossRef] [PubMed]
6. Aspelund, A.; Antila, S.; Proulx, S.T.; Karlsen, T.V.; Karaman, S.; Detmar, M.; Wiig, H.; Alitalo, K. A dural lymphatic vascular system that drains brain interstitial fluid and macromolecules. *J. Exp. Med.* **2015**, *212*, 991–999. [CrossRef] [PubMed]
7. Nikolenko, V.N.; Oganesyan, M.V.; Yakhno, N.N.; Orlov, E.A.; Porubayeva, E.E.; Popova, E.Y. The brain'sglymphatic system: Physiological anatomy and clinical perspectives. *Neurol. Neuropsychiatry Psychosom.* **2018**, *10*, 94–100. (In Russian) [CrossRef]
8. Munn, L.L.; Padera, T.P. Imaging the lymphatic system. *Microvasc. Res.* **2014**, *96*, 55–63. [CrossRef]
9. Sharma, R.; Wendt, J.A.; Rasmussen, J.C.; Adams, K.E.; Marshall, M.V.; Sevick-Muraca, E.M. New Horizons for Imaging Lymphatic Function. *Ann. N. Y. Acad. Sci.* **2008**, *1131*, 13–36. [CrossRef]
10. Liu, N.F.; Lu, Q.; Jiang, Z.H.; Wang, C.G.; Zhou, J.G. Anatomic and functional evaluation of the lymphatics and lymph nodes in diagnosis of lymphatic circulation disorders with contrast magnetic resonance lymphangiography. *J. Vasc. Surg.* **2009**, *49*, 980–987. [CrossRef]
11. Sevick-Muraca, E.M.; Sharma, R.; Rasmussen, J.C.; Marshall, M.V.; Wendt, J.A.; Pham, H.Q.; Bonefas, E.; Houston, J.P.; Sampath, L.; Adams, K.E.; et al. Imaging of Lymph Flow in Breast Cancer Patients after Microdose Administration of a Near-Infrared Fluorophore: Feasibility Study. *Radiology* **2008**, *246*, 734–741. [CrossRef] [PubMed]
12. Kwon, S.; Sevick-Muraca, E.M. Noninvasive Quantitative Imaging of Lymph Function in Mice. *Lymphat. Res. Biol.* **2007**, *5*, 219–232. [CrossRef] [PubMed]
13. Sharma, R.; Wang, W.; Rasmussen, J.C.; Joshi, A.; Houston, J.P.; Adams, K.E.; Cameron, A.; Ke, S.; Kwon, S.; Mawad, M.E.; et al. Quantitative imaging of lymph function. *Am. J. Physiol. Heart Circ. Physiol.* **2007**, *292*, H3109–H3118. [CrossRef] [PubMed]
14. Dixon, J.B.; Zawieja, D.C.; Gashev, A.A.; Cote, G.L. Measuring microlymphatic flow using fast video microscopy. *J. Biomed. Opt.* **2005**, *10*, 064016. [CrossRef]
15. Zawieja, S.D.; Castorena-Gonzalez, J.A.; Dixon, B.; Davis, M.J. Experimental Models Used to Assess Lymphatic Contractile Function. *Lymphat. Res. Biol.* **2017**, *15*, 331–342. [CrossRef]
16. Blatter, C.; Meijer, E.F.J.; Nam, A.S.; Jones, D.; Bouma, B.E.; Padera, T.P.; Vakoc, B.J. In vivo label-free measurement of lymph flow velocity and volumetric flow rates using Doppler optical coherence tomography. *Sci. Rep.* **2016**, *6*. [CrossRef]
17. Quick, C.M.; Venugopal, A.M.; Gashev, A.A.; Zawieja, D.C.; Stewart, R.H. Intrinsic pump-conduit behavior of lymphangions. *Am. J. Physiol. Regul. Integr. Comp. Physiol.* **2007**, *292*, R1510–R1518. [CrossRef]
18. Macdonald, A.J.; Arkill, K.P.; Tabor, G.R.; McHale, N.G.; Winlove, C.P. Modeling flow in collecting lymphatic vessels: one-dimensional flow through a series of contractile elements. *Am. J. Physiol. Heart Circ. Physiol.* **2008**, *295*, H305–H313. [CrossRef]
19. Lobov, G.I.; Pan'kova, M.N.; Abdreshov, S.N. Phase and tonic contractions of lymphatic vessels and nodes under the action of atrial natriuretic peptide. *Reg. Blood Circ. Microcirc.* **2015**, *14*, 72–77. (In Russian) [CrossRef]
20. Ohhashi, T.; Mizuno, R.; Ikomi, F.; Kawai, Y. Current topics of physiology and pharmacology in the lymphatic system. *Pharmacol. Ther.* **2005**, *105*, 165–188. [CrossRef]

21. Venugopal, A.M.; Stewart, R.H.; Laine, G.A.; Dongaonkar, R.M.; Quick, C.M. Lymphangion coordination minimally affects mean flow in lymphatic vessels. *Am. J. Physiol. Heart Circ. Physiol.* **2007**, *293*, H1183–H1189. [CrossRef]
22. Bertram, C.D.; Macaskill, C.; Moore, J.E. Simulation of a Chain of Collapsible Contracting Lymphangions With Progressive Valve Closure. *J. Biomech. Eng.* **2010**, *133*. [CrossRef] [PubMed]
23. Margaris, K.N.; Black, R.A. Modelling the lymphatic system: challenges and opportunities. *J. R. Soc. Interface* **2012**, *9*, 601–612. [CrossRef] [PubMed]
24. Roose, T.; Tabor, G. Multiscale Modelling of Lymphatic Drainage. In *Multiscale Computer Modeling in Biomechanics and Biomedical Engineering*; Springer: Berlin/Heidelberg, Germany, 2012; pp. 149–176. [CrossRef]
25. Guyton, A.C.; Hall, J.E. *Textbook of Medical Physiology*; Logosphere: Moscow, Russia, 2008; p. 1296. (In Russian)
26. Schmidt, R.; Thews, G. *Human Physiology*; Mir: Moscow, Russia, 2005; Volume 2, p. 314. (In Russian)
27. McKinley, M.; O'Loughlin, V.D. *Human Anatomy*; McGraw-Hill: New York, NY, USA, 2012; p. 966.
28. Moore, J.E.; Bertram, C.D. Lymphatic System Flows. *Ann. Rev. Fluid Mech.* **2018**, *50*, 459–482. [CrossRef]
29. Borzyak, E.; Bocharov, V.; Sapin, M. *Human Anatomy*; Medicine: Moscow, Russia, 1993; Volume 2, p. 560. (In Russian)
30. Sinelnikov, R.; Sinelnikov, Y. *Atlas of Human Anatomy. The Doctrine of the Vessels*; Medicine: Moscow, Russia, 1996; Volume 3, p. 219. (In Russian)
31. Petrenko, V.M. *Functional Morphology of Lymphatic Vessels*; DEAN: Saint Petersburg, Russia, 2008; p. 400. (In Russian)
32. Lobov, G.I.; Pan'kova, M.N. Lymph transport in lymphatic nodes: mechanisms of regulation. *Ross. Fiziol. Zhurnal Im. I.M. Sechenova* **2013**, *98*, 1350–1361. (In Russian)
33. Zawieja, D.C. Contractile Physiology of Lymphatics. *Lymphat. Res. Biol.* **2009**, *7*, 87–96. [CrossRef] [PubMed]
34. Quéré, I. Description anatomique et histologique, physiologie du système lymphatique. *La Presse Médicale* **2010**, *39*, 1269–1278. [CrossRef] [PubMed]
35. Schulte-Merker, S.; Sabine, A.; Petrova, T.V. Lymphatic vascular morphogenesis in development, physiology, and disease. *J. Cell Biol.* **2011**, *193*, 607–618. [CrossRef] [PubMed]
36. Krstic, R. *Human Microscopic Anatomy: An Atlas for Students of Medicine and Biology*; World and Education: Moscow, Russia, 2010; p. 608. (In Russian)
37. Reddy, N.P.; Krouskop, T.A.; Paul, H.; Newell, j. Biomechanics of a Lymphatic Vessel. *J. Vasc. Res.* **1975**, *12*, 261–278. [CrossRef]
38. Gajani, G.S.; Boschetti, F.; Negrini, D.; Martellaccio, R.; Milanese, G.; Bizzarri, F.; Brambilla, A. A lumped model of lymphatic systems suitable for large scale simulations. In Proceedings of the 2015 European Conference on Circuit Theory and Design (ECCTD), Trondheim, Norway, 24–26 August 2015. [CrossRef]
39. Contarino, C.; Toro, E.F. A one-dimensional mathematical model of collecting lymphatics coupled with an electro-fluid-mechanical contraction model and valve dynamics. *Biomech. Model. Mechanobiol.* **2018**, *17*, 1687–1714. [CrossRef]
40. Mozokhina, A.S.; Mukhin, S.I. Pressure Gradient Influence on Global Lymph Flow. In *Trends in Biomathematics: Modeling, Optimization and Computational Problems*; Springer International Publishing: Berlin/Heidelberg, Germany, 2018; pp. 325–334. [CrossRef]
41. Tretyakova, R.M.; Lobov, G.I.; Bocharov, G.A. Modelling lymph flow in the lymphatic system: From 0D to 1D spatial resolution. *Math. Model. Nat. Phenom.* **2018**, *13*, 45. [CrossRef]
42. Milišić, V.; Quarteroni, A. Analysis of lumped parameter models for blood flow simulations and their relation with 1D models. *ESAIM Math. Model. Numer. Anal.* **2004**, *38*, 613–632. [CrossRef]
43. Kokalari, I.; Karaja, T.; Guerrisi, M. Review on lumped parameter method for modeling the blood flow in systemic arteries. *J. Biomed. Sci. Eng.* **2013**, *6*, 92–99. [CrossRef]
44. Rahbar, E.; Moore, J.E.J. A model of a radially expanding and contracting lymphangion. *J. Biomech.* **2011**, *44*, 1001–1007. [CrossRef]
45. Reddy, N.P.; Krouskop, T.A.; Newell, P.H. A computer model of the lymphatic system. *Comput. Biol. Med.* **1977**, *7*, 181–197. [CrossRef]
46. Sherwin, S.; Franke, V.; Peiró, J.; Parker, K. One-dimensional modelling of a vascular network in space-time variables. *J. Eng. Math.* **2003**, *47*, 217–250. [CrossRef]

47. Bunicheva, A.Y.; Mukhin, S.I.; Sosnin, N.V.; Khrulenko, A.B. Mathematical modeling of quasi-one-dimensional hemodynamics. *Comput. Math. Math. Phys.* **2015**, *55*, 1381–1392. [CrossRef]
48. Barnard, A.L.; Hunt, W.; Timlake, W.; Varley, E. A Theory of Fluid Flow in Compliant Tubes. *Biophys. J.* **1966**, *6*, 717–724. [CrossRef]
49. Phionik, O. Clinical and Morpho-Functional Bases for Diagnostic and Treatment of Lymphedema of Low Limbs. Ph.D. Thesis, St Petersburg University, Saint Petersburg, Russia, 2008. (In Russian)
50. Zweifach, B.W. Micropressure measurements in the terminal lymphatics. *Bibl. Anat.* **1973**, *12*, 361–365.
51. Mynard, J.P.; Davidson, M.R.; Penny, D.J.; Smolich, J.J. A simple, versatile valve model for use in lumped parameter and one-dimensional cardiovascular models. *Int. J. Numer. Methods Biomed. Eng.* **2011**, *28*, 626–641. [CrossRef]
52. Simakov, S.; Gamilov, T.; Soe, Y.N. Computational study of blood flow in lower extremities under intense physical load. *Russ. J. Numer. Anal. Math. Model.* **2013**, *28*. [CrossRef]
53. Mozokhina, A.S.; Mukhin, S.I. Quasi-One-Dimensional Flow of a Fluid with Anisotropic Viscosity in a Pulsating Vessel. *Differ. Equ.* **2018**, *54*, 938–944. [CrossRef]
54. Mozokhina, A.; Mukhin, S.; Koshelev, V. *Quasi-Onedimensional Approach for Modeling the Lymph Flow in the Lymphatic System*; MAKS Press: Moscow, Russia, 2017; p. 20. (In Russian)
55. Jamalian, S.; Bertram, C.D.; Richardson, W.J.; Moore, J.E. Parameter sensitivity analysis of a lumped-parameter model of a chain of lymphangions in series. *Am. J. Physiol. Heart Circ. Physiol.* **2013**, *305*, H1709–H1717. [CrossRef] [PubMed]
56. Quick, C.M.; Ngo, B.L.; Venugopal, A.M.; Stewart, R.H. Lymphatic pump-conduit duality: contraction of postnodal lymphatic vessels inhibits passive flow. *Am. J. Physiol. Heart Circ. Physiol.* **2009**, *296*, H662–H668. [CrossRef] [PubMed]
57. Vassilevski, Y.V.; Salamatova, V.Y.; Simakov, S.S. On the elasticity of blood vessels in one-dimensional problems of hemodynamics. *Comput. Math. Math. Phys.* **2015**, *55*, 1567–1578. [CrossRef]
58. Bertram, C.; Macaskill, C.; Moore, J. Pump function curve shape for a model lymphatic vessel. *Med. Eng. Phys.* **2016**, *38*, 656–663. [CrossRef]
59. Absi, R. Revisiting the pressure-area relation for the flow in elastic tubes: Application to arterial vessels. *Ser. Biomech.* **2018**, *32*, 47–59.
60. Macdonald, A.J. The Computational Modelling of Collecting Lymphatic Vessels. Ph.D. Thesis, University of Exeter, England, UK, 2008.
61. Mozokhina, A.; Lobov, G. Simulation of lymph flow with consideration of natural gravity force influence. *ITM Web Conf.* **2020**, *31*, 01003. [CrossRef]
62. Reddy, N.; Patel, K. A mathematical model of flow through the terminal lymphatics. *Med. Eng. Phys.* **1995**, *17*, 134–140. [CrossRef]
63. Mozokhina, A.S.; Mukhin, S.I.; Lobov, G.I. Pump efficiency of lymphatic vessels: numeric estimation. *Russ. J. Numer. Anal. Math. Model.* **2019**, *34*, 261–268. [CrossRef]
64. Franzone, P.C.; Pavarino, L.F.; Scacchi, S. *Mathematical Cardiac Electrophysiology*; Springer International Publishing: Berlin/Heidelberg, Germany, 2014. [CrossRef]
65. Tretyakova, R.; Savinkov, R.; Lobov, G.; Bocharov, G. Developing Computational Geometry and Network Graph Models of Human Lymphatic System. *Computation* **2017**, *6*, 1. [CrossRef]
66. Plasticboy. Plasticboy Pictures 2009 CC. Available online: http://www.plasticboy.co.uk/store/Human_Lymphatic_System_no_textures.html (accessed on 21 April 2020).
67. Mendoza, E.; Schmid-Schonbein, G.W. A Model for Mechanics of Primary Lymphatic Valves. *J. Biomech. Eng.* **2003**, *125*, 407–414. [CrossRef] [PubMed]
68. Heppell, C.; Richardson, G.; Roose, T. A Model for Fluid Drainage by the Lymphatic System. *Bull. Math. Biol.* **2012**, *75*, 49–81. [CrossRef] [PubMed]
69. Galie, P.; Spilker, R.L. A Two-Dimensional Computational Model of Lymph Transport Across Primary Lymphatic Valves. *J. Biomech. Eng.* **2009**, *131*. [CrossRef]
70. Novkovic, M.; Onder, L.; Cheng, H.W.; Bocharov, G.; Ludewig, B. Integrative Computational Modeling of the Lymph Node Stromal Cell Landscape. *Front. Immunol.* **2018**, *9*. [CrossRef]
71. Mueller, S.N.; Germain, R.N. Stromal cell contributions to the homeostasis and functionality of the immune system. *Nat. Rev. Immunol.* **2009**, *9*, 618–629. [CrossRef]

72. Turley, S.J.; Fletcher, A.L.; Elpek, K.G. The stromal and haematopoietic antigen-presenting cells that reside in secondary lymphoid organs. *Nat. Rev. Immunol.* **2010**, *10*, 813–825. [CrossRef]
73. Link, A.; Vogt, T.K.; Favre, S.; Britschgi, M.R.; Acha-Orbea, H.; Hinz, B.; Cyster, J.G.; Luther, S.A. Fibroblastic reticular cells in lymph nodes regulate the homeostasis of naive T cells. *Nat. Immunol.* **2007**, *8*, 1255–1265. [CrossRef]
74. Alberts, B.; Johnson, A.; Lewis, J.; Raff, M.; Roberts, K.; Walter, P. *Molecular Biology of the Cell*, 4th ed.; Garland Science: New York, NY, USA, 2002.
75. Bachmann, M.F.; Oxenius, A. Interleukin 2: from immunostimulation to immunoregulation and back again. *EMBO Rep.* **2007**, *8*, 1142–1148. [CrossRef]
76. Hochrein, H.; Shortman, K.; Vremec, D.; Scott, B.; Hertzog, P.; O'Keeffe, M. Differential Production of IL-12, IFN-α, and IFN-γ by Mouse Dendritic Cell Subsets. *J. Immunol.* **2001**, *166*, 5448–5455. [CrossRef]
77. Cooper, L.J.; Heppell, J.P.; Clough, G.F.; Ganapathisubramani, B.; Roose, T. An Image-Based Model of Fluid Flow Through Lymph Nodes. *Bull. Math. Biol.* **2015**, *78*, 52–71. [CrossRef] [PubMed]
78. Jafarnejad, M.; Woodruff, M.C.; Zawieja, D.C.; Carroll, M.C.; Moore, J. Modeling Lymph Flow and Fluid Exchange with Blood Vessels in Lymph Nodes. *Lymphat. Res. Biol.* **2015**, *13*, 234–247. [CrossRef]
79. Kislitsyn, A.; Savinkov, R.; Novkovic, M.; Onder, L.; Bocharov, G. Computational Approach to 3D Modeling of the Lymph Node Geometry. *Computation* **2015**, *3*, 222–234. [CrossRef]
80. Savinkov, R.; Kislitsyn, A.; Watson, D.J.; van Loon, R.; Sazonov, I.; Novkovic, M.; Onder, L.; Bocharov, G. Data-driven modelling of the FRC network for studying the fluid flow in the conduit system. *Eng. Appl. Artif. Intell.* **2017**, *62*, 341–349. [CrossRef]
81. Donovan, G.M.; Lythe, G. T cell and reticular network co-dependence in HIV infection. *J. Theor. Biol.* **2016**, *395*, 211–220. [CrossRef] [PubMed]
82. Pepper, M.S.; Tille, J.C.; Nisato, R.; Skobe, M. Lymphangiogenesis and tumor metastasis. *Cell Tissue Res.* **2003**, *314*, 167–177. [CrossRef] [PubMed]
83. Eymard, N.; Volpert, V.; Quere, I.; Lajoinie, A.; Nony, P.; Cornu, C. A 2D Computational Model of Lymphedema and of its Management with Compression Device. *Math. Model. Nat. Phenom.* **2017**, *12*, 180–195. [CrossRef]

© 2020 by the authors. Licensee MDPI, Basel, Switzerland. This article is an open access article distributed under the terms and conditions of the Creative Commons Attribution (CC BY) license (http://creativecommons.org/licenses/by/4.0/).

Article

Drift of Scroll Waves in a Mathematical Model of a Heterogeneous Human Heart Left Ventricle

Sergey Pravdin [1,2,†], Pavel Konovalov [3,†], Hans Dierckx [4] and Olga Solovyova [2,3] and Alexander V. Panfilov [2,5,*]

1. Krasovskii Institute of Mathematics and Mechanics, 620990 Ekaterinburg, Russia; sfpravdin@imm.uran.ru
2. Laboratory of Computational Biology and Medicine, Ural Federal University, 620075 Ekaterinburg, Russia; o.solovyova@iip.uran.ru
3. Institute of Immunology and Physiology UB RAS, 620049 Ekaterinburg, Russia; p.konovalov@iip.uran.ru
4. KULeuven Campus KULAK, 8500 Kortrijk, Belgium; h.dierckx@kuleuven.be
5. Department of Physics and Astronomy, Ghent University, 9000 Ghent, Belgium
* Correspondence: Alexander.Panfilov@ugent.be
† These authors contributed equally to this work.

Received: 24 March 2020; Accepted: 6 May 2020; Published: 12 May 2020

Abstract: Rotating spiral waves of electrical excitation underlie many dangerous cardiac arrhythmias. The heterogeneity of myocardium is one of the factors that affects the dynamics of such waves. In this paper, we present results of our simulations for scroll wave dynamics in a heterogeneous model of the human left ventricle with analytical anatomically based representation of the geometry and anisotropy. We used a set of 18 coupled differential equations developed by ten Tusscher and Panfilov (TP06 model) which describes human ventricular cells based on their measured biophysical properties. We found that apicobasal heterogeneity dramatically changes the scroll wave dynamics. In the homogeneous model, the scroll wave annihilates at the base, but the moderate heterogeneity causes the wave to move to the apex and then continuously rotates around it. The rotation speed increased with the degree of the heterogeneity. However, for large heterogeneity, we observed formation of additional wavebreaks and the onset of complex spatio-temporal patterns. Transmural heterogeneity did not change the dynamics and decreased the lifetime of the scroll wave with an increase in heterogeneity. Results of our numerical experiments show that the apex may be a preferable location of the scroll wave, which may be important for development of clinical interventions.

Keywords: spiral wave; heterogeneity; heart modeling; myocardium; left ventricle

---

1. Introduction

Vortices in excitable medium, which are called spiral waves in 2D or scroll waves in 3D, were found in many physical, chemical, and biological systems [1]. They play an important role in dynamics of these systems. For example, formation of such vortices in the heart causes cardiac arrhythmias, which remain the largest cause of death in the industrialized countries [2]. The dynamics of such vortex in the heart determines the type of cardiac arrhythmia. For example, the drift of a scroll wave in the ventricles of the heart may result in the onset of an arrhythmia called polymorphic ventricular tachycardia [3] and a breakup of the vortex into complex spatio-temporal patterns results in the onset of ventricular fibrillation, which causes cardiac arrest and sudden cardiac death [4]. Therefore, the factors responsible for drift of scroll waves are of great interest.

In a very general sense, the factors which can cause drift of a scroll wave include geometrical factors, anisotropy, and tissue heterogeneity. Previous research on generic models showed that, in homogeneous isotropic medium with high excitability, scroll waves drift to the regions where the thickness of the domain is minimal [5,6]. This result was recently confirmed in studies using detailed models of cardiac tissue [7]. Anisotropy of cardiac tissue also causes drift of vortices in 2D [8] and 3D [9]. Another important factor is the tissue heterogeneity, which will be a main focus of this paper.

The human heart, as in other mammals, has two main types of normal heterogeneity: transmural [10–12] and apicobasal [13,14]. The transmural heterogeneity in the human left ventricle (LV) involves an increase in action potential duration (APD) from subendocardial to subepicardial regions [15]. At the same time, data on the apicobasal heterogeneity are controversial: some papers report longer APD [13,14] while other show shorter APD [16] at the LV apex in comparison with the LV base. In a normal heart, this heterogeneity minimizes the dispersion of repolarization of cardiomyocytes from different regions and, as a consequence, makes the heart contraction maximally effective [17]. However, in pathological conditions, the heterogeneity amplifies and plays an important role during cardiac arrhythmias [18,19].

The effect of heterogeneity on vortex dynamics has mainly been studied for 2D vortices, known as spiral waves. It was shown in generic [20] and in ionic models [21] of cardiac tissue that heterogeneity in APD causes drift of a spiral wave to the regions of longer period. The effect of transmural heterogeneity on scroll waves was studied only in generic models [22,23], and it was shown that small heterogeneity transversal to the plane of rotation of a scroll wave results in the onset of a twisted scroll wave, i.e., the spiral waves have different rotation phase in different sections [22]. However, a large heterogeneity can cause a breakup of a scroll wave into a complex spatio-temporal pattern [23].

Because a real heart has regions with different thickness, anisotropy, and heterogeneity, all of these factors can potentially affect dynamics of scroll waves during cardiac arrhythmia. Recently, studies of effects of some of these factors on scroll wave dynamics in a model human ventricle were performed. A three-dimensional model describes the shape and anisotropy of the left ventricle of the human heart using a small number of parameters and has been verified against histological and DT-MRI data [24]. Using this geometrical model, the effect of the thickness of myocardial wall and anisotropy of cardiac tissue [7,25] has been studied. It was shown that the scroll wave usually stabilizes at a certain region of the ventricle; this location is mainly affected by the thickness of the wall and anisotropy.

Recently [26], we studied the effects of apicobasal heterogeneity using a two-variable model of cardiac tissue by Aliev and Panfilov [27]. This model belongs to a class of low-dimensional models of cardiac tissue which describe generic properties for waves propagating in the heart, but do not reproduce a detailed biophysical mechanism for cardiac excitation. We found that heterogeneity has some effect on the location of the scroll wave, but only if it was extremely large, 200 ms or more. Although such low-dimensional model reproduces the generic dynamics of waves in the heart [28], to obtain more qualitative estimates, one needs to use ionic models which describe the underlying mechanisms of action potential generation. Recently, one such study using a detailed ionic model for human ventricular cell TP06 was performed [29]. It was shown that the use of the ionic model substantially affected the observed dynamics of the vortices. In particular, it was found that, in most of the cases, we do not see any stabilization as we did in the AP model. On the contrary, the scroll wave drifts to the base of the heart and disappears. The main aim of this paper is to perform a study of scroll wave dynamics in a realistic ionic model of human ventricular tissue which includes not only correct shape and anisotropy, but also the heterogeneity of the heart ventricles.

2. Materials and Methods

2.1. Baseline Homogeneous Model. Numerical Approach and Software

The numerical computation of electrophysiological activity requires models at three levels: cell, tissue, and organ.

We used the LV anatomical model described in [24]. This model is axisymmetric and represents an average ventricle of a healthy adult human. The LV model was represented as a body of revolution with the shape fitted to experimental data [30]. The rotation axis is vertical axis Oz. An example of the

LV model is shown in Figure 1C. The epicardium is the colored surface, and the endocardium is the mesh surface.

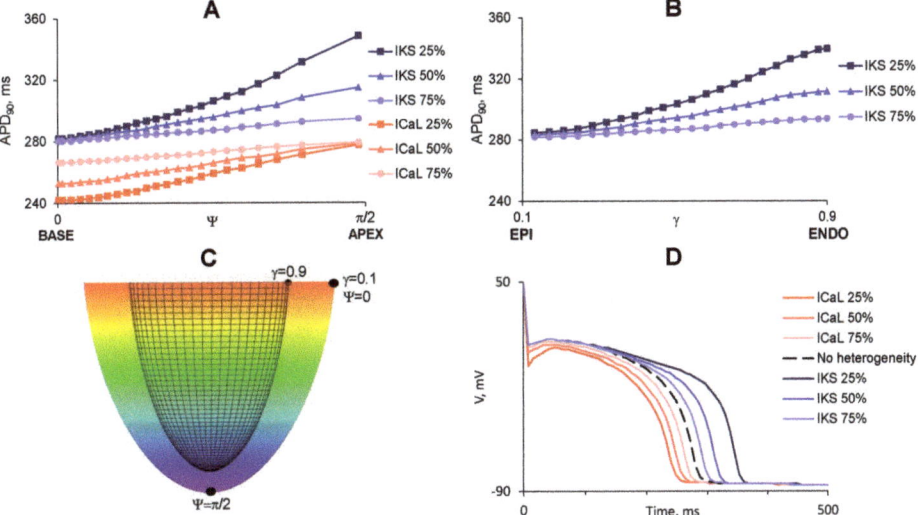

Figure 1. Distribution of APD_{90} (shown by color, from blue, 353 ms, to red, 248 ms) in the considered heterogeneous models of the LV. (**A**) apicobasal heterogeneity is set by a decrease of IKs at the apex (shades of blue) or ICaL at the base (shades of red). APD90 increases with increase of Ψ from 0 (base) to $\pi/2$ (apex); (**B**) Transmural heterogeneity is set by decrease of IKs at the subendocardium. $\gamma = 0.1$ means subepicardium cells, $\gamma = 0.9$, subendocardial; (**C**) points with maximal difference of APD90 in cases of apicobasal ($\Psi = 0$, $\Psi = \pi/2$) and transmural ($\gamma = 0.1$, $\gamma = 0.9$) heterogeneity. Color shows the z-coordinate; (**D**) plot of the action potential against time in LV points with maximal heterogeneity.

The model has the following list of parameters. R_b is the LV outer radius on the LV equator; Z_b is the LV height including the apical wall thickness; L is the LV wall thickness at the LV equator; h is the LV wall thickness at the LV apex; and $\epsilon \in [0,1]$ is a dimensionless parameter influencing the conicity–ellipticity characteristic of the LV shape.

A spherical-like local coordinate system (γ, ψ, ϕ) is connected with the LV model. Its coordinate γ gives position in the wall from the endocardium $\gamma = \gamma_0$ to the epicardium $\gamma = \gamma_1$; ψ is an analogue of geographical longitude with the base, $\psi = 0$, and the apex, $\psi = \pi/2$; $\phi \in [0, 2\pi)$.

The local coordinates are linked with the cylindrical coordinates (ρ, φ, z) by the following formulae:

$$\rho = (R_b - \gamma L)\left(\epsilon \cos \psi + (1 - \epsilon)(1 - \sin \psi)\right),$$
$$\varphi = \phi,$$
$$z = (Z_b - \gamma h)(1 - \sin \psi) + (1 - \gamma)h.$$

In every LV point, a vector of myofibre direction is defined as described in [24].

The parameters of the anatomical model were kept constant between all our simulations: outer (epicardial) radius at the base $R_b = 33$ mm, basal thickness $L = 12$ mm, full height $Z_b = 60$ mm, and ellipticity coefficient $\epsilon = 0.85$.

A set of parameters was taken as reference. The reference ventricle had anisotropy coefficients in ratio 9:1, transmural fibre rotation angle (FRA) of 147°, and apical LV thickness $h = 6$ mm.

The LV model allows us to easily change LV geometrical parameters—for example, to vary its thickness at the base and apex. These parameters are important for studying scroll waves drift which goes, accordingly with scroll waves theory, to a thin place if filament tension is positive.

For the tissue level, we used a monodomain approach [31] and anisotropic medium. Reaction–diffusion equations describe processes in the modelled body and have the following form:

$$\frac{\partial u}{\partial t} = \text{div}(D \text{ grad } u) - \frac{I_{ion}}{C_m}, \quad (1)$$

$$I_{ion} = I_{Kr} + I_{Ks} + I_{K1} + I_{to} + I_{Na} + I_{bNa} + I_{CaL} + I_{bCa} + I_{NaK} + I_{NaCa} + I_{pCa} + I_{pK}. \quad (2)$$

Here, u = u(r, t) is the transmembrane potential; the intracellular processes are captured by $I_{ion} = I_{ion}(r,t)$, which is the sum of the ionic transmembrane currents. C_m is the capacitance of the cell membrane. In a general sense, this equation shows that the potential difference V_m over the cell membrane changes due to currents (I) which flow through the membrane. These currents are conveyed by different ions (Na, K, Ca) and via various biophysical processes. In modern computational cardiac electrophysiology, the properties of all such currents are fitted to their experimentally measured values and their dynamics is fitted using additional differential equations. Each of the currents typically depends on V_m and time, and the time-dependency is normally given by an exponential relaxation equation for the gating variable(s) g. For example, consider a hypothetical current I_* which conveys ion '*':

$$I_* = G_* g_*^\alpha g_\bullet^\beta (V_m - V_*) \quad (3)$$

$$\frac{dg_i}{dt} = \frac{g_i^\infty(V_m) - g_i}{\tau_i(V_m)}, \quad i = *, \bullet, \quad (4)$$

It has a maximal conductivity of $G_* = const$ and is zero for V_*, the so-called Nernst potential for ion '*'. This Nernst potential can be easily computed from the concentrations of specific ions outside and inside the cell membrane. The time dynamics of this current is given by two gating variables $g_*, g_\bullet$ to the power α, β. The variables $g_*, g_\bullet$ approach their voltage-dependent steady state values $g_i^\infty(V_m)$ with characteristic time $\tau_i(V_m)$ (4). All parameters and functions are chosen here to fit experimentally measured properties of the specific ionic current. Most of ionic currents have one or two gating variables, with $\alpha = \beta = 1$. In our simulations we use equations derived and fitted in [32], which has 18 state variables. Note that a proper description of ionic currents is very important as most cardiac drugs affect the maximal conductivities of specific ion channels. Furthermore, cardiac tissue is heterogeneous with respect to expression levels of these channels. Thus, using Equations (2)–(4), one can model the effect of pharmacological preparations on cardiac tissue, and, as we do in this paper, study effects of heterogeneity of the heart on cardiac arrhythmias. One of the standard ways to do so is to make the maximal conductivity of a specific current dependent on space in a manner based on experimental data. As described in the next subsection, the most important currents accounting for the heterogeneity are I_{CaL} and I_{Ks}.

As in [33], the diffusion matrix $D = (D^{ij})$ was computed from the unit vectors in fibre direction v using the formula

$$D^{ij} = D_2 \delta_{i,j} + (D_1 D_2) v_i v_j, \quad (5)$$

where D_1 and D_2 are the diffusion coefficients along and across the fibres and $\delta_{i,j}$ is the Kronecker symbol.

Ionic model parameters were taken from [32]; they correspond to physiological characteristics of cardiomyocytes.

A scroll wave was initiated using a standard protocol S1S2 [29] and rotated counterclockwise. We used a standard boundary condition – zero flux of potential through the LV surface. A uniform grid with spatial step $dr = 0.28$ mm was set in Cartesian coordinates.

To solve the differential Equations (1)–(3), we used a finite difference approach. To approximate the diffusion term, we used a stencil of 18 points for 3D using the following equation:

$$L(i,j,k) = \sum_{l=0}^{18} w_l V_m(l) \qquad (6)$$

where l is an index over the 18 neighbors of the point (i,j,k) including itself, and w_l is the weight of the voltage of a particular neighboring grid point which accounts for its contribution to the Laplacian. The weights w_l were computed based on the conduction tensor D_{ij} and location of the point with respect to the local boundary. All points outside the heart geometry have weights $w_l = 0$. The weights were precomputed for each geometry used and allowed an efficient evaluation of the Laplacian during the simulations. The gating variables in the TP06 model were integrated using the Rush and Larsen approach [34]. To speed up calculations, voltage-dependent functions in gating variables equations were precomputed and placed into look-up tables.

Additional details on the integration procedure can be found, for example, in [33].

We studied scroll wave dynamics depending on LV wall thickness at the apex and degree of heterogeneity. In our simulations, LV basal thickness was $L = 12$ mm, while apical thickness was varied to $h = 6, 12, 18$ mm. We considered the physiological value of the anisotropy ratio $D_1/D_2 = 9$.

Our procedure of postprocessing filament coordinates was thoroughly described in [29].

All calculations were performed on a C program on clusters "URAN" (IMM, Ural Branch of RAS) and "UrFU" of Ural Federal University (Ekaterinburg). The program uses CUDA for GPU parallelization and was compiled with a Nvidia C Compiler "nvcc". Computational nodes have graphical cards Tesla K40m0.

2.2. Heterogeneity Representation

The heterogeneity in the ventricular myocardium is a result of changes of balance between inward currents (mainly ICaL) and outward potassium currents. Although in reality these currents are changing in combination, the exact degree of modifiication is not fully quantified. In order to compare effects of the heterogeneities in the same conditions, we decided to study a separate effect of ICaL and one of the most important potassium currents IKs.

We considered LV models with the following types of apicobasal heterogeneity:

1. Apicobasal heterogeneity caused by decrease of ICaL current with $APD_{base} < APD_{apex}$. We reduced ICaL current at the LV base to 75%, 50%, and 25% of its original value, which resulted in the gradients of 14, 28, and 38 ms. We denote these cases as ICaL-75, ICaL-50 and ICaL-25.
2. Apicobasal heterogeneity caused by decrease of IKs current with $APD_{base} < APD_{apex}$. We reduced IKs at the apex to 75%, 50%, and 25% of its original value, which resulted in the APD gradients of 14, 34, and 68 ms. We denote these cases as IKs-75, IKs-50, and IKs-25.

To represent transmural (TM) heterogeneity, we decreased the IK current because IK is the main current affecting APD, and it is responsible for transmural heterogeneity [35]. We followed the work [36] and made models with $APD_{epi} < APD_{endo}$ by multiplying IKs conductivity by 75%, 50%, and 25%, which resulted in the APD gradients of 12, 31, and 58 ms. We denote these cases as IKs-75-TM, IKs-50-TM, and IKs-25-TM.

The effects of the changes of the currents on AP shape are shown in Figure 1D. Figure 1A shows distribution of APD from apex to base for each type of apicobasal heterogeneity and Figure 1B shows the three types of transmural heterogeneity.

3. Results

Below, we present results of our simulations for scroll wave dynamics in ventricles of different geometries and various types of heterogeneity.

We considered three geometries with the same anisotropy but different in the thickness of the apex $h = 6$, $h = 12$, and $h = 18$ mm. In all these geometries in homogeneous ventricles, the scroll wave drifted to the base and disappeared. We studied the dynamics of the scroll waves with the same initial conditions, but in the presence of the heterogeneity.

We considered two types of regional heterogeneity: apicobasal and transmural.

3.1. Apicobasal Heterogeneity

Figure 2A,B show a typical trajectory of the scroll wave drift in the absence of heterogeneity. The scroll was initially located between apex and base, then drifted upwards and disappeared after collision with the boundary at the base. We found that the presence of apicobasal heterogeneity substantially changed the scroll wave dynamics. Figure 2C shows an example of scroll wave drift from the same initial conditions, but for a small apicobasal heterogeneity of 14 ms, IKs-75. We see that the scroll wave drifts to the apex, stabilizes at 8 mm from it, and continues its circumferential rotation with a speed of 0.19 mm/s.

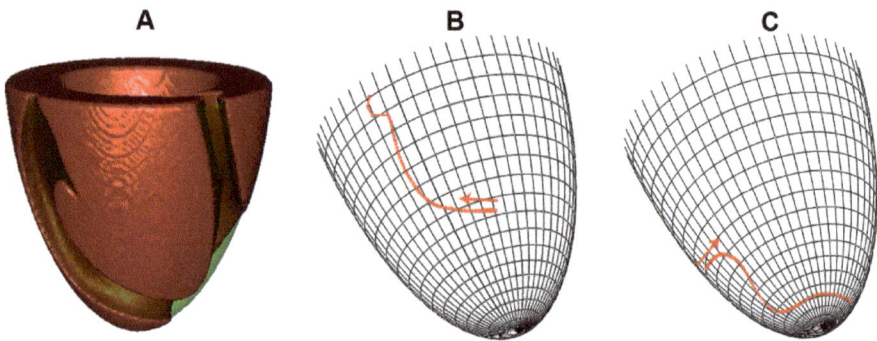

Figure 2. Examples of average filament trajectory. (**A**) typical view of the scroll wave in the absence of heterogeneity; (**B**) no heterogeneity; drift to the LV base, where the scroll wave disappeared; (**C**) apicobasal heterogeneity of 14 ms, IKs-75; drift to the apex, where filament approached a dynamic attractor with vertical coordinate $\Psi = 1.24$. The apical thickness h was 6 mm.

We performed 18 computations for six types of the heterogeneity in three LVs of different geometry. The results are presented in Figure 3, where we mark the type of observed dynamics (e.g., annihillaiton at the base and breakup) and in case of stabilisation at attractor we show the coordinates of the attractor, and the drift speed along it.

Our findings were the following. From the 18 cases studied, in only one case (IKs-75 and $h = 18$ mm), we observed the same dynamics as in the homogeneous case: annihilation at the base.

In 12 out of 18 cases, we observed a dramatic change of dynamics: instead of annihilation at the base, the scroll wave stabilized at the apex and then continued moving around it along the circle. One example of such dynamics was already shown in Figure 2C, where the filament stabilized at 8 mm from the apex and rotated with a speed of 0.19 mm/s. The apex-base heterogeneity in that case was 14 ms (the corresponding grey columns in Figure 3A). For the same geometry of $h = 6$ mm and a larger heterogeneity (IKs-25, $DR_{AB} = 68$ ms), the scroll wave filament stabilized also at approximately 8 mm from the apex ($\psi = 1.24$) and was moving along the circle with speed 2.27 mm/s. The speed of the motion along the attractor increased with the increase in the heterogeneity. Note that the location of attractor in most of the cases did not depend on the degree of heterogeneity. However, for $h = 18$ mm, we saw that the location of the attractor for ICaL-50 was closer to the base than for ICaL-75.

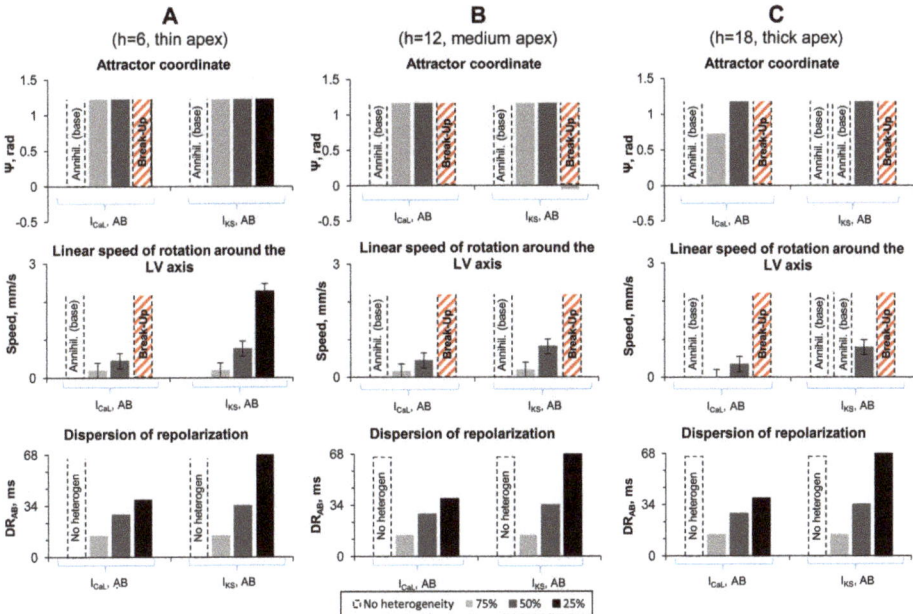

Figure 3. Results of numerical experiments on LV models with apicobasal heterogeneity. (**A–C**) LV models with apical thickness 6, 12, and 18 mm, respectively. Top row: Attractor coordinate. Dotted bars mean that scroll wave disappeared at the LV base. Red dotted bars show the cases with wave break up. Middle row: Linear speed of rotation around the LV axis. Bottom row: Apicobasal dispersion of repolarization (DR_{AB}) for points with coordinates $\Psi = 0$ and $\Psi = \pi/2$ (see Figure 1).

In the five other cases, all with large heterogeneity, we observed a more complex dynamics. Here, the initial scroll wave generated waves with a high frequency. Such waves were propagating toward the apex where the APD and refractory period were the longest and tissue was not able to recover. As a result, we observed formation of the wave breaks. Figure 4 shows an example of such process. Here, the scroll wave, whose tip is located at the other side of the LV, generates two wavebreaks (Figure 4B) which evolve to a complex spatiotemporal pattern (Figure 4C). This occurred in cases IKs-25 and ICaL-25.

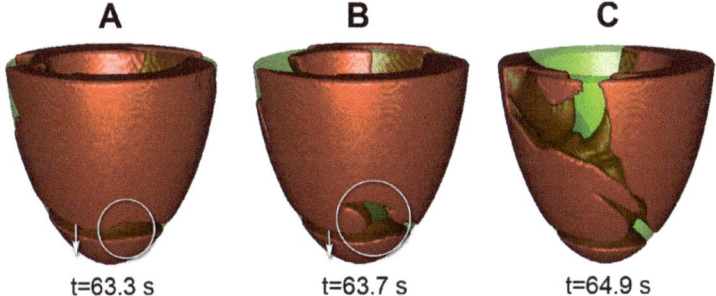

Figure 4. Example of wave break on the LV with apical thickness 6 mm, apicobasal heterogeneity ICaL-25. (**A–C**) show wave patterns at t = 63.3 s; t = 63.7 s.; t = 64.9 s.

Interestingly, long filaments with both ends at the epicardial surface sometimes emerged (Figure 5). Such filament may have a complex behavior which can lead to annihilation of the pattern or formation of new filaments. However, they never stabilized to attractors as in the previous 12 cases.

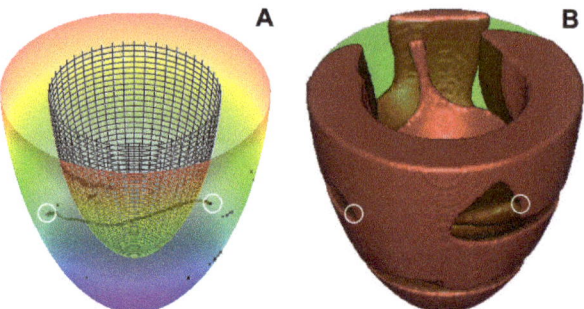

Figure 5. Scroll wave in the LV can have a filament with both ends at the epicardium. (**A**) the filament shape at t = 25.4 s for the LV with apical thickness h = 6 mm, apicobasal heterogeneity, IKs-25. We see that the filament is located between the LV epicardium (colorful surface) and endocardium (meshy surface); (**B**) the scroll wave at the same time (red area shows excited cells). Ends of the filament are marked with white circles.

Overall, we can say that, while a moderate increase in heterogeneity resulted in stabilization of filament towards the region of longer APD, a larger degree of heterogeneity resulted in the formation of additional wavebreaks and complex spatio-temporal patterns.

3.2. Transmural Heterogeneity

To investigate the effect of transmural heterogeneity, we performed nine simulations: three LVs with different geometry with three degrees of heterogeneity, where we changed IKs. This is because IK is the main ionic current responsible for the transmural heterogeneity. The results of these simulations are shown in Figure 6.

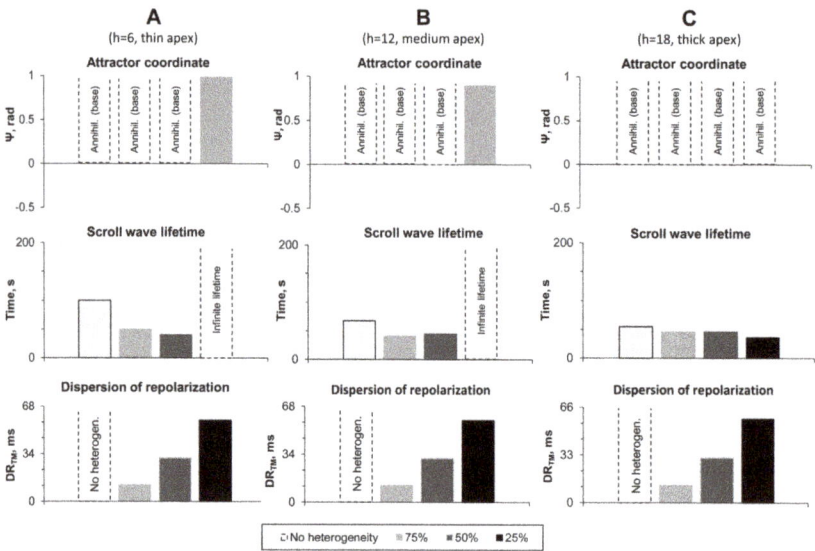

Figure 6. Results of numerical experiments on LV models with transmural heterogeneity. (**A–C**): LV models with apical thickness 6, 12, and 18 mm, respectively. Top row: Attractor coordinates. Dotted bars mean that scroll wave disappeared at the LV base. Middle row: Scroll wave lifetime (200 means it survived until the end of the simulation). Bottom row: Transmural dispersion of repolarization (DR_{TM}) for points with coordinates $\gamma = 0.1$ and $\gamma = 0.9$ (see Figure 1C).

In seven out of nine cases, we observed the same type of dynamics as in the absence of the heterogeneity. The scroll wave drifted to base and annihilated there. We saw, however, that the lifetime of the scroll wave decreased with increase in heterogeneity. In two cases (IKs-25-TM; apex thickness 6 and 12 mm), the scroll wave stabilized at the apex and either rotated with a velocity of 0.19 mm/s ($h = 6$ mm) or did not move at all ($h = 12$ mm).

4. Discussion

In this paper, we study the effects of heterogeneity on scroll wave dynamics in a model of the left ventricles of the human heart. We used a detailed ionic model for human ventricular cell and analytical representation of the geometry and anisotropy. We considered two types of heterogeneity: apicobasal and transmural, and we represent them by changing ICaL and IK currents.

The apicobasal heterogeneity was found to change the dynamics of the scroll wave. Instead of annihilation at the base, the scroll wave in most of the cases stabilized at the apex. This result can be explained by the following known dynamics of the spiral wave in cardiac tissue. In [20,21], it was shown that spiral waves drift to the region of longer APD. Because in our case APD was longest at the apex, we can assume that stabilization at the apex is a result of such dynamics. Note that for $h = 6$ mm and 12 mm the location of the attractor did not change with the degree of the heterogeneity. However, for $h = 18$ mm, we had a gradual change of the location of the attractor $\Psi = 0.72$ for ICaL-75, while $\Psi = 1.2$ for ICaL-50. In addition, for $h = 18$ mm and IKs-75, the scroll wave still drifted to the base and annihilated there. The mechanisms of the observed phenomenon can be the following. In case $h = 18$ mm, the apex is substantially thicker than the base. Because in the absence of heterogeneities the scroll wave drifts toward the region of minimal thickness, such gradient attempts to push the scroll wave towards the base. On the other hand, the apicobasal heterogeneity thrusts the scroll wave toward the apex. As a result of interaction of these opposite processes, we observe a stabilization of the scroll closer to the apex for larger heterogeneity in ICaL. In addition, the small heterogeneity IKs-75 was not sufficient to stabilize the scroll and it annihilated at the base.

Another important type of the dynamics is a breakup. Basically in all cases of large heterogeneity, some form of breakup was observed. These results can also be explained in the following way. The formation of wavebreak at the regions with a longer refractory period (APD) is one of the most classical mechanisms of generation of spiral waves in 2D [37]. As in our case, the breakup occurred via formation of wavebreak at the apex, where the APD is longest, we think that the mechanism [37] is also responsible for the onset of breakup in our case.

For the transmural heterogeneity, we did not observe substantial effects on the filament dynamics. It did not change the location of the attractor: in almost all cases as in the control case, the scroll wave drifted towards the base and annihilated there. We saw a slight decrease in the life time of the scroll wave, but the effect was small and not significant. In a few cases of largest heterogeneity, we saw stabilization at the apex. Unfortunately, the mechanisms of such behaviour are not clear. In [22,23], it was shown that transmural heterogeneity affects the period of rotation of a scroll wave, and in extreme cases can cause its breakup. However, studies there were perfomed in a simplified model of cardiac tissue. In [38], a sproing (elongation) of the filament was also observed in a simplified model of cardiac tissue. However, we did not observe such phenomena and found effects that require additional study.

It should be noted that filaments have complex dynamics; however, not all complexity of the dynamics comes from the heterogeneity. As pointed out by Papadimitriou [39], the complexity does not necessarily require heterogeneity and that is definitely true for filaments. For example, even in completely homogeneous cardiac tissue, for certain parameter values, one can have a breakup of filaments due to negative filament tension [40], or dynamic instabilities [41]. In addition, anisotropy by itself can produce complex filament shapes [42]. It would be interesting to further characterize complexity of filament dynamics in the heart in a wider range of parameters and study if heterogeneity will be a determinant for their dynamics as it was in our study.

In a previous paper [26], we studied scroll wave drift on the Aliev–Panfilov model and symmetrical LV model with isotropy and anisotropy 9:1 in homogeneous and heterogeneous myocardium. In that paper, we also found a tendency of scroll wave to move towards the apex and an increase of the rotation speed of waves with an increase in the heterogeneity. The drift speed in the AP model was smaller than one found in this paper, which is a result of using of a more accurate model for cardiac excitation.

Overall, these results show that the scroll wave drifts and stabilizes at the apex of the left ventricle in the presence of moderate apicobasal heterogeneity. Thus, independently of their initial location, the scroll waves are after some transient processes more likely to be observed at the apex of the heart. Therefore, the apex may be a preferable location of the scroll wave. Identification of the location of a source of cardiac arrhythmias is an important question in practical cardiology. One of the most effective ways to treat cardiac arrhythmia is a cardiac ablation procedure, during which flexible catheters are passed into the heart through a blood vessel. Once located near the target region, the catheter is activated to locally destroy or modify the arrhythmia substrate by delivering radio-frequency energy or extreme cold. Thus, locating the arrhythmia source is a problem of great importance. In this paper we show that the source of cardiac arrhythmia is more likely to be located at the apex of the heart. Hence, the modification of the apex by some sort of ablation, which makes rotation of scroll wave impossible, can eliminate the cardiac arrhythmia. Recently, a non-invasive method of ablation was proposed, which enables modifying the tissue not only on the surface but also inside the myocardial wall by noninvasive delivery of radiation with stereotactic body radiation [43]. For such methodology, the location of the arrhythmia source is again crucial. Note, however, that ablation in 3D ventricle to eliminate a scroll wave is a non-trivial issue and needs to be further studied, e.g., using a modeling methodology applied in our paper.

A limitation of our approach is that we considered only spiral waves with one chirality. The chirality may substantially affect the dynamics of scroll waves [29]. In this paper, we chose that particular case for counterclockwise rotation as without heterogeneity we had always one regime (annihilation at the base). It was thus straightforward to study the effect of heterogeneity in that situation. The effect of chirality in the presence of heterogeneity should be investigated in a subsequent study.

We studied transmural and apicobasal heterogeneity separately. It would be also interesting to study their combined effect on scroll wave dynamics.

In this paper, we performed our studies only in a limited parameter range. This is because such modeling is extremely challenging, as the studied effects can be observed only in anatomical models of the heart on a long time scale (of the order of minutes) while typical simulations in anatomical models of the heart are normally performed for a few seconds (see e.g., [44]). It would be interesting in the future to extend studies for a larger parameter range and also model direct clinical interventions which can stop arrhythmias caused by scroll waves, such as catheter ablation.

5. Conclusions

We performed a study of filament dynamics in the left ventricle of the heart using a state-of-art ionic model for cardiac tissue and combining all the most important factors which can be responsible for the filament drift: the shape of the ventricle, variation in thickness, anisotropy, and heterogeneity. Under those conditions, we find that, in spite of the importance of all factors, the apicobasal heterogeneity was the most important in determining the final position of the filament. For moderate heterogeneity, the filament was stabilized at the apex of the ventricle, while for a large heterogeneity a break-up into a complex spatio-temporal pattern was observed. The transmural heterogeneity did not substantially affect the filament dynamics. As predicting of location of the arrhythmia sources is important for clinical interventions, our results can help in developing clinical procedures to remove arrhythmias organized by filaments.

Author Contributions: Conceptualization, A.V.P. and O.S.; methodology, A.V.P. and S.P.; software S.P., H.D., and P.K.; formal analysis, S.P.; investigation, P.K.; writing—original draft preparation, P.K. and S.P.; writing—review and editing, P.K., S.P., A.V.P., H.D., and O.S.; supervision, A.V.P. and O.S. All authors have read and agreed to the published version of the manuscript.

Funding: P.K., S.P., O.S., and A.V.P. were funded by the Russian Science Foundation (project 14-35-00005). A.V.P., P.K., and O.S. were funded by the Russian Foundation for Basic Research (#18-29-13008). A.V.P. and O.S. were funded by RF Government Act #211 of 16 March 2013 (agreement 02.A03.21.0006). P.K. and O.S. work was carried out within the framework of the IIF UrB RAS theme No. AAAA-A18-118020590031-8. A.V.P. and H.D. were partially funded by BOF Ghent University.

Acknowledgments: Simulations were performed at the supercomputer Uran of Institute of Mathematics and Mechanics (Ekaterinburg, Russia) and at the supercomputer of Ural Federal University (Ekaterinburg, Russia).

Conflicts of Interest: The authors declare no conflict of interest.

References

1. Winfree, A.T.; Strogatz, S.H. Organizing centres for three-dimensional chemical waves. *Nature* **1984**, *311*, 611–615. [CrossRef] [PubMed]
2. Mehra, R. Global public health problem of sudden cardiac death. *J. Electrocardiol.* **2007**, *40*, S118–S122. [CrossRef] [PubMed]
3. Gray, R.A.; Jalife, J.; Panfilov, A.; Baxter, W.T.; Cabo, C.; Davidenko, J.M.; Pertsov, A.M. Nonstationary Vortexlike Reentrant Activity as a Mechanism of Polymorphic Ventricular Tachycardia in the Isolated Rabbit Heart. *Circulation* **1995**, *91*, 2454–2469. [CrossRef] [PubMed]
4. Katz, A.M. *Physiology of the Heart*; Lippincott Williams & Wilkins: Philadelphia, PA, USA, 2010.
5. Panfilov, A.; Rudenko, A.; Krinsky, V. Scroll rings in three-dimensional active medium with two component diffusion. *Biofizika* **1986**, *31*, 850–854.
6. Panfilov, A.; Aliev, R.; Mushinsky, A. An integral invariant for scroll rings in a reaction-diffusion system. *Phys. D Nonlinear Phenom.* **1989**, *36*, 181–188. [CrossRef]
7. Pravdin, S.; Dierckx, H.; Markhasin, V.S.; Panfilov, A.V. Drift of Scroll Wave Filaments in an Anisotropic Model of the Left Ventricle of the Human Heart. *BioMed Res. Int.* **2015**, *2015*. [CrossRef]
8. Rogers, J.M.; McCulloch, A.D. Nonuniform muscle fiber orientation causes spiral wave drift in a finite element model of cardiac action potential propagation. *J. Cardiovasc. Electrophysiol.* **1994**, *5*, 496–509. [CrossRef]
9. Dierckx, H.; Bernus, O.; Verschelde, H. A geometric theory for scroll wave filaments in anisotropic excitable media. *Phys. D Nonlinear Phenom.* **2009**, *238*, 941–950. [CrossRef]
10. Nabauer, M.; Beuckelmann, D.J.; Uberfuhr, P.; Steinbeck, G. Regional Differences in Current Density and Rate-Dependent Properties of the Transient Outward Current in Subepicardial and Subendocardial Myocytes of Human Left Ventricle. *Circulation* **1996**, *93*, 168–177. [CrossRef]
11. Glukhov, A.V.; Fedorov, V.V.; Lou, Q.; Ravikumar, V.K.; Kalish, P.W.; Schuessler, R.B.; Moazami, N.; Efimov, I.R. Transmural Dispersion of Repolarization in Failing and Nonfailing Human Ventricle. *Circ. Res.* **2010**, *106*, 981–991. [CrossRef]
12. Opthof, T.; Janse, M.J.; Meijborg, V.M.; Cinca, J.; Rosen, M.R.; Coronel, R. Dispersion in ventricular repolarization in the human, canine and porcine heart. *Prog. Biophys. Mol. Biol.* **2016**, *120*, 222–235. [CrossRef] [PubMed]
13. Franz, M.R.; Bargheer, K.; Rafflenbeul, W.; Haverich, A.; Lichtlen, P.R. Monophasic action potential mapping in human subjects with normal electrocardiograms: Direct evidence for the genesis of the T wave. *Circulation* **1987**, *75*, 379–386. [CrossRef] [PubMed]
14. Janse, M.J.; Coronel, R.; Opthof, T.; Sosunov, E.A.; Anyukhovsky, E.P.; Rosen, M.R. Repolarization gradients in the intact heart: Transmural or apico-basal? *Prog. Biophys. Mol. Biol.* **2012**, *109*, 6–15. [CrossRef]
15. Boukens, B.J.; Sulkin, M.S.; Gloschat, C.R.; Ng, F.S.; Vigmond, E.J.; Efimov, I.R. Transmural APD gradient synchronizes repolarization in the human left ventricular wall. *Cardiovasc. Res.* **2015**, *108*, 188–196. [CrossRef] [PubMed]
16. Szentadrassy, N.; Banyasz, T.; Biro, T.; Szabo, G.; Toth, B.I.; Magyar, J.; Lazar, J.; Varro, A.; Kovacs, L.; Nanasi, P.P. Apico-basal inhomogeneity in distribution of ion channels in canine and human ventricular myocardium. *Cardiovasc. Res.* **2005**, *65*, 851–860. [CrossRef]

17. Solovyova, O.; Katsnelson, L.; Konovalov, P.; Kursanov, A.; Vikulova, N.; Kohl, P.; Markhasin, V. The cardiac muscle duplex as a method to study myocardial heterogeneity. *Prog. Biophys. Mol. Biol.* **2014**, *115*, 115–128. [CrossRef]
18. Antzelevitch, C. Heterogeneity and cardiac arrhythmias: An overview. *Heart Rhythm* **2007**, *4*, 964–972. [CrossRef]
19. Keldermann, R.H.; ten Tusscher, K.H.; Nash, M.P.; Hren, R.; Taggart, P.; Panfilov, A.V. Effect of heterogeneous APD restitution on VF organization in a model of the human ventricles. *Am. J. Physiol.-Heart Circ. Physiol.* **2008**, *294*, H764–H774. [CrossRef]
20. Rudenko, A.; Panfilov, A. Drift and interaction of vortices in two-dimensional heterogeneous active medium. *Stud. Biophys.* **1983**, *98*, 183–188.
21. Ten Tusscher, K.; Panfilov, A.V. Reentry in heterogeneous cardiac tissue described by the Luo–Rudy ventricular action potential model. *Am. J. Physiol.-Heart Circ. Physiol.* **2003**, *284*, H542–H548. [CrossRef]
22. Panfilov, A.V.; Rudenko, A.N.; Pertsov, A.M. Twisted scroll waves in three- dimensional active media. *Dokl. Akad. Nauk SSSR* **1984**, *279*, 1000–1002.
23. Panfilov, A.V.; Keener, J.P. Twisted scroll waves in heterogeneous excitable media. *Int. J. Bifurc. Chaos* **1993**, *3*, 445–450. [CrossRef]
24. Pravdin, S.F.; Berdyshev, V.I.; Panfilov, A.V.; Katsnelson, L.B.; Solovyova, O.; Markhasin, V.S. Mathematical model of the anatomy and fibre orientation field of the left ventricle of the heart. *Biomed. Eng. Online* **2013**, *12*, 54. [CrossRef]
25. Pravdin, S.F.; Dierckx, H.; Katsnelson, L.B.; Solovyova, O.; Markhasin, V.S.; Panfilov, A.V. Electrical wave propagation in an anisotropic model of the left ventricle based on analytical description of cardiac architecture. *PLoS ONE* **2014**, *9*. [CrossRef]
26. Konovalov, P.V.; Pravdin, S.F.; Solovyova, O.E.; Panfilov, A.V. Scroll Wave dynamics in a model of the heterogeneous heart. *JETP Lett.* **2016**, *104*, 821–821. [CrossRef]
27. Aliev, R.R.; Panfilov, A.V. A simple two-variable model of cardiac excitation. *Chaos Solitons Fractals* **1996**, *7*, 293–301. [CrossRef]
28. Panfilov, A. Three-dimensional organization of electrical turbulence in the heart. *Phys. Rev. E* **1999**, *59*, R6251–R6254. [CrossRef]
29. Pravdin, S.F.; Dierckx, H.; Panfilov, A.V. Drift of scroll waves in a generic axisymmetric model of the cardiac left ventricle. *Chaos Solitons Fractals* **2019**, *120*, 222–233. [CrossRef]
30. Streeter, D., Jr. The Heart. In *Handbook of physiology*; Section 2; American Physiological Society: Bethesda, MD, USA, 1979; Chapter Gross Morphology and Fiber Geometry of the Heart; Volume I, pp. 61–112.
31. Panfilov, A.; Holden, A. Computer simulation of re-entry sources in myocardium in two and three dimensions. *J. Theor. Biol.* **1993**, *161*, 271–285. [CrossRef]
32. ten Tusscher, K.; Panfilov, A.V. Alternans and spiral breakup in a human ventricular tissue model. *Am. J. Physiol.-Heart Circ. Physiol.* **2006**, *291*, H1088–H1100. [CrossRef]
33. Kazbanov, I.V.; Clayton, R.H.; Nash, M.P.; Bradley, C.P.; Paterson, D.J.; Hayward, M.P.; Taggart, P.; Panfilov, A.V. Effect of Global Cardiac Ischemia on Human Ventricular Fibrillation: Insights from a Multi-scale Mechanistic Model of the Human Heart. *PLoS Comput. Biol.* **2014**, *10*, e1003891. [CrossRef] [PubMed]
34. Rush, S.; Larsen, H. A practical algorithm for solving dynamic membrane equations. *IEEE Trans. Biomed. Eng.* **1978**, *4*, 389–392. [CrossRef] [PubMed]
35. Pereon, Y.; Demolombe, S.; Baro, I.; Drouin, E.; Charpentier, F.; Escande, D. Differential expression of KvLQT1 isoforms across the human ventricular wall. *Am. J. Physiol.-Heart Circ. Physiol.* **2000**, *278*, H1908–H1915. [CrossRef]
36. ten Tusscher, K.; Noble, D.; Noble, P.J.; Panfilov, A.V. A model for human ventricular tissue. *Am. J. Physiol.-Heart Circ. Physiol.* **2004**, *286*, H1573–H1589. [CrossRef] [PubMed]
37. Krinskii, V. Spread of excitation in an inhomogeneous medium (state similar to cardiac fibrillation). *Biophysics-USSR* **1966**, *11*, 676–683.
38. Henze, C.; Lugosi, E.; Winfree, A. Helical organizing centers in excitable media. *Can. J. Phys.* **1990**, *68*, 683–710. [CrossRef]
39. Papadimitriou, F. Geo-mathematical modeling of spatial-ecological complex systems: An evaluation. *Geogr. Environ. Sustain.* **2010**, *3*, 67–80. [CrossRef]

40. Biktashev, V.; Holden, A.; Zhang, H. Tension of organizing filaments of scroll waves. *Philos. Trans. R. Soc. London. Ser. A Phys. Eng. Sci.* **1994**, *347*, 611–630.
41. Panfilov, A.V.; Hogeweg, P. Mechanisms of cardiac fibrillation. *Science* **1995**, *270*, 1223–1224.
42. Verschelde, H.; Dierckx, H.; Bernus, O. Covariant stringlike dynamics of scroll wave filaments in anisotropic cardiac tissue. *Phys. Rev. Lett.* **2007**, *99*, 168104. [CrossRef]
43. Cuculich, P.S.; Schill, M.R.; Kashani, R.; Mutic, S.; Lang, A.; Cooper, D.; Faddis, M.; Gleva, M.; Noheria, A.; Smith, T.W.; et al. Noninvasive cardiac radiation for ablation of ventricular tachycardia. *N. Engl. J. Med.* **2017**, *377*, 2325–2336. [CrossRef] [PubMed]
44. Arevalo, H.J.; Vadakkumpadan, F.; Guallar, E.; Jebb, A.; Malamas, P.; Wu, K.C.; Trayanova, N.A. Arrhythmia risk stratification of patients after myocardial infarction using personalized heart models. *Nat. Commun.* **2016**, *7*, 11437. [CrossRef] [PubMed]

© 2020 by the authors. Licensee MDPI, Basel, Switzerland. This article is an open access article distributed under the terms and conditions of the Creative Commons Attribution (CC BY) license (http://creativecommons.org/licenses/by/4.0/).

Article

Myocardial Fibrosis in a 3D Model: Effect of Texture on Wave Propagation

Arsenii Dokuchaev [1], Alexander V. Panfilov [2,3,*] and Olga Solovyova [1,3,*]

1 Institute of Immunology and Physiology, Ural Branch of Russian Academy of Sciences, Ekaterinburg 620049, Russia; a.dokuchaev@iip.uran.ru
2 Department of Physics and Astronomy, Ghent University, Krijgslaan 281, 9000 Gent, Belgium
3 Laboratory of Computational Biology and Medicine, Ural Federal University, Ekaterinburg 620075, Russia
* Correspondence: alexander.panfilov@ugent.be (A.V.P.); o.solovyova@iip.uran.ru (O.S.)

Received: 21 July 2020; Accepted: 7 August 2020; Published: 12 August 2020

Abstract: Non-linear electrical waves propagate through the heart and control cardiac contraction. Abnormal wave propagation causes various forms of the heart disease and can be lethal. One of the main causes of abnormality is a condition of cardiac fibrosis, which, from mathematical point of view, is the presence of multiple non-conducting obstacles for wave propagation. The fibrosis can have different texture which varies from diffuse (e.g., small randomly distributed obstacles), patchy (e.g., elongated interstitional stria), and focal (e.g., post-infarct scars) forms. Recently, Nezlobinsky et al. (2020) used 2D biophysical models to quantify the effects of elongation of obstacles (fibrosis texture) and showed that longitudinal and transversal propagation differently depends on the obstacle length resulting in anisotropy for wave propagation. In this paper, we extend these studies to 3D tissue models. We show that 3D consideration brings essential new effects; for the same obstacle length in 3D systems, anisotropy is about two times smaller compared to 2D, however, wave propagation is more stable with percolation threshold of about 60% (compared to 35% in 2D). The percolation threshold increases with the obstacle length for the longitudinal propagation, while it decreases for the transversal propagation. Further, in 3D, the dependency of velocity on the obstacle length for the transversal propagation disappears.

Keywords: cardiac fibrosis; excitable media; wave break; elongated obstacle

1. Introduction

Non-linear waves of electrical excitation propagate through the heart and initiate cardiac contraction. Myocardial tissue consists of inter-connected excitable cells—cardiomyocytes forming so-called myocardial fibers ensuring prevailing conducting pathways in the tissue. In normal myocardium, the electrical wave propagation is about 2–3 times faster in the longitudinal direction of the myocardial fibers than in the transversal directions. From a mathematical point of view, the electrical waves belong to a large class of waves in the Reaction–Diffusion systems which are widely studied in applied mathematics [1]. The structural properties of the myocardium are accounted in the models as electro-diffusion anisotropy of the medium with higher diffusion coefficient along the myofiber direction. In many forms of heart disease, normal wave propagation is disturbed by an excessive growth of the fraction of connective tissue within myocardium including inexitable cells—fibroblasts and myofibroblasts. This pathological condition is called cardiac fibrosis.

Cardiac fibrosis is a common attribute of myocardial diseases of different etiology such as ischaemic cardiomyopathy and myocardial infarction, dilated cardiomyopathy, aortic stenosis, hypertrophic cardiomyopathy, and myocarditis [2]. Fibrosis is shown to be a predictor of adverse outcomes, including heart failure, ventricular arrhythmias, sudden cardiac death and all-cause mortality [3]. Fibrosis affects excitation propagation in the tissue creating inexitable obstacles, and zones of increasing heterogeneity and reduced conduction velocity in the myocardium. This may essentially increase the risk of arrhythmia induction and sustaining [4]. Experimental studies show that the frequency of ventricular arrhythmias linearly increases with fibrosis density [5]. On the other hand, fibrotic connective tissue is much stiffer than myocardium that reduces contractile performance of the tissue [6].

The pattern and distribution of fibrosis differ between conditions and vary from reversible diffuse forms to irreversible replacement fibrosis [7]. There are several types of fibrosis with varying texture between diffuse, patchy, and focal forms [4]. In diffuse fibrosis, small obstacles are randomly distributed between cardiomyocytes. In interstitial and patchy fibrosis, there are more structured non-conducting (low-conducting) obstacles elongated in form of stria or patches separating myocardial fibers. The latter fibrosis textures were shown to be even more arrhythmogenic than compact fibrosis due to the zig-zag conduction allowing micro-reentry to arise in the tissue [8].

Mathematical models were used to study effects of fibrosis on the myocardial excitation, and its role in the arrhythmia onset [9–16]. As any mathematical model is an idealisation of reality, representation of fibrosis in such studies was also simplified. The most important effect of fibrosis is that fibroblasts are inexcitable cells. Thus most of the studies considered cardiac tissue in the presence of fibrosis as a mixture of excitable cells (myocytes) described by non-linear differential equations coupled to each other by the diffusion operator describing electrical current between coupled myocytes. The fibrotic cells were mainly considered as obstacles [17,18], i.e., cells through which electrical currents from myocytes cannot flow and thus these cells were described as internal boundaries with no-flux boundary conditions [19].

Most modelling research has been devoted to the effects of diffuse or compact fibrosis. Effects of the diffuse fibrosis in the myocardium were considered by A. Panfilov's group on simplified models of cardiac tissue [17,18], as well as on modern ionic models [9]. In these studies, the domain representing cardiac tissue consisted of rectangular (squared or cubic) cells which with some probability were assigned to be excitable myocytes or inexcitable obstacles simulating fibroblasts. It was shown that two-dimensional biophysical models of myocardial tissue, in spite of their simplicity, allow one to find out main scenarios of fibrosis consequences, namely, the progressive slowing down of the wave propagation velocity, fractalization of its front and more irregular propagation of the excitation wave, and even predict the conditions for the break-up of spiral waves, i.e., in the settings of intact myocardium—the transition from tachycardia to life-treating fibrillation. It was found that if external impulse stimulation was applied to the fibrous tissue, the tendency to arrhythmias increased with an increase in the degree of tissue fibrosis [18]. Further studies showed that one of the most important factors influencing the occurrence of arrhythmias was the spatial heterogeneity of fibrosis [10].

Local (compact) fibrosis, as a rule, is associated with development of a myocardial post-infarction scar that replaces cardiomyocytes. The role in the onset and dynamics of arrhythmias of post-infarction scar simulated as a compact inexcitable area in myocardial tissue and its border zone simulated as a surrounding excitable tissue with impaired properties (reduced tissue conductivity and/or cellular excitability) was considered in the framework of computer modeling [20–22]. A case of compact fibrosis represented by a small number of large inexcitable regions in myocardial tissue was studied in paper [23].

Another approach used to study cardiac fibrosis in computational models is to simulate the electrical coupling between myocytes and fibroblasts. Works using this approach were performed in papers [24–26] showing emergence of complex dynamics of spiral waves and increase in the likelihood of arrhythmias depending on the fibroblasts properties and degree of coupling between myocytes and fibroblasts. A series of works was performed by N. Trayanova's group, which demonstrates the effect of myofibroblast density on the probability of arrhythmia (see, for example, [27]). The role of increasing amount of fibroblasts coupled to cardiomyocytes in induction of atrial fibrillations was studied in [21].

Effects of more complex structures of fibrosis formed of interstitial stria surrounding myofibers either randomly or being organized in patchy areas that may replace a large fraction of myocardial tissue have not been sufficiently studied neither in experiment nor in simulations. Role of the texture and type of fibrosis in the rhythm disturbances and the risk of their complications is not quantitatively evaluated. One of the first works demonstrating the importance of the texture and the characteristic dimensions of unexcitable obstacles in the tissue for the onset of spiral waves was the work of A. Pertsov's group (see [28]). In our recent study, [29], we made a first attempt to study the effects of fibrotic texture representing some features of interstitial fibrosis with primarily linear accumulations of collagen separating bundles of myocytes with little alteration in the alignment of myocytes or bundles [30]. One of main features of this pattern is that inexcitable inclusions to cardiac tissue here have some directional dependency laying between myofibers. In idealised representation they can be seen as thin elongated obstacles directed along cardiac fibers. In [29], we considered case of parallel fibers, thus texture of linear inexcitable elements which all have the same orientation and we studied propagation in such 2D models of myocardium for wave along this direction (longitudinal) and perpendicular to it (transversal) propagation. Main findings of this study revealed an opposite change in the conduction velocity (CV) with the obstacle length depending on the direction of the wavefront propagation. The CV increases with obstacle length if the excitation propagates along the obstacle strips, while the CV decreases under transverse wavefront propagation. This causes increasing tissue anisotropy with obstacle elongation [28]. This effect was explained by zig-zag propagation in the tissue with more short excitation path along the obstacles than across. The 2D model we used has natural limitations not allowing the wave to propagate in the third spacial direction transverse to the longitudinal coordinate of the obstacles. In this article, we extend our study of structured fibrosis to 3D myocardial tissue models and analyze if the third spatial dimension and the depth of the 3D tissue volume affect the characteristics of the excitation wave and the arrhythmogenic power of the fibrosis.

In particular, the aim of this paper is to study in detail effects of randomly generated textures of fibrosis with various lengths and percentages of obstacles on wave propagation in cardiac tissue slab of various depth.

2. Materials and Methods

2.1. Electrophysiology Model

To describe propagation of the excitation wave in the myocardial tissue we used a 3-dimensional monodomain formulation:

$$C_m \frac{\partial V}{\partial t} = \mathbf{D}\nabla^2 V - I_{ion}, \tag{1}$$

where V is transmembrane potential, $\mathbf{D}$—electro-diffusion matrix for anisotropic tissue, in this study we considered D constant, I_{ion}—sum of all transmembrane ionic currents, described with biophysically detailed cardiac action potential model TP06 [31].

Initial conditions were set as the rest potential $V = V_{rest}$ for the cardiac tissue. Boundary conditions were formulated as the no flux through the boundaries:

$$\vec{n}\nabla V = 0, \tag{2}$$

where $\vec{n}$—is the normal to the boundary.

The problem was solved in a 3D cuboid simulating myocardial slab with inclusions of structured fibrosis. Fibrosis elements of the tissue were simulated as non-conducting inexcitable obstacles and considered as the boundaries (no flux) for the myocardial elements.

Numerical Methods

To solve the problem (1) and (2) we used a finite-difference method with 18-point stencil discretization scheme as described in [32] with 0.25 mm for the spatial step and 0.02 ms for the time step.

Rush-Larsen formalism [33] was used for TP06 gating variables integration.

2.2. Fibrosis Pattern Generation

We considered 3D finite-difference mesh as matrix $X \times Y \times Z$ (where X, Y, Z are the numbers of elements in **x, y** and **z** direction) with each element to be either excitable myocardial tissue or inexcitable fibrosis. In general we used matrices $200 \times 200 \times 10$ elements (or 50 mm $\times$ 50 mm $\times$ 2.5 mm) in size, but we also considered matrices with $200 \times 200 \times Z$ elements, where Z might be 1, 2, 3 or 5.

Similarly to Nezlobinsky et al. [29], we uniformly distributed fibrosis elements and varied length of elements (in x-axis direction) and density of elements distribution. To distribute fibrosis elements through the mesh we implemented two approaches. Fibrosis pattern was generated in the same manner as described in [29], namely, we subdivided each $X \times Y \times Z$ matrix in **x** or **y** direction into $Y \cdot Z$ or $X \cdot Z$ rows with size $1 \times 200 \times 1$ or $200 \times 1 \times 1$ respectively. Then we subdivided each row into blocks of length n and assign it as fibrotic with a probability p. This approach allowed us to set fibrosis percentage precisely close to desired value.

Examples of wave propagation in a myocardial slab with fibrosis of different texture are shown in Figure 3. Upper panels A-B show an example of elongated fibrosis texture with linear $3 \times 1 \times 1$ elements of 3 x-node length occupying 60% of the tissue volume. Panel A shows the wave propagation in the longitudinal direction of the fibrotic obstacles, panel B shows the wave in the transversal direction. Bottom panel C shows an example of the diffuse fibrosis texture of the same density. Details on the wave dynamics are described in the Results section.

2.3. Shortest Path Calculation

The shortest possible path of excitation propagation between the opposite sides of the myocardial cuboid was calculated to demonstrate the zig-zag propagation through the fibrotic texture with elongated obstacles. We used iterative multi-dimensional binary dilation routine. Figure 1 shows and 2D example of our algorithm. Firstly, we mark an initial side of the mesh (all cells in left lateral surface except fibrosis) as "activated" and applied 3D binary dilatation operation until at least one cell on the opposite side of mesh becomes "activated" (see Figure 1A). Then, starting from this early approached cell on the opposite side (if there were several points we randomly picked one single point), we repeated this procedure in backward direction. The number of dilation iterations was taken as the length of shortest possible path for excitation propagation (see Figure 1B). Very first activated opposite cells for both steps were then used to calculate the trace of the shortest path.

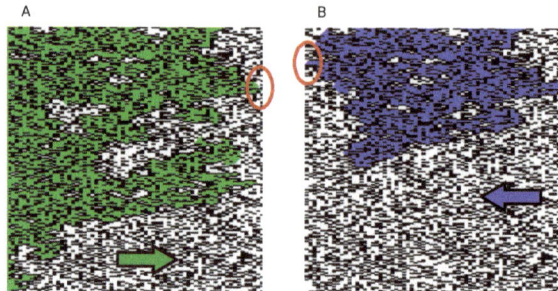

Figure 1. An example of the shortest path calculation routine. (**A**) Left lateral side of the myocardial slab is marked as "activated" and multiple dilation operations are performed. Wave propagates from left to right. Green color indicates activated cells, white and black indicate non-activated cells and fibrosis elements respectively. Red ellipse indicates the very first activated cell on the right side. (**B**) Starting from the right side activated cell, a number of the dilation operations is performed until any of left side cells is activated. This number N of dilation iterations is taken as the length of the shortest propagation path for the given fibrosis texture. Here, blue color shows activated cells. Red ellipse indicates the very first activated cell on the left side.

To trace the shortest possible path we build an undirected graph connecting neighbouring "active" cells in the fibrosis matrix and then breadth-first search algorithm with cells obtained on previous step assigned to be source and target nodes was used. An example of the shortest possible path trace is shown in Figure 2.

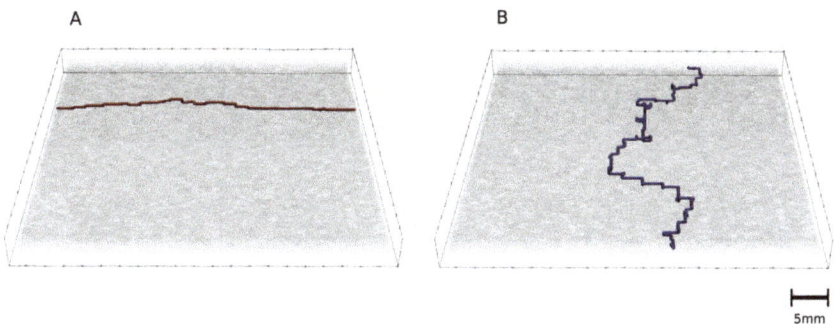

Figure 2. Traces of the shortest possible path of excitation propagation in the longitudinal (**A**) and transversal (**B**) directions in the model of 60% fibrosis with 4-node length elements.

3. Results

We evaluated properties of the excitation wave propagating either in the longitudinal direction of the fibrosis stria (along the long face of the strips in the direction of horizontal axis x) and across them (transversal, along axis y). We varied several parameters: (1) the depth of the myocardial slab from monolayer, which is equivalent 2D tissue to a 3D multilayer slab (2, 3, 5 and 10 layers); (2) the length of fibrosis stria from random 0.25 mm cubic elements simulating diffuse fibrosis (number of nodes $n = 1$ on axis x) to $1.0 \times 0.25 \times 0.25$ mm^3 strips ($n = 4$) elongated along axis x; (3) the density of the fibrosis from 10 to 60%.

3.1. Wavefront Velocity Depends on Fibrosis Fraction and Propagation Direction

Representative simulations of the wave propagation in the longitudinal and transverse directions in the 3D myocardial volume with structural fibrosis of 60% density formed of linear $3 \times 1 \times 1$ elements of 3 x-node length are shown in Figure 3. We compare propagation of the wave initiated by the stimulation on the left lateral surface and spreading along the obstacle elongation (see panel A) or on the upper lateral surface and travelling across the obstacles (panel B). We see that the wavefront has a complex shape due to interaction with obstacles. We also see that the wave propagates along the obstacles about two times faster than across them. For example, 50 ms after stimulation the wave propagated at a distance of approximately 25 mm for the longitudinal direction. However, for the transversal propagation the wavefront covered about 16 mm of the tissue. Panel C shows wave propagation in the model with diffuse fibrosis of the same density initiated on the upper surface of the volume. Here, average dynamics of the wave does not depend on the direction as the fibrotic elements are cubic. The conduction velocity at diffuse fibrosis is much slower than that at the longitudinal wavefront direction for the elongated obstacles, and is near to that for the transversal wavefront direction (CV = 0.29 mm/ms in panel C versus 0.48 mm/ms in panel A, and 0.24 mm/ms in panel B).

Figure 3. Electrical wave propagation in models with structural (panels **A**,**B**) and diffuse (panel **C**) fibrosis. Panels **A**,**B** show a myocardial slab with elongated fibrosis texture of linear $3 \times 1 \times 1$ elements of 3 x-node length. Wave propagation in the 3D myocardial slab was initiated on the left lateral surface of the cuboid for the longitudinal propagation (panel **A**) and on the upper lateral surface for the transversal propagation (panel **B**). The red color shows activated cells, the gray color indicates non-activated cells and transparent cells within the gray zone indicates fibrosis. Fibrosis density is 60%. Snapshots of excitation propagation are shown at 50 (left panels) and 100 ms (right panels) after excitation onset. The figure illustrates faster excitation propagation for the wavefront co-directed along the elongation of the fibrosis elements. In this case (panel **A**, longitudinal wavefront), about two times larger part of the myocardial volume is activated at the certain time moments as compared to the wave propagation across the fibrosis elongation (panel **B**, transversal wavefront). Panel **C** shows wave propagation in the model with 60% diffuse fibrosis. Here, excitation wave is initiated at the upper surface, and is independent of the wavefront direction. The conduction velocity at diffuse fibrosis is much slower than that at the longitudinal wavefront direction for the elongated obstacles, and is near to that for the transversal wavefront direction (CV = 0.29 mm/ms in panel **C** versus 0.48 mm/ms in panel **A**, and 0.24 mm/ms in panel **B**).

Figure 4 shows the dependence of the conduction velocity (CV) on the percentage of the fibrosis for textures with 0.25 to 1 mm long elements (n = 1, 2, 3, 4 x-nodes). Upper panels A, B show the results we obtained in a 3D monolayer of the tissue (0.25 mm depth, 1 z-layer). As this case is equivalent to 2D model, we use it as a reference for comparison of 2D and 3D models, in particular with the results reported earlier [29]. We see that the velocity significantly decreases with an increase in the fibrosis percentage. For the diffuse fibrosis ($n = 1$) the velocity does not depend on the direction of activation (see red lines in Figure 4) and is used as the reference to compare with results for structural fibrosis. In the monolayer model, elongation of obstacles has a pronounced opposite effect on the longitudinal and transversal CV with increasing longitudinal velocity with obstacle length but decreasing transversal velocity.

Middle C, D panels in Figure 4 show the results obtained in the multilayer tissue of the 2.5 mm depth (10 z-layers). It is seen that qualitative effects of fibrosis density and obstacle length on the CV are similar to that of the monolayer model. At the same time, the tissue depth allowing excitation to propagate around obstacles in the additional transversal direction (z-axis). It differently affect the longitudinal and transversal propagation. We see a significant effect of the obstacle length on the longitudinal CV similar to monolayer case, while the transversal CV in the ten-layer tissue model is almost independent of the obstacle length in contrast to the monolayer model.

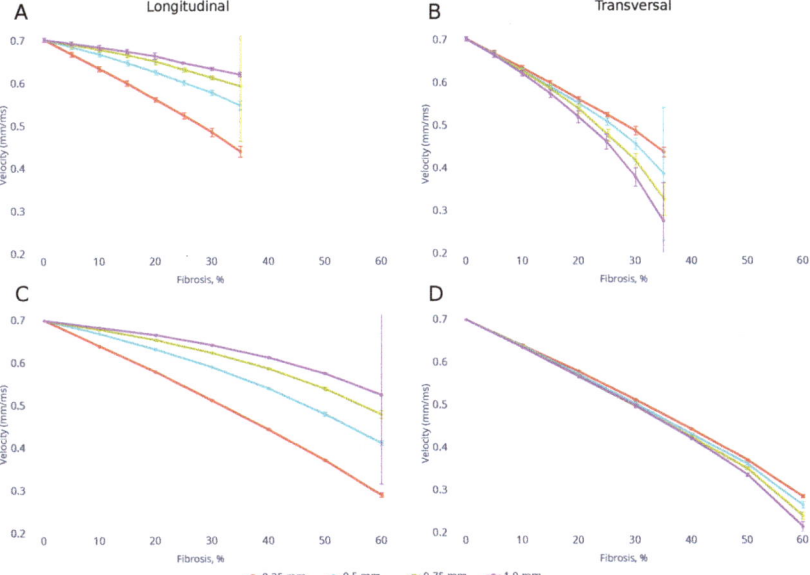

Figure 4. Conduction velocity (CV) depending on the fibrosis area in the longitudinal (left) and transversal (right) wavefront directions in a 3D monolayer of 0.25 mm depth (upper, **A,B**) and a multilayer myocardial slab of 2.50 mm depth (middle, **C,D**). Error bars show the standard deviation (STD) of the CV in 20 fibrosis pattern samples for each obstacle length. E-F Average wavefront velocities normalized by the values at the diffuse fibrosis of the same percentage for the 3D monolayer (dashed lines) and multilayer models (solid lines).

Therefore, the texture of fibrosis produces anisotropic propagation with increasing longitudinal/transversal anisotropy ratio with the density of fibrosis and obstacle length. This is illustrated in Figure 5. The anisotropy ratio in multilayer model reaches 2:1 with increase in the fibrosis density, the dependency is steeper at longer obstacles and more steep in the monolayer vs. multilayer models. The dependency is non-linear at higher than 25% fibrosis percentage.

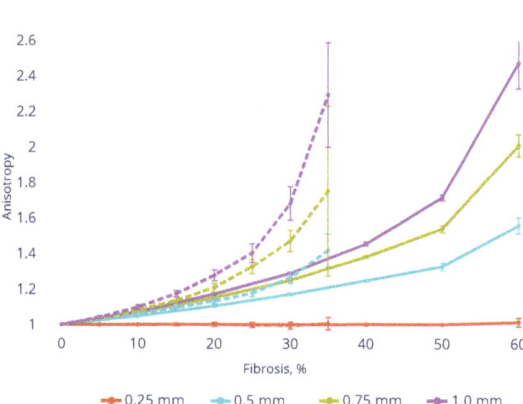

Figure 5. Ratio in the conduction velocity between the longitudinal and transversal wavefront direction as a characteristic of the fibrotic texture anisotropy. Dotted lines show the ratio in the monolayer models, solid lines—in the multilayer models. Error bars show STD for the same model groups as in Figure 4.

3.2. Wave Propagation Stopping Depends on the Tissue Depth, Fibrosis Percentage and Wavefront Direction

The CV dependencies in Figure 4 are spanned for different ranges of fibrosis percentage for monolayer models (panels A,B) and multilayer models (panels C,D), indicating different percolation threshold for 2D and 3D cases. The fibrosis percentage in the monolayer models is limited by 35%, after which the wave cannot propagate anymore, while in the multilayer models the wave propagates for the fibrosis density up to 55%. Close to percolation threshold we observed a steep increase in the variability of CV for both models (see increasing STD whiskers on the curves in Figure 4). For example, in the monolayer models with 35% fibrosis and three-node long obstacles, coefficient of variability in CV is of 21.8% for the longitudinal direction and 11.5% for the transverse direction. In the multilayer models with 60% fibrosis and three-node long obstacles the coefficients of variability in CV are 1.8% and 3.6% respectively.

Figure 6 characterizes the percolation threshold statistically: it shows the fraction of models with failed wave propagation depending on the fibrosis percentage in the mono- (upper panels) and multilayer (lower panels) model. Here, we used the binary dilation method to analyze if the wave is able to approach the opposite side of the tissue volume (see Methods section for detail) and counted a number of models with propagation failing. The dependencies are approximated by logistic Hill curves.

It is seen that the curves are right-shifted for longitudinal wavefront (left panels) compared to transversal wavefront (right panels) and in the multilayer models (lower panels) compared to monolayer models (upper panels). This means that the probability of excitation failure is higher for for the transversal versus longitudinal wavefront propagation and in the monolayer against multilayer tissue. Moreover, the curves shift in the opposite directions for the longitudinal and transversal propagation when the obstacle length is increasing.

We see that for the longitudinal propagation the probability of excitation failure decreases while for the transversal propagation the probability of excitation failure increases with increase in the obstacle length.

To clarify these results further we show the dependencies of the percolation threshold FP_{95} (i.e., fibrosis percentage where propagation failure occurs for 95% of models) on the obstacle length for the longitudinal and transversal propagation in the mono- and multilayer models (Figure 7). It is seen that FP_{95} is much higher in the multilayer (upper lines) compared to the monolayer (lower lines) models, and FP$_{95}$ is higher for the longitudinal wavefront direction (solid lines) compared to the transversal (dotted lines). For longitudinal wave direction the minimal FP$_{95}$ is observed for the diffuse fibrosis and FP$_{95}$ is about 43% for the monolayer models and FP$_{95}$ and is about 65% for the multilayer models. For transversal propagation the minimal FP$_{95}$ is lower than the minimal FP$_{95}$ for the longitudinal propagation. It is observed for the fibrosis of 4-node long and FP$_{95}$ is about 35% for the monolayer models and FP$_{95}$ and is about 60% for the multilayer models. Note, that in Figures 4 and 5 we considered the latter fibrosis percentages as upper limits for the mono- and multilayer models.

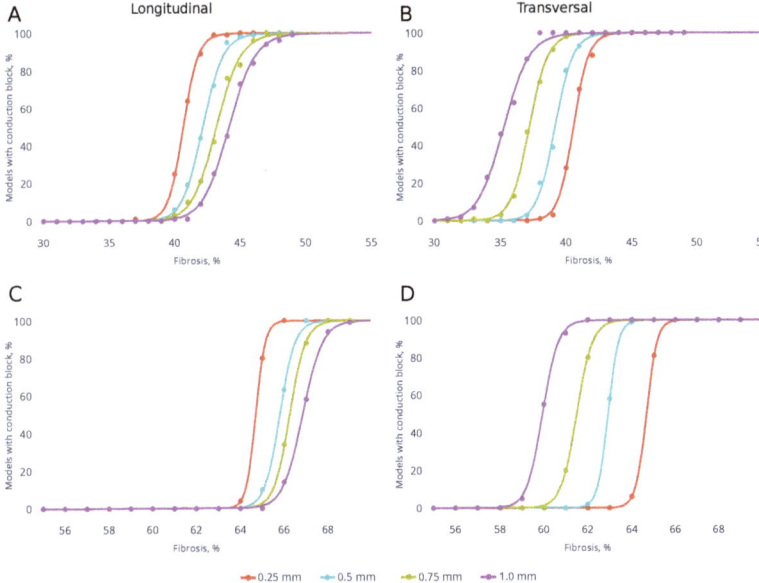

Figure 6. Fraction of models with failed wave propagation depending on the fibrosis percentage in the mono- (upper panels) and multilayer (lower panels) tissue for longitudinal (panels **A**,**C**) and transversal (panels **B**,**D**) wavefront direction. Dots show percentage of models, solid lines show approximation by logistic curves.

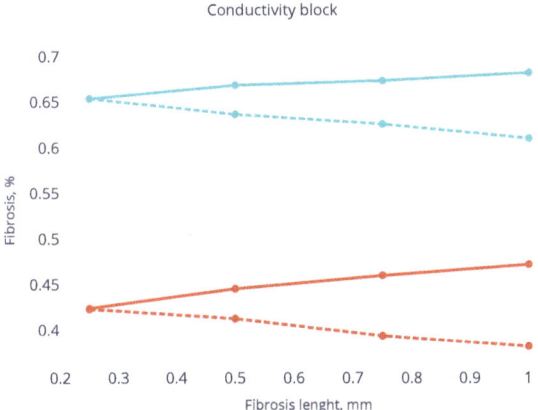

Figure 7. A threshold fibrosis percentage (FP$_{95}$) for 95% of models with conduction block depending on the fibrosis obstacle length for longitudinal (solid lines) and transversal (dotted lines) wavefront direction in the monolayer (lower red lines) and multilayer (upper blue lines) models.

3.3. Effects of Fibrosis on the Conduction Velocity Depends on the Tissue Depth

In Figure 8 we compare effects of obstacle length on the CV in the ten-layer models (red lines) with that in the mono-layer models at 35% fibrosis. We see that the length of the obstacle substantially affects longitudinal propagation and the dependency in 3D is similar to that in 2D. However, for the transversal propagation the velocity is almost independent on the obstacle length in 3D.

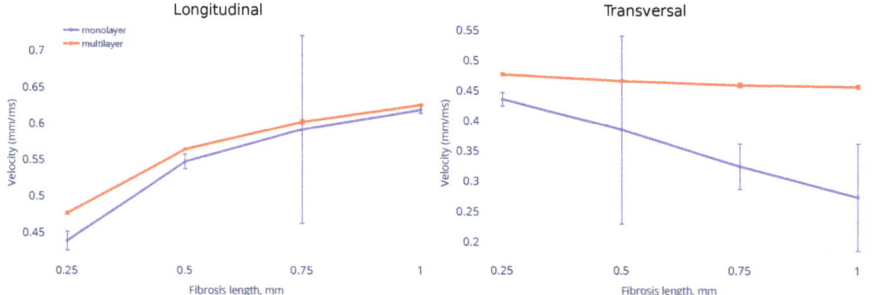

Figure 8. Dependence of the conduction velocity on the obstacle length in mono- (n = 1, blue lines) and multilayer (n = 10, red lines) models at 35% fibrosis for longitudinal (**Left**) and transversal (**Right**) wavefront direction.

To quantify further effects of the tissue depth on the degree of CV change with the length of fibrosis obstacles, we demonstrate the ratio of the average CV for the models with 1 mm (n = 4) obstacle length to the model of diffuse fibrosis of the same percentage (Figure 9) at different thickness of the tissue (number of layers is 1, 2, 3, 5, and 10). We see big change in velocity between 1 and 2 layers. However starting from 3-layer thick tissue, the curves become independent of the number of layers. So, for both wavefront directions the thickness when 3D effects become dominating is about 0.5 mm.

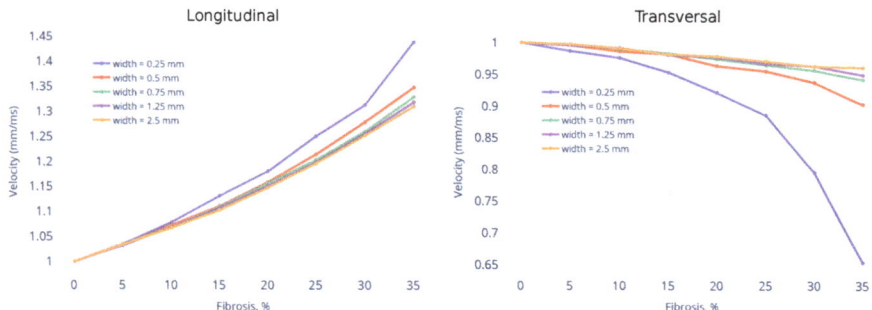

Figure 9. Ratio in the conduction velocities between the average CV for the model with fibrosis of 1 mm (n = 4) obstacle length and the model of diffuse fibrosis (n = 1) of the same percentage. (**Left**) longitudinal wavefront direction. (**Right**) transversal direction. The curves show dependences of the ratio on the fibrosis percentage at varying tissue depth from 0.25 to 2.5 mm (1 to 10 z-layers).

3.4. Shortest Path for Wave Propagation Depends on the Obstacle 0 and the Tissue Depth

In paper [29], we explained an opposite effect of increasing obstacle length on the longitudinal and transverse propagation velocity by a zig-zig propagation [8]. It was characterized by estimating the shortest path length between opposite boundaries of the medium. Here, we use the similar approach and calculate the shortest path from one to the opposite face of the 3D myocardial slab in the longitudinal and transverse direction (see Methods section in detail). As shown in Figure 2 the path is more straight and short in the longitudinal direction and it is more complex and long in the transverse direction. The distribution of the shortest path length for the 3D mono- (upper panels) and multilayer models (lower panels) for the same fibrosis density of 30% for obstacle length $n = 4$ in comparison with diffuse fibrosis with $n = 1$ are shown in Figure 10. It is seen that the average path is much shorter for the longitudinal wavefront direction than that for the diffuse fibrosis and the latter is shorter than the path for the transversal wavefront direction. Interestingly, for a particular texture shown in Figure 2 the ratio of conduction velocities in the longitudinal and transversal directions is 2.47, which is reasonably close to the ratio of average path lengths (2.23). Note that the average path length almost does not change with the tissue depth for the longitudinal wavefront direction. In contrast, the average path for the transverse direction essentially shortens in the multilayer tissue (116.71 for transverse fibrosis of $n = 4$ node length and 30% density. see Figure 10) coming closer to the value for the average value for the diffuse fibrosis models.

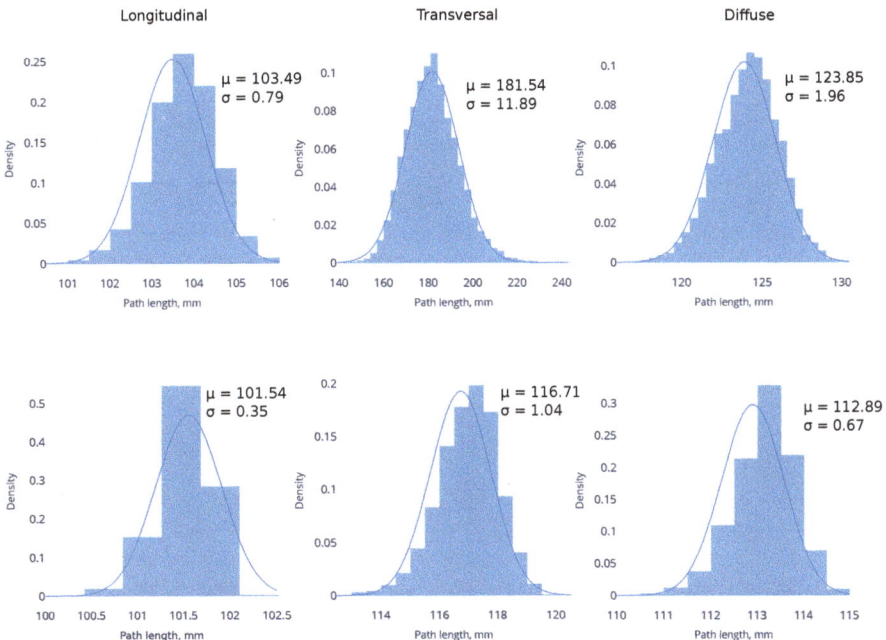

Figure 10. Normalized histograms of the shortest path for the longitudinal propagation and transversal propagation for $n = 4$ node length fibrosis and for $n = 1$ diffuse fibrosis of the same 30% density in the monolayer (**upper**) and multilayer (**lower**) models. 5000 shortest possible trajectories calculated in 5000 fibrosis matrices are used for each histogram.

4. Discussion

In this paper, we extended our previous study on the effects of fibrosis texture on wave propagation in cardiac tissue by transition from 2D to 3D models with varying depth of the myocardial slab. First, we checked if the effects we observed in the 2D models are reproduced in the 3D formulations of a monolayer model with 1 node depth of the tissue. Then, we increased the number of layers in the slab to evaluate effects of the third spatial dimension on the properties of the wave in the fibrosis textures of various densities and obstacle elongation. The simple model of straight elongated obstacle stria in the myocardial tissue were considered as a simplified representation of interstitial fibrosis [8] localized between myocardial fibers replacing part of cardiomyocytes. To evaluate effects of obstacle elongation in comparison with more rounded small elements of the diffuse fibrosis, we varied the obstacle length considering the texture with fibrosis strips of equal length. Then, we varied fibrosis percentage in a rather wide range to approach such density where excitation wave stopped due to obstacle texture blocking the propagation at a certain distance from the initially stimulated area.

We show that the elongation of obstacles has a substantial effect on the waves. It results in anisotropic wave propagation in which the velocity of the longitudinal propagation increases with increasing elongation, while the velocity of the transversal direction decreases (Figure 4). The extent of the effect increases with the fibrosis percentage and with the the obstacle length.

We found that qualitative effects of the fibrosis texture are similar in the 3D and 2D models of the tissue, which validates main predictions based on the 2D models and allows one to translate them into the more adequate 3D settings. At the same time we showed that the effects of fibrosis in the 3D tissue are less expressed due to additional spacial dimension transversal to the obstacle elongation that allows the wave to find more ways to bypass the fibrotic obstacles. The effects of the third tissue dimension are more pronounced for the transversal wavefront direction increasing the conduction velocity in the 3D multilayer models compared to monolayer models and making almost negligible difference between the CV at various obstacle lengths at the same fibrosis percentage (Figure 4).

We showed that in 3D tissue the dependencies of the tissue anisotropy on the fibrosis density are much less steep (Figure 5), so the higher fibrosis percentage is needed to approach the same extent of anisotropy as produced by the 2D models. For instance, an anisotropy of 2:1 was approached at about 30% fibrosis in the monolayer models while the same anisotropy was demonstrated in the models of 60% fibrosis in ten-layer deep models.

We showed that threshold fibrosis density for conduction block is also about twice higher in the 3D multilayer models as compared to the monolayer models (see Figure 7). Of note, the threshold fibrosis density increases with obstacle elongation for the longitudinal wavefront direction, while decreases for the transversal direction that makes much less percentage of fibrosis to be more dangerous.

The geometric approach we used here for determining the shortest path of the wave is the main approach which was used in most of the papers on fibrosis [12,13,19,34]. In real myocardial tissue, the path of the wave is governed by the dynamic electrotonic load (source-sink relationship), which can affect the propagation up to producing local blocks in spite of the geometrical connectivity [3]. In our previous work [29] we specially analyzed if the geometry consideration only is sufficient to estimate the path and to discriminate the cases of different elongation of the fibrosis elements. We performed additional simulations initiating a propagation wave by point stimulation at one boundary and calculating its shortest path to the opposite boundary. We found that the source-sink relationship slightly increases the path of zig-zag propagation as compared to percolation path for both the longitudinal and transversal wavefront direction, and an increase in the transversal path was about 15%, while for the longitudinal only about 3%. We also found that the ratio of the average path lengths and the ratio of the conduction velocities (anisotropy) in the longitudinal and transversal directions were almost the same for a particular fibrosis texture. Thus, we showed that the percolation path provided a good estimate for the real path of the wave and the difference in the geometry-based paths at fibrosis of different textures reflects adequately the real difference rather underestimating it quantitatively.

These simulations suggest that the severity of the fibrosis effects on the myocardial tissue is essentially dependent on the local density of the fibrosis and its orientation relative to the myocardial fibers that can vary depending on the tissue depth and orientation of the myofibers within the heart chambers.

Our research is a computation prediction on some 3D effects of fibrosis texture on wave propagation. It would be interesting to compare it with available experimental data. There is a lot of experimental data which qualitatively confirms the results of our study. In many papers and in various preparations it was shown that fibrosis decreases the conduction velocity of the wave propagation (see reviews [3,4]). Further in [35] it was shown that increased interstitial fibrosis affects transverse conduction much more than the longitudinal conduction. Further, one of our results is a complex shape of the wave in the presence fibrosis (Figure 3) is also qualitatively confirmed in [36], where it was shown that increased fibrosis has also been associated with heterogeneioty of conduction. It was shown in patients with dilated cardiomyopathy that the amount of myocardial fibrosis correlated with the severity of abnormal propagation [30].

However, currently quantitative comparison is impossible. This is because it is challenging to create cardiac tissues with controlled textures of fibrosis in which length interstitial stria gradually changes. However, we think that using modern technologies of cardiac cell co-cultures [37] and/or optogenetics [38] it will be possible to mimic multiple elongated obstacles with controlled length and study wave propagation there. There are also recent developments which allow construction of such 3D preparations [39]. It would be interesting to use these technologies to study wave propagation in such experimental systems and compare it with the predictions of our modelling.

Our approach has several limitations.

The major limitation is that we studied models of idealised fibrosis formed by linear inexcitable obstacles of the same length. The geometry and texture of fibrosis in real tissue is much more complex on many aspects. Fibrosis can be formed not only by fibroblasts, but also by myofibroblasts, which are much larger in size. Substantial amounts of collagen are present in fibrotic cardiac tissue and have different space scale and organisation than fibroblasts. Further, direction, size and shape of fibrotic inclusions are not so regular as we assumed in our paper. It would be interesting to perform studies on realistic textures of cardiac fibrosis in the human heart similar to those from [40] obtained for 3D tissue [41] using computational approaches presented in study [42,43]. Further, a novel approach of modelling which takes into account shapes of cardiac cells can be an interesting tool to better represent fibrosis textures [26]. In this paper we studied effects of fibrosis texture on the wave propagation. It would be also interesting to study effects of fibrosis texture on the onset and properties of cardiac arrhythmias which either occur due to re-entry mechanism of abnormal propagation related to the structural abnormality of the tissue [44] or due to ectopic activity related to the abnormal depolarization-repolarization of cardiac cells [45].

We considered 3D isotropic tissue without taking into account pre-excisting anysotrophy of myocardium due to myofiber structure with different electro-diffusion along and across the fibers. This is also the task for our future studies on realistic textures of cardiac fibrosis.

Numerical Implications for Fibrosis Modelling

Currently two main approaches are used to represent fibrotic remodeling: (1) modification of the diffusion tensor in the regions where fibrosis is present [20] and (2) introduction of real obstacles or internal boundary conditions (this study, [19]). Although, in our view approach (2) can potentially better represent texture of the tissue and better reproduce mechanisms of initiation of arrhythmias, its numerical implementation is much more challenging. In our paper we used approach (2), however, our approach can also be applied for simulations performed using approach (1) in combination with patient specific data in the following way. For some patients with cardiac myopathy a biopsy of endocardial tissue is often taken. The bioptic tissue can be studied using histological methods and a patient specific texture of fibrosis for this given patient is available. This texture can be input to computer and fibrotic tissue can be modelled as obstacles, as we did in our paper. Then we can compute the velocities of propagation for wave in three orthogonal directions and based on that calculate diffusion coefficients as well as a rule for their modification. We illustrate it with a specific example. We generated a fibrotic texture in a small 3D patch of cardiac tissue (Supplementary Materials Figure S1). The preexisting anisotropy was 1:0.5:0.5 with respect to velocity. From studies of wave propagation in 3 orthogonal direction we found that for given fibrosis texture (50% fibrosis produced by obstacles of 0.75 mm $\times$ 0.25 mm) decrease in velocities can be reproduced by decrease in diffusion coefficient by 1.75, 2.08, 2.08 for the longitudinal and 2 transversal directions. Thus if one wants to introduce effect of such fibrosis to the large scale model, the diffusion coefficients in fibrotic regions should be reduced by such factors, e.g., instead of 1:0.25 :0.25 one should use 0.57:0.12:0.12.

In conclusion, in 3D idealised models for human myocardial tissue, we showed that inexcitable fibrotic elements produce anisotropy of cardiac tissue that depends on the fibrosis percentage, obstacle elongation and tissue depth. The anisotropy results from zig-zag propagation depending on the elongation of fibrosis elements. We showed that a threshold fibrosis density to stop excitation propagation depends on the obstacle length and tissue depth.

Supplementary Materials: The following are available online at http://www.mdpi.com/2227-7390/8/8/1352/s1, Figure S1: Electrical wave propagation in a model with structural fibrosis and a reduced model simulating propagation timing in the fibrosis tissue. Table S1: Electro-diffusion coefficients (D), anisotropy (longitudinal to transverse ratio) and conduction velocity (CV) in the models with structural fibrosis and corresponding reduced model.

Author Contributions: Conceptualization, A.V.P. and O.S.; methodology, A.D. and A.V.P.; software, A.D.; simulations, A.D.; analysis, A.D., A.V.P. and O.S.; writing—original draft preparation, A.D., A.V.P., O.S.; writing—review and editing, A.V.P. and O.S.; visualization, A.D. All authors have read and agreed to the published version of the manuscript.

Funding: A.D., A.V.P. and O.S. were funded by the Russian Foundation for Basic Research (#18-29-13008). A.V.P. and O.S. were funded by RF Government Act #211 of 16 March 2013 (agreement 02.A03.21.0006). A.D. and O.S. work was carried out within the framework of the IIP UB RAS theme No. AAAA-A18-118020590031-8. A.V.P. was partially funded by BOF Ghent University.

Acknowledgments: We are thankful to Timur Nezlobinsky for help in setting up of computations.

Conflicts of Interest: The authors declare no conflict of interest.

Abbreviations

The following abbreviations are used in this manuscript:
CV   Conduction velocity

References

1. Volpert, A.; Volpert, V.; Volpert, V. *Traveling Wave Solutions of Parabolic Systems: Translations of Mathematical Monographs*; American Mathematical Society: Providence, RI, USA, 1994; Volume 140.
2. Bing, R.; Dweck, M.R. Myocardial fibrosis: Why image, how to image and clinical implications. *Heart* **2019**, *105*, 1832–1840. [CrossRef] [PubMed]
3. Nguyen, T.P.; Qu, Z.; Weiss, J.N. Cardiac fibrosis and arrhythmogenesis: The road to repair is paved with perils. *J. Mol. Cell. Cardiol.* **2014**, *70*, 83–91. [CrossRef] [PubMed]
4. De Jong, S.; Van Veen, T.A.; Van Rijen, H.V.; De Bakker, J.M. Fibrosis and cardiac arrhythmias. *J. Cardiovasc. Pharmacol.* **2011**, *57*, 630–638. [CrossRef]
5. Stein, M.; Noorman, M.; van Veen, T.A.; Herold, E.; Engelen, M.A.; Boulaksil, M.; Antoons, G.; Jansen, J.A.; van Oosterhout, M.F.; Hauer, R.N.; et al. Dominant arrhythmia vulnerability of the right ventricle in senescent mice. *Heart Rhythm* **2008**, *5*, 438–448. [CrossRef] [PubMed]
6. Jugdutt, B.I. Ventricular remodeling after infarction and the extracellular collagen matrix. *Circulation* **2003**, *108*, 1395–1403. [CrossRef]
7. Travers, J.G.; Kamal, F.A.; Robbins, J.; Yutzey, K.E.; Blaxall, B.C. Cardiac fibrosis. *Circ. Res.* **2016**, *118*, 1021–1040. [CrossRef]
8. De Bakker, J.M.; Van Capelle, F.J.; Janse, M.J.; Tasseron, S.; Vermeulen, J.T.; De Jonge, N.; Lahpor, J.R. Slow conduction in the infarcted human heart: 'Zigzag' course of activation. *Circulation* **1993**, *88*, 915–926. [CrossRef]
9. Ten Tusscher, K.H.; Panfilov, A.V. Influence of diffuse fibrosis on wave propagation in human ventricular tissue. *Europace* **2007**, *9* (Suppl. 6), 38–45. [CrossRef]
10. Kazbanov, I.V.; Ten Tusscher, K.H.; Panfilov, A.V. Effects of heterogeneous diffuse fibrosis on arrhythmia dynamics and mechanism. *Sci. Rep.* **2016**, *6*, 1–14. [CrossRef]
11. Alonso, S.; Bär, M. Reentry near the percolation threshold in a heterogeneous discrete model for cardiac tissue. *Phys. Rev. Lett.* **2013**, *110*, 158101. [CrossRef]

12. Alonso, S.; dos Santos, R.W.; Bär, M. Reentry and ectopic pacemakers emerge in a three-dimensional model for a slab of cardiac tissue with diffuse microfibrosis near the percolation threshold. *PLoS ONE* **2016**, *11*, e0166972. [CrossRef] [PubMed]
13. Vigmond, E.; Pashaei, A.; Amraoui, S.; Cochet, H.; Hassaguerre, M. Percolation as a mechanism to explain atrial fractionated electrograms and reentry in a fibrosis model based on imaging data. *Heart Rhythm* **2016**, *13*, 1536–1543. [CrossRef] [PubMed]
14. McDowell, K.S.; Vadakkumpadan, F.; Blake, R.; Blauer, J.; Plank, G.; MacLeod, R.S.; Trayanova, N.A. Mechanistic inquiry into the role of tissue remodeling in fibrotic lesions in human atrial fibrillation. *Biophys. J.* **2013**, *104*, 2764–2773. [CrossRef] [PubMed]
15. Krueger, M.W.; Rhode, K.S.; O'Neill, M.D.; Rinaldi, C.A.; Gill, J.; Razavi, R.; Seemann, G.; Doessel, O. Patient-specific modeling of atrial fibrosis increases the accuracy of sinus rhythm simulations and may explain maintenance of atrial fibrillation. *J. Electrocardiol.* **2014**, *47*, 324–328. [CrossRef]
16. Varela, M.; Colman, M.A.; Hancox, J.C.; Aslanidi, O.V. Atrial heterogeneity generates re-entrant substrate during atrial fibrillation and anti-arrhythmic drug action: mechanistic insights from canine atrial models. *PLoS Comput. Biol.* **2016**, *12*, e1005245. [CrossRef]
17. ten Tusscher, K.H.; Panfilov, A.V. Influence of nonexcitable cells on spiral breakup in two-dimensional and three-dimensional excitable media. *Phys. Rev. E Stat. Phys. Plasmas Fluids Relat. Interdiscip. Top.* **2003**, *68*, 2–5. [CrossRef]
18. ten Tusscher, K.H.W.J.; Panfilov, A.V. Wave propagation in excitable media with randomly distributed obstacles. *Multiscale Model. Simul.* **2005**, *3*, 265–282. [CrossRef]
19. Mendonca Costa, C.; Prassl, A.J.; Weber, R.; Campos, F.O.; Prassl, A.J.; Dos Santos, R.W.; Sanchez-Quintana, D.; Ahammer, H.; Hofer, E.; Plank, G.; et al. An efficient finite element approach for modeling fibrotic clefts in the heart. *IEEE Trans. Biomed. Eng.* **2014**, *61*, 900–910. [CrossRef]
20. Arevalo, H.J.; Vadakkumpadan, F.; Guallar, E.; Jebb, A.; Malamas, P.; Wu, K.C.; Trayanova, N.A. Arrhythmia risk stratification of patients after myocardial infarction using personalized heart models. *Nat. Commun.* **2016**, *7*, 11437. [CrossRef]
21. Gao, Y.; Gong, Y.; Xia, L. Simulation of atrial fibrosis using coupled myocyte-fibroblast cellular and human atrial models. *Comput. Math. Methods Med.* **2017**, *2017*. [CrossRef]
22. Mangion, K.; Gao, H.; Husmeier, D.; Luo, X.; Berry, C. Advances in computational modelling for personalised medicine after myocardial infarction. *Heart* **2018**, *104*, 550–557. [CrossRef] [PubMed]
23. Majumder, R.; Pandit, R.; Panfilov, A.V. Turbulent electrical activity at sharp-edged inexcitable obstacles in a model for human cardiac tissue. *Am. J. Physiol. Heart Circ. Physiol.* **2014**, *307*, H1024–H1035. [CrossRef] [PubMed]
24. Nayak, A.R.; Shajahan, T.K.; Panfilov, A.V.; Pandit, R. Spiral-wave dynamics in a mathematical model of human ventricular tissue with myocytes and fibroblasts. *PLoS ONE* **2013**, *8*. [CrossRef] [PubMed]
25. Sridhar, S.; Vandersickel, N.; Panfilov, A.V. Effect of myocyte-fibroblast coupling on the onset of pathological dynamics in a model of ventricular tissue. *Sci. Rep.* **2017**, *7*, 1–12. [CrossRef]
26. Kudryashova, N.; Nizamieva, A.; Tsvelaya, V.; Panfilov, A.V.; Agladze, K.I. Self-organization of conducting pathways explains electrical wave propagation in cardiac tissues with high fraction of non-conducting cells. *PLoS Comput. Biol.* **2019**, *15*, e1006597. [CrossRef]
27. McDowell, K.S.; Arevalo, H.J.; Maleckar, M.M.; Trayanova, N.A. Susceptibility to arrhythmia in the infarcted heart depends on myofibroblast density. *Biophys. J.* **2011**, *101*, 1307–1315. [CrossRef]
28. Pertsov, A.M. Scale of geometric structures responsible for discontinuous propagation in myocardial tissue. In *Discontinuous Conduction in the Heart*; Futura Publishing Company: Armonk, NY, USA, 1997.
29. Nezlobinsky, T.; Solovyova, O.; Panfilov, A.V. Anisotropic conduction in the myocardium due to fibrosis: The effect of texture on wave propagation. *Sci. Rep.* **2020**, *10*, 764. [CrossRef]
30. Anderson, K.P.; Walker, R.; Une, P.; Ershler, P.R.; Lux, R.L.; Karwandee, S.V. Myocardial electrical propagation in patients with idiopathic dilated cardiomyopathy. *J. Clin. Investig.* **1993**, *92*, 122–140. [CrossRef]
31. ten Tusscher, K.H.W.J. Alternans and spiral breakup in a human ventricular tissue model. *AJP Heart Circ. Physiol.* **2006**, *291*, H1088–H1100. [CrossRef]
32. Tusscher, K.H.W.J.T.; Panfilov, A.V. Modelling of the ventricular conduction system. *Prog. Biophys. Mol. Biol.* **2008**, *96*, 152–170. [CrossRef]

33. Rush, S.; Larsen, H. A practical algorithm for solving dynamic membrane equations. *IEEE Trans. Biomed. Eng.* **1978**, *BME-25*, 389–392. [CrossRef] [PubMed]
34. Bub, G.; Shrier, A.; Glass, L. Spiral wave generation in heterogeneous excitable media. *Phys. Rev. Lett.* **2002**, *88*, 058101. [CrossRef] [PubMed]
35. Spach, M.S.; Miller, W.T.; Dolber, P.C.; Kootsey, J.M.; Sommer, J.R.; Mosher, C.E. The functional role of structural complexities in the propagation of depolarization in the atrium of the dog. Cardiac conduction disturbances due to discontinuities of effective axial resistivity. *Circ. Res.* **1982**, *50*, 175–191. [CrossRef] [PubMed]
36. Li, D.; Fareh, S.; Leung, T.K.; Nattel, S. Promotion of atrial fibrillation by heart failure in dogs. *Circulation* **1999**, *100*, 87–95. [CrossRef] [PubMed]
37. Zlochiver, S.; Muñoz, V.; Vikstrom, K.L.; Taffet, S.M.; Berenfeld, O.; Jalife, J. Electrotonic myofibroblast-to-myocyte coupling increases propensity to reentrant arrhythmias in two-dimensional cardiac monolayers. *Biophys. J.* **2008**, *95*, 4469–4480. [CrossRef]
38. Majumder, R.; Feola, I.; Teplenin, A.S.; de Vries, A.A.; Panfilov, A.V.; Pijnappels, D.A. Optogenetics enables real-time spatiotemporal control over spiral wave dynamics in an excitable cardiac system. *eLife* **2018**, *7*, 1–17. [CrossRef]
39. Zuppinger, C. 3D cardiac cell culture: A critical review of current technologies and applications. *Front. Cardiovasc. Med.* **2019**, *6*, 1–9. [CrossRef]
40. Pope, A.J.; Sands, G.B.; Smaill, B.H.; LeGrice, I.J. Three-dimensional transmural organization of perimysial collagen in the heart. *Am. J. Physiol. Heart Circ. Physiol.* **2008**, *295*, H1243–H1252. [CrossRef]
41. Glashan, C.A.; Androulakis, A.F.; Tao, Q.; Glashan, R.N.; Wisse, L.J.; Ebert, M.; De Ruiter, M.C.; Van Meer, B.J.; Brouwer, C.; Dekkers, O.M.; et al. Whole human heart histology to validate electroanatomical voltage mapping in patients with non-ischaemic cardiomyopathy and ventricular tachycardia. *Eur. Heart J.* **2018**, *39*, 2867–2875. [CrossRef]
42. Keldermann, R.H.; Ten Tusscher, K.H.; Nash, M.P.; Hren, R.; Taggart, P.; Panfilov, A.V. Effect of heterogeneous APD restitution on VF organization in a model of the human ventricles. *Am. J. Physiol. Heart Circ. Physiol.* **2008**, *294*, 764–774. [CrossRef]
43. Keldermann, R.H.; Ten Tusscher, K.H.; Nash, M.P.; Bradley, C.P.; Hren, R.; Taggart, P.; Panfilov, A.V. A computational study of mother rotor VF in the human ventricles. *Am. J. Physiol. Heart Circ. Physiol.* **2009**, *296*. [CrossRef] [PubMed]
44. Panfilov, A.; Keener, J. Re-entry in an anatomical model of the heart. *Chaos Solitons Fractals* **1995**, *5*, 681–689. [CrossRef]
45. Vandersickel, N.; Kazbanov, I.V.; Nuitermans, A.; Weise, L.D.; Pandit, R.; Panfilov, A.V. A study of early afterdepolarizations in a model for human ventricular tissue. *PLoS ONE* **2014**, *9*, e84595. [CrossRef]

© 2020 by the authors. Licensee MDPI, Basel, Switzerland. This article is an open access article distributed under the terms and conditions of the Creative Commons Attribution (CC BY) license (http://creativecommons.org/licenses/by/4.0/).

Article

Hemodynamic Effects of Alpha-Tropomyosin Mutations Associated with Inherited Cardiomyopathies: Multiscale Simulation

Fyodor Syomin [1,2,*], Albina Khabibullina [3], Anna Osepyan [1,3] and Andrey Tsaturyan [1]

1. Institute of Mechanics, Lomonosov Moscow State University, 119192 Moscow, Russia
2. Peoples' Friendship University of Russia (RUDN University), 117198 Moscow, Russia
3. Mathematics and Mechanics Department, Lomonosov Moscow State University, 119991 Moscow, Russia
* Correspondence: f.syomin@imec.msu.ru

Received: 17 June 2020; Accepted: 14 July 2020; Published: 16 July 2020

Abstract: The effects of two cardiomyopathy-associated mutations in regulatory sarcomere protein tropomyosin (Tpm) on heart function were studied with a new multiscale model of the cardiovascular system (CVS). They were a Tpm mutation, Ile284Val, associated with hypertrophic cardiomyopathy (HCM), and an Asp230Asn one associated with dilated cardiomyopathy (DCM). When the molecular and cell-level changes in the Ca^{2+} regulation of cardiac muscle caused by these mutations were introduced into the myocardial model of the left ventricle (LV) while the LV shape remained the same as in the model of the normal heart, the cardiac output and arterial blood pressure reduced. Simulations of LV hypertrophy in the case of the Ile284Val mutation and LV dilatation in the case of the Asp230Asn mutation demonstrated that the LV remodeling partially recovered the stroke volume and arterial blood pressure, confirming that both hypertrophy and dilatation help to preserve the LV function. The possible effects of changes in passive myocardial stiffness in the model according to data reported for HCM and DCM hearts were also simulated. The results of the simulations showed that the end-systolic pressure–volume relation that is often used to characterize heart contractility strongly depends on heart geometry and cannot be used as a characteristic of myocardial contractility.

Keywords: mathematical modeling; cardiac mechanics; multiscale simulation; cardiomyopathies; left ventricle remodeling

1. Introduction

The inherited cardiac diseases, hypertrophic (HCM) and dilated (DCM) cardiomyopathies, can be caused by mutations in genes encoding sarcomere proteins expressed in heart muscle. At least 31 mutations in the TPM1 gene encoding regulatory protein tropomyosin (Tpm) are associated with HCM, DCM, or, more rear, left ventricular non-compaction [1–3]. HCM is characterized by a thickening of the left ventricular (LV) wall occurring in the absence of other diseases. This remodeling often results in a decrease in the volume of LV cavity and obstruction of the LV outflow tract. DCM is characterized by an increased volume of the LV cavity and a reduced ejection fraction in the absence of coronary artery diseases. The remodeling of the LV geometry is believed to play an adaptive role in the protection of the heart function despite impaired myocardial mechanics [4]. Changes in passive mechanical properties of LV myocardium upon HCM and DCM were also found [5–7].

Tpm is a coiled-coil dimer of parallel α-helices that serves as a gatekeeper in Ca^{2+} regulation of the actin–myosin interaction in sarcomeres of striated muscles. The Tpm molecules bind to each other in a head-to-tail manner and form a continuous strand located in a helical groove on the surface of an actin filament. The strand controls the availability of actin sites for the binding of motor domains of myosin molecules—myosin heads. Another regulatory protein, troponin (Tn), binds Tpm in a 1:1 stoichiometry and forms, together with the fibrillary actin and Tpm, a regulated thin filament. Tn controls Tpm movement with respect to the axis of an actin filament in a Ca^{2+}-dependent manner [8,9]. The regulation of muscle contraction is highly cooperative: relatively small changes in intracellular Ca^{2+} concentration cause large changes in force and actin–myosin ATPase rate. Modeling [10–12] suggests that the local movements of a stiff Tpm strand (caused by Ca^{2+} binding to Tn or myosin binding to actin) are transmitted to neighbor parts of the strand, providing high cooperativity.

Tpm mutation Asp230Asn is associated with DCM, while the Ile284Val one is associated with HCM. Both significantly change the Ca^{2+} regulation of the myosin–actin interaction measured in the in vitro motility assay or experiments with single cardiomyocytes [13–15]. These molecular and cellular level changes are believed to underlie the impairment of the heart function. Two questions remain unanswered: (1) how the changes in the actin–myosin interaction at the molecular and cellular levels caused by the Tpm mutations affect the pumping function of the heart and the LV particularly; and (2) how does the remodeling of the LV wall associated with the cardiomyopathies change the heart function? To address these questions, we performed a computer simulation of the heart mechanics using a recently developed multiscale LV model [16] incorporated into a simple lumped parameter model of circulation [17]. A model of myocardial mechanics used in the multiscale model was described previously [18]. It describes all major mechanical features of cardiac muscle including the force–velocity and stiffness–velocity relations, tension responses to step-like and ramp changes in muscle length, high cooperativity of Ca^{2+}-force relation, and its length-dependence, etc. Here, we used available experimental data to modify model parameters to account for the changes in the Ca^{2+} regulation of myocardial mechanics at the molecular and cellular levels caused by the cardiomyopathy-associated Tpm mutations. Then we simulated the effect of these changes on the movement of the LV wall during heartbeats. We also simulated the effects of remodeling of the left ventricle—hypertrophy and stenosis of the outflow tract for the Ile284Val mutation and the dilatation for the Asp230Asn one—on the calculated cardiac output. The effects of changes in the passive myocardial stiffness associated with HCM and DCM [5–7] were also estimated.

2. Materials and Methods

2.1. Cardiac Muscle Mechanics and Regulation

A model of cardiac muscle mechanics was described in detail previously [18]. The myocardium was treated as an anisotropic incompressible material with passive elastic and active stress components. Passive elastic stress was a sum of an isotropic hyper-elastic part and a part caused by the tension of titin filaments in sarcomeres. The overall passive stiffness was highly non-linear and anisotropic. The active tension was essentially one-dimensional acting along the axis of muscle fibers. It depended on two molecular variables that characterize the actin–myosin interaction: the fraction of myosin heads bound to actin n, and their ensemble averaged distortion δ. These cell-level variables were defined by a system of kinetic ordinary differential equations (ODE) specifying the interactions of contractile and regulatory proteins. Regulation of the contraction was determined by kinetic variables A_1 and A_2, which represent the fractions of available myosin binding sites on actin in the overlap zone and outside this zone, respectively. The kinetic equations for the variables accounted for the Ca^{2+} regulation of the thin filaments. The balance equation for Ca^{2+} concentration in cytoplasm included the terms of Ca^{2+} influx (set as a given function of time), Ca^{2+} binding to troponin and to other cytoplasmic proteins, and Ca^{2+} uptake from cytoplasm. Variation of the average micro-distortion depended on the sliding velocity of the myosin and actin filaments, thus being defined by the macroscopic strain rate tensor. The full set of equations describing the passive and active stress components in terms of continuum mechanics and the kinetic equations for the interaction of contractile and regulatory proteins and Ca^{2+} dynamics are given in [18] and Appendix A. For the values of the model parameters, see Supplementary Materials Table S1.

2.2. Geometry of the Left Ventricle

The LV was approximated by a body of rotation with a shape and distribution of the fiber orientation similar to those observed in human hearts. Despite the axial symmetry of the LV model, all three strain components—radial, axial and angular—were present, so that the ventricle was able to expand, contract, and twist during a heartbeat cycle. The fiber angle with respect to a plane perpendicular to the axis of symmetry changed linearly from +80° at the endocardium to −55° at the epicardium. To describe an increase in ventricular stiffness near its base and the peculiarities of ventricular anatomy near the apex, an additional anisotropic elastic stress of circumferential collagen fibers was added to the stress tensor in the base, and an isotropic elastic stress term depended on the number of bound myosin heads was added in the apex to account for the fiber disorder. The LV model used here was described in detail previously [16,17] (see also Appendix A and Supplementary Materials Table S2).

2.3. Model of Circulation

A lumped parameter model of systemic and pulmonary circulations that included the LV ventricle model treated the atria and the right ventricle as viscoelastic reservoirs with non-linear passive and time-varied active stiffness and viscosity [17]. Pressures and blood flows in different parts of the vascular bed were described by a system of ODEs. The compliances, hydraulic and inertial resistances of aorta, large systemic and pulmonary arteries and veins, and the resistances of the systemic and pulmonary microcirculation were parameters of these equations. These parameters were set up to describe the characteristic values and the time course of the pressures, volumes, and blood flows in different parts of the CVS of healthy humans. The model also accounted for changes in the inertial and hydraulic resistance upon constrictions of the left ventricular outflow tract and was capable of describing the changes in pressures caused by aortic valve stenosis of different severity. The model was described in detail previously [17] (see also Appendix A and Supplementary Materials Table S3).

2.4. Numerical Simulation and the Model Validation

The finite element (FE) method formulated in small increments was implemented to solve the problem, as described in detail [16,17]. Triangular FEs with linear displacement interpolation were used. The incompressibility equations were set up and solved for every two triangles connected into a quadrilateral element.

The CVS model used here was validated thoroughly. Our cell-level model of myocardium reproduced a large set of uniaxial experiments describing tension time courses and calcium transients in isometric twitches and load-dependent relaxation in mixed isometric/isotonic modes of contraction correctly. Validation of the model was presented in detail [18]. Our CVS model, the choice of the parameters for the hemodynamics block of the model and the comparison of hemodynamic values (ventricle and atrial pressures and volumes, pressures in different vessels, blood flow through the mitral valve) was discussed and validated [17]. Not only did the model reproduce typical time-courses of hemodynamical values in healthy humans correctly, but it also was successfully used to simulate the aortic and mitral valve stenosis and insufficiency. The results of the numerical research matched the data of clinical guidelines for the classification of the valves pathologies quantitatively. The numerical methods implemented here are commonly used and have been validated for the convergence, which was checked by variation of the time-step and the size and number of the FEs [16,17]. Local strains of our simulated axisymmetric left ventricle fit clinical data [16].

2.5. Modeling Cell-Level Effects of Two Cardiomyopathy-Associated Mutations

We assumed that the Asp230Asn and Ile284Val Tpm mutations did not affect the kinetics, the unitary force, and the unitary myosin displacement during its interaction with actin. The kinetics of Ca^{2+} release and uptake in myocardial cells was also assumed to be the same as in normal myocardium.

Only the parameters of the equations that describe Ca^{2+} binding to Tn, strain-dependence of the binding, and the number of myosin filaments per cross-section area in the myocardial model were changed to simulate the effects of the mutations.

Sequeira et al. [15] studied the Ca^{2+}–force relationship in single permeabilized myocardial cells from the LV of a patient with HCM-associated Ile284Val Tpm mutation and healthy donors. They had found that maximal tension at saturating Ca^{2+} concentration decreased by approximately 55% upon the Ile284Val Tpm mutation, while the Ca^{2+} concentration required for half-maximal activation decreased by a third. Besides, the length-dependent activation [19] in the cells from the HCM patient was significantly reduced as compared to that in healthy donors [15]. The maximal sliding velocity of reconstructed thin filaments containing Ile284Val Tpm in vitro was the same as of those with wild-type (WT) Tpm [20]. To simulate these experimentally observed changes, we decreased the density of myosin filaments per cross-section area (leading to a decrease in the maximal active tension) and the parameter of length-dependency of the activation while increasing the parameter of the Tn affinity for Ca^{2+} (thus increasing Ca^{2+} sensitivity). The model parameters and changes introduced to simulate the effects of the Tpm mutations listed above are described in Supplementary Materials Table S4. The resulting dependencies of isometric tension and the activation level in the overlap zone A_1 on dimensionless Ca^{2+} concentration are shown in Figure 1 (red).

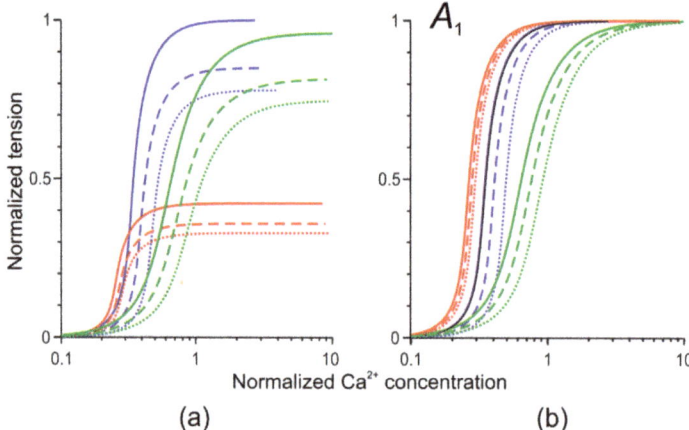

Figure 1. The dependence of the normalized isometric tension (**a**) and thin filament activation A_1 (**b**) on the dimensionless concentration of Ca^{2+} ions for the simulation of normal cardiac muscle (blue) and of those with the Ile284Val (red) and Asp230Asn (green) tropomyosin (Tpm) mutations. Continuous, dashed, and dotted lines correspond to sarcomere lengths of 2.2 µm, 2.0 µm, and 1.8 µm, respectively.

We could not find in the literature detailed characteristics of changes in the Ca^{2+} regulation and mechanical properties of human cardiac muscle caused by the Asp230Asn Tpm mutation. Several individuals from two families carrying the mutation have shown a mild or severe heart failure with the ejection fraction reduced down to 20% and significantly increased end-diastolic diameter of the left ventricle [13]. In vitro studies of the effects of the Asp230Asn Tpm mutation on Ca^{2+} regulation of myosin ATPase in the presence of regulated thin filaments and Ca^{2+} binding to thin filaments have shown a reduced Ca^{2+} sensitivity and decreased Ca^{2+} affinity for the thin filaments [13]. LV mechanics in vivo and in situ at the cellular and molecular levels were studied in transgenic mice carrying the Asp230Asn mutation [14]. A significantly reduced Ca^{2+} sensitivity and the cooperativity of Ca^{2+} regulation were found in vitro in the presence of the Asp230Asn Tpm mutation compared to WT Tpm [14,21]. To simulate the experimentally observed changes, we varied two model parameters in the regulation block of our model: the cooperativity parameter and the Tpm affinity for Ca^{2+}. The details are described in Supplementary Materials Table S4. The effects of these changes on the calculated dependencies of isometric tension and the activation level on Ca^{2+} in the overlap zone A_1 are shown in Figure 1 (green).

2.6. Modeling LV remodeling for HCM and DCM-Associated Tpm Mutations

The approximating geometry for the left ventricle shape was set up by the expressions from [22]. A detailed description of the approximation and the algorithm used to search for the unloaded initial configuration were described previously [16].

Sequeira et al. [15] reported that the maximal end-diastolic thickness of the septal wall of the hypertrophic left ventricle of the patient with the Ile284Val Tpm mutation was about 16 mm, while in healthy donors, it was 13 mm. The average end-diastolic thickness of the ventricle wall in a large population of healthy people was approximately 7 mm in men and 6 mm in women [23] with maximal septal wall thickness being, on average, 8.2 mm in men and 6.9 mm in women. It was also reported [24,25] that the end-systolic volume of the hypertrophic LV and the short radius of its cavity are, generally, slightly decreased.

In order to simulate the remodeling accompanying HCM according to the above cited data, the inner and outer end-systolic basal radii and the thickness of the ventricular apex were changed as described in Supplementary Materials (Table S5, HCM). These changes resulted in the increase in the end-diastolic LV wall thickness to 11 mm from 6 mm in the model of the normal LV.

To simulate ventricular dilatation caused by the Asp230Asn Tpm mutation, we changed the near-end-systolic geometry in accordance with the data reported for the transgenic mice [14]. In particular, the parameter choice was based on the relative value of the increase in the end-diastolic size and volume of the LV cavity of the mice. The changes in the model parameters made to simulate DCM caused by the mutation are specified in the Supplementary Materials (Table S5, DCM).

2.7. Modeling Changes in Passive Myocardial Stiffness Accompanying HCM and DCM

In the simulation of the LV remodeling described above (Section 2.6), no changes in the passive myocardial stiffness accompanying HCM or DCM were taken into account. There are no data on the passive properties of the LV myocardium associated with the Tmp mutations simulated here. However, there are clinical research data for similar HCM and DCM cases. A decrease in the titin-based stiffness in patients with the end-stage heart failure caused by nonischemic DCM was found [5]. These changes correlated with an increased expression of the long N2BA titin isoform in DCM myocardium as compared to its expression in normal myocardium, where the shorter N2B isoform was mainly expressed. Interestingly, no changes in the strain–stress diagram were found after removing the titin stress component by high ionic strength solutions. To reproduce these DCM data in our model, we increased the contour titin length [26] from 0.35 μm in normal myocardium to 0.725 μm in DCM myocardium, leaving the isotropic stress–strain relation the same as in normal myocardium.

In clinical research, the myocardial stiffness estimated by the measurement of shear wave velocity was 2–3 times higher in HCM patients than in healthy volunteers [7]. In rats with LV hypertrophy caused by aortic banding, titin stiffness increased significantly, while non-titin components showed only the slightest increase [6]. As the data are controversial, we tested the effects of an increase in each of two components of passive myocardial stiffness: a twofold increase in the isotropic extracellular stiffness or a decrease in titin contour length (0.24 μm instead of 0.35 μm in normal myocardium model).

3. Results

3.1. Simulation of Hemodynamic Changes Caused by the Asp230Asn and Ile284Val Tpm Mutations Without the LV Remodeling

To understand how the cell-level changes in mechanical properties of cardiac muscle caused by the Tpm mutations might affect the heart function, we simulated steady-state heartbeats at a rate of 60 heartbeats per min by the model with the default 'normal' LV size and shape [16,17], while the model parameters of Ca^{2+} regulation were changed as described in Section 2.5 (see also Supplementary Materials, Table S4) and shown in Figure 1. All other parameters of the model were the same as those for the normal heart model. The simulations performed with the model parameters corresponding to the normal right ventricle showed blood redistribution from the pulmonary circulation to the systemic one and a significant LV overfill. To overcome this problem in the absence of available data regarding changes in the right ventricle geometry and function caused by the mutations, we decreased the parameter of maximal isovolumetric pressure for the right ventricle to obtain the same end-diastolic LV volume as that in the normal heart model. The parameter was decreased from 85 to 70 mm Hg for the simulation of the Ile284Val Tpm mutation and to 60 mm Hg for the case of the Asp230Asn one. The duration of the right ventricle systole remained unchanged for simplicity. The hemodynamic variables obtained during the heartbeat simulations are shown in Figure 2.

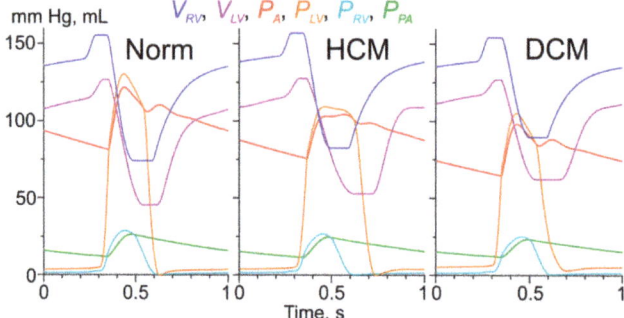

Figure 2. The results of the simulations of heartbeats for model of the normal (Norm) left ventricle (LV) cardiac muscle and those with the Ile284Val (hypertrophic cardiomyopathy, HCM) and Asp230Asn (dilated cardiomyopathy, DCM) Tpm mutations. V_{RV}, V_{LV} are volumes of the right and left ventricles, respectively; P_A, P_{LV}, P_{RV}, and P_{PA} are pressures in aorta, left and right ventricles and pulmonary artery, respectively. The color code is shown on top of the plots.

The end-diastolic and end-systolic LV volumes for the model of the normal heart were 127 mL and 45 mL, respectively, while the ejection fraction was 65%. These hemodynamic characteristics as well as the systolic (121 mm Hg) and diastolic (81 mm Hg) aortic pressures were close to those reported for healthy humans. The simulation of heartbeats in the presence of the Ile284Val Tpm mutation associated with HCM showed a mild reduction in the heart performance: the stroke volume and the rejection fraction reduced to 75 mL and 59%, respectively, while the systolic and diastolic aortic pressure decreased to 105 mm Hg and 75 mm Hg, respectively (Figure 2, HCM). The simulations of the effect of the Ile284Val Tpm mutation also showed a prolongation of LV systole from 189 ms in the normal heart model to 253 ms. Simulation of the effects of the Asp230Asn Tpm mutation resulted in more severe hemodynamic changes. The stroke volume and the LV ejection fraction reduced to 65 mL and 51%, respectively. The aortic blood pressure was also decreased compared to that in the simulations with the normal LV and was equal to 98/65 mm Hg (Figure 2, DCM).

3.2. Changes in Ca^{2+} Transients Caused by the Tpm Mutations

The changes in the cell-level model parameters caused by the Tpm mutations affected the time course of the intracellular variables describing myocardial activation, as shown in Figure 3.

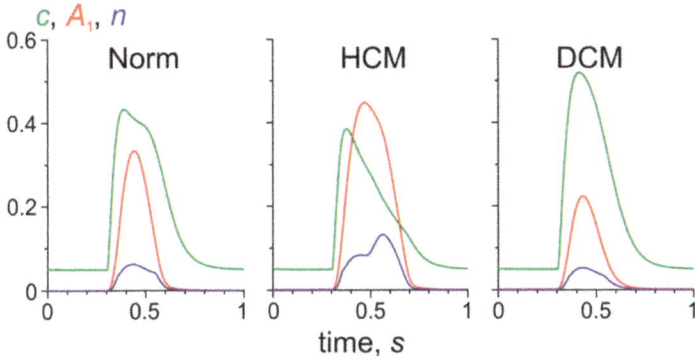

Figure 3. The time course of the normalized Ca^{2+} concentration in the cytosol (c), the level of activation of the Tpm–Tn (troponin) system in the overlap zone of sarcomeres (A_1), and the fraction of myosin heads bound to actin (n) in a mid-wall finite element located near the LV equator. The results obtained in the normal LV model (Norm) and those with normal LV geometry and cell-level parameters characteristic for the HCM and for DCM Tpm mutations are shown.

In the HCM model, the peak systolic values of A_1 and n increased compared to their normal values, while the Ca^{2+} peak decreased (Figure 3, HCM). These changes are caused by the shift of the force-Ca^{2+} toward lower Ca^{2+} concentration (Figure 1) and by a reduction of the free intracellular Ca^{2+} concentration due to its binding to Tn. In contrast, in the DCM model, the peaks of A_1 and n were reduced, while the peak of the free Ca^{2+} concentration was enhanced (Figure 3, DCM).

3.3. Simulation of Hemodynamic Changes Caused by the Tpm Mutations and the LV Remodeling

The end-diastolic and end systolic shapes of the normal model LV and those with DCM and HCM are shown in Figure 4.

The model of dilated LV with the Asp230Asn Tpm mutation showed higher end-diastolic and end-systolic volumes than those of the normal LV (Figure 4). The fiber strain in the DCM LV model was noticeably smaller than in the normal LV model. In contrast to DCM that caused more uniform distribution of sarcomere length than in the normal LV model, the model of HCM LV was characterized by a higher transmural difference in sarcomere length. This resulted in very short sarcomeres at the end of systole in the subepicardium.

The twist of the LV apex with respect to the base for the model of normal LV between systole and diastole was 16.3°. It decreased to 6.4° for the DCM model and increased to 37.9° for the HCM model. The higher twist may be responsible for the sarcomere length heterogeneity observed in the HCM simulation (Figure 4). We have also measured the global longitudinal strains (GLS) of the simulated LVs with the standard procedure used in 2D echocardiography. In simulations, GLS was decreased moderately in HCM LV (−15.1%) and drastically in DCM LV (−11.1%) compared to the value of approximately 18.7% for normal LV. The animations showing the changes in the LV shape during a heartbeat are given in the Supplementary Materials (Video S1).

The results of the simulation of the effects of the Tpm mutation and the LV remodeling (dilatation for the Asp230Asn mutation and hypertrophy for the Ile284Val one) on systemic and pulmonary hemodynamics are shown in Figure 5.

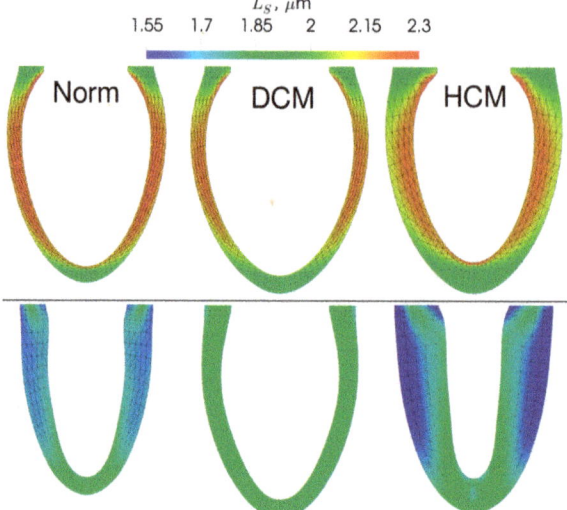

Figure 4. The end-diastolic (top) and end-systolic (bottom) LV geometry for the models of the normal (Norm), dilated (DCM), and hypertrophic (HCM) LV obtained during the simulations. The color code shows sarcomere length.

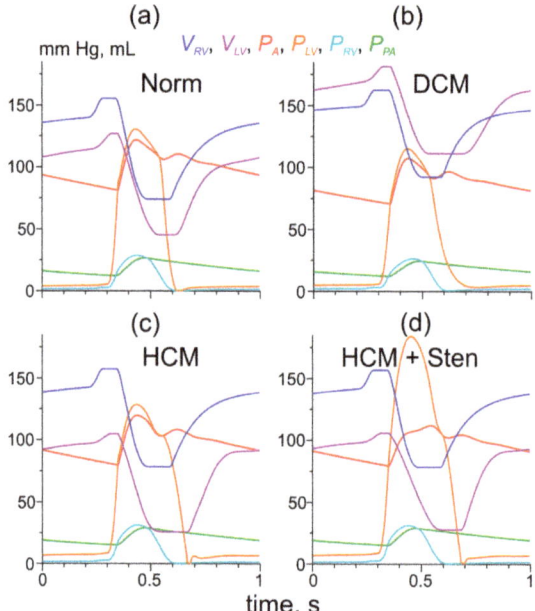

Figure 5. The results of simulations of hemodynamic variables during a heartbeat for the model of normal LV myocardium and normal LV geometry (**a**) and those with the DCM- and HCM-associated Tpm mutations and LV remodeling (**b**,**c**,**d**). (**b**) The cell level effects of the Asp230Asn TPM mutation were combined with LV dilatation as described in Methods; (**c**) the cell-level effects of the Ile284Val Tpm mutation were combined with LV hypertrophy as described in Methods; (**d**) the same as (**c**) plus the stenosis of the LV outflow tract. The color codes for the pressures and volumes are shown.

An increase in the volume of the LV cavity (dilatation) in the model with the Asp230Asn Tpm mutation led to partial compensation of a decrease in the LV performance caused by the changes in Ca^{2+} sensitivity (Figure 5b). The stroke volume and the aortic blood pressure reduced, although remaining close to those in the normal heart model (70 mL, 107/71 mm Hg versus 82 mL and 121/81 mm Hg, respectively) while the ejection fraction was reduced significantly to 39% (compared to 65%).

The LV hypertrophy was accompanied by an increase in passive LV stiffness: at a 'normal' end-diastolic LV pressure (EDLVP) of 5.3 mm Hg, the LV volume was only 92 mL. Our simulation of the remodeled LV was performed with the 'normal' initial values of the hemodynamic variables (excluding the LV volume) and the 'healthy' right ventricle (default values of E_{3RV}, Table S3). Such conditions resulted in an increased LV preload of 8.6 mm Hg and the end-diastolic volume of 105 mL. Simulation of the effect of LV hypertrophy for the HCM-associated Ile284Val Tpm mutation also resulted in a more complete recovery of the hemodynamic parameters (Figure 5c). The aortic pressure of 120/80 mm Hg and the stroke volume of 79 mL were close to those calculated for the normal heart model (121/81 mm Hg, 82 mL), while the ejection fraction increased to 75%. When the stenosis of the LV output tract (maximal orifice area 1 cm^2 at aortic area of 3 cm^2) was taken into account in the model of HCM caused by the Tpm mutation, the peak LV systolic pressure increased up to 184 mm Hg, while the aortic pressure (112/79 mm Hg) and the stroke volume (78 mL) reduced slightly compared to the HCM model without the stenosis (Figure 5d). The peak pressure gradient (72 mm Hg) was similar to that in a patient with this mutation [15].

As one would expect, a decrease in titin stiffness in the LV myocardium at DCM led to a slight increase in the end-diastolic LV volume at fixed values of end-diastolic pressure and to an increase in the stroke volume and arterial blood pressure. An increase in any component of the passive myocardial stiffness in the LV myocardium at HCM led to a further impairment of LV diastolic function and decreased LV performance (Table 1).

Table 1. The effects of passive myocardial stiffness on the model LV performance at HCM and DCM.

LV Hemodynamic Characteristics	HCM + Normal Passive Stiffness	HCM + Stiff Titin Component	HCM + Stiff Isotropic Component	DCM + Normal Passive Stiffness	DCM + Soft Titin Component
		Low Preload			
End-diastolic pressure, mm Hg		6.4		4.3	
Peak pressure, mm Hg	115.2	112.6	92.5	105.4	108.5
End-diastolic volume, mL	92.2	88.3	73.2	157.5	165.1
Stroke volume, mL	67.7	64.1	48.3	64	66.1
Ejection fraction, %	73	73	66	41	40
		Average Preload			
End-diastolic pressure, mm Hg		8.7		5.1	
Peak pressure, mm Hg	128.4	123.1	99.3	110.7	113.2
End-diastolic volume, mL	104.9	99.2	80	169.8	176.6
Stroke volume, mL	79	74	54.8	67.3	68.9
Ejection fraction, %	75	75	69	40	40

Table 1. Cont.

LV Hemodynamic Characteristics	HCM + Normal Passive Stiffness	HCM + Stiff Titin Component	HCM + Stiff Isotropic Component	DCM + Normal Passive Stiffness	DCM + Soft Titin Component
		High Preload			
End-diastolic pressure, mm Hg		11.5			5.8
Peak pressure, mm Hg	141.1	133	108.6	115.2	118.3
End-diastolic volume, mL	117.9	109.5	88.8	181.5	191
Stroke volume, mL	89.9	83.1	63	70.2	72.1
Ejection fraction, %	76	76	71	39	38

3.4. Simulation of the Effects of the Tpm Mutations and the LV Remodeling on the Pressure-Volume Loops

The LV pressure–volume loops (PV-loops) obtained at different preloads are often used to estimate the systolic and diastolic functions of the heart chambers. In particular, the slope of the line plotted through the end-systolic point of the loops, the so-called end-systolic pressure–volume relationship (ESPVR) is believed to characterize the LV contractility [27]. Figure 6 shows the PV-loops obtained in the simulations of normal and cardiomyopathic LVs. To probe the LV performance at various preloads, initial blood pressures in systemic and pulmonary veins were varied.

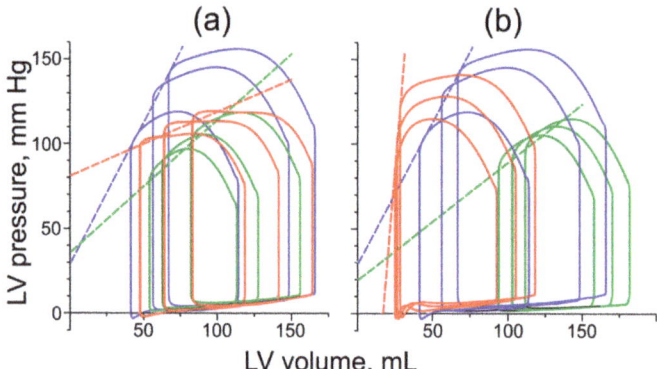

Figure 6. The LV pressure–volume loops obtained from the simulation of the LV with normal cardiac muscle and normal geometry (blue) and the simulations of DCM (green) and HCM (red). Different loops were obtained at different preloads. (**a**) PV loops obtained with normal LV geometry; (**b**) PV loops obtained for the remodeled LVs with DCM and HCM at default passive myocardial stiffness. The dashed straight lines are the ESPVR lines plotted for each simulation case.

Both HCM and DCM resulted in a decrease in the slope of ESPVR if the LV remodeling was not considered, and only changes in myocardial properties caused by the Tpm mutations were taken into account (Figure 6a. This agrees with the idea that the slope characterizes myocardial contractility [28]: the DCM-associated mutation decreases the Ca^{2+} sensitivity of active tension, while the HCM-associated mutation decreases the maximal active tension and length-dependent activation (Figure 1).

LV dilatation led to a further decrease in the ESPVR slope compared to the effect of the Asp230Asn Tpm mutation alone (Figure 6b, green). The slope of ESPVR in the DCM simulations decreased by a factor of approximately 2.5 compared to the normal LV model. On the contrary, the LV hypertrophy (Figure 6b, red) dramatically increased the ESPVR slope compared to that in the absence of hypertrophic remodeling and even compared to the slope in the simulations of the normal heartbeat, despite impaired myocardial properties.

4. Discussion

4.1. Cell-Level Changes in Ca^{2+} Regulation and Cardiac Function

When the molecular-level effects of the Asp230Asn Tpm mutation [13,14] were introduced into our myocardial model, the amplitude of Ca^{2+} transient increased as was observed in transgenic mice [14], while the activation level (A_1) and the fractions of actin-bound myosin heads (n) decreased compared to the model of normal heart (Figure 3, DCM). When changes in Ca^{2+} regulation found in myocardial cells with the HCM-associated Ile284Val Tpm mutation [15] were simulated, the amplitude of Ca^{2+} transient decreased, while the fraction of actin bound myosin heads increased (Figure 3, HCM). These effects result from a combination of several factors: change in Ca^{2+} sensitivity of the thin filaments caused by the mutations (Figure 1) and Ca^{2+} binding to Tn affecting the free Ca^{2+} concentration. Despite the difference, changes in the cell-level model parameters corresponding to those caused by both the HCM and DCM Tpm mutations decreased arterial blood pressure, stroke volume, and ejection fraction compared to the model of normal LV (Figure 2).

4.2. Effects of LV Remodeling

To simulate the effect of the remodeling accompanying DCM or HCM, we changed the LV geometry in the model according to available data for the Asp230Asn and Ile284Val Tpm mutations, respectively.

The LV dilatation (Figure 4, DCM) partially, although not completely, compensated for the loss of the stroke volume caused by the Asp230Asn Tpm mutation (Figure 5b). In this case, a partial restoration of the heart function was accompanied by a further reduction in the ejection fraction to a value close to those found in members of two families with the mutation [13]. Compensation of the stroke volume at a decreased ejection fraction was also observed in transgenic mice [14]. We suppose that the compensation occurred due to the increase in the end-diastolic volume of the LV, which provided the normal stroke volume despite the Starling's law of the heart estimated by the PV loops being violated (Figure 6b green).

A decrease in passive myocardial stiffness due to an increase in the contour length of titin in the model had only a slight effect on the heart performance, enhancing the compensatory effect of LV dilation a little (Table 1). The results of the simulation show that the changes in cardiac titin observed in myocardial samples from DCM patients [5] might be a part of the LV remodeling that helps preserve its function.

When the LV hypertrophy that accompanied the Ile284Val Tpm mutation was included in the LV model, it resulted in nearly full compensation of the stroke volume and the aortic blood pressure at preserved initial hemodynamic conditions, which provided an increased preload of the LV. At the 'normal' values of the preload, no compensation was observed. When stenosis of the outflow tract, which accompanied the LV hypertrophy in a patient with the Ile284Val Tpm mutation [15], was accounted for, the stroke volume and the ejection fraction were close to their values for the model of normal LV at elevated preload (Figure 5c,d). The results of our simulations suggest that the wall thickening is able to compensate for the reduction of active force and the impaired length-dependent Ca^{2+} activation caused by the Ile284Val Tpm mutation. However, an increase in the isotropic or anisotropic (titin) component of passive myocardial stiffness along with the wall thickening led to an even more severe impairment of LV diastolic and systolic function (Table 1). The hypertrophy also led to heterogeneity in sarcomere length distribution across the LV wall (Figure 4) and to an increase in the LV twist.

Significant values of the decrease in the LV twist and absolute values of GLS in our simulation of the DCM LV were similar to those observed in DCM patients [29–31]. The decrease in the LV GLS obtained here was also close to that in DCM patients [30–32]. An increase in the LV twist compared to control was observed in patients with HCM. The observed increase varied, being either significantly lower than that calculated here [33,34] or close to it [30]. The decrease in the calculated GLS was also similar to that in HCM patients [30,34–36]. An increase in the isotropic (extracellular) or anisotropic (titin) component of the passive myocardial stiffness in the HCM LV model worsened the LV function by impairing its diastolic function (Table 1). The increase in passive stiffness together with the thickening of the LV wall itself are the factors impairing LV diastolic function.

4.3. PV Loops Depend on Not Only on Myocardial Contractility but Also LV Geometry

Our simulation of PV loops for the normal and cardiomyopathic LVs resulted in the following conclusions and suggestions.

Firstly, the ESPVR slope does characterize the contractile properties of cardiac muscle at fixed LV geometry, as was initially suggested by Suga and Sagawa [27]. The simulated ESPVRs had lower slopes for the cases of both HCM-and DCM-associated mutations in not remodeled LV compared to those for the normal LV model (Figure 6a). The decrease in the ESPVR slope was caused by the impaired properties of the muscle: the decreased Ca^{2+} sensitivity for the DCM mutation and the decreased maximal force with a simultaneous reduction in the length-dependence of Ca^{2+} activation for the HCM mutation.

Secondly, in the simulations of the dilated LVs with the Asp230Asn Tpm mutation, the ESPVR slope was even lower than in the LV with the same mutation and normal geometry. The decrease in the ESPVR slope by a factor of 2.5 in the simulations of the DCM LVs compared to the model of normal LV is similar to that observed [14] for WT and transgenic mice. A possible explanation for the further reduction in ESPVR is as follows. An enlargement of the LV cavity at a normal wall thickness leads to a decrease in the end-systolic myocardial stress at the same end-systolic blood pressure.

Thirdly, the LV hypertrophy that accompanies the Ile284Val Tpm mutation (in contrast to the effects of the LV dilatation) led to an even steeper ESPVR slope than that obtained for the normal LV model (Figure 6b. We were not able to find any PV-loop data for this particular mutation, but an increase in the ESPVR slope for the patients with heart failure, concentric HCM, and non-impaired ejection fraction was reported [37,38]. We suppose that the increase in the ESPVR slope for the HCM LV, compared to that with the Ile284Val Tpm mutation and normal geometry, was caused by an increase in the thickness of the LV wall and, possibly, by the enhanced Ca^{2+} sensitivity of cardiac muscle. From our results and published clinical data, we can suggest that the ESPVR slope is strongly affected by changes both in the contractile properties of the myocardium and the LV geometry and cannot be used as a characteristic of myocardial contractility for hypertrophic LVs.

4.4. Relation to Previous Works

A number of models were suggested to simulate heart work in health and disease. A simple 0D lumped parameter model [39] was used to simulate the heartbeats of normal and DCM hearts. In this model, the myocardial dysfunction was described by a decrease in the end-systolic LV elastance, which is a value that is difficult to relate to the specific changes in cardiac muscle characteristics underlying DCM. A multiscale model of the CVS containing an accurate cell-level description of myocardial electromechanics and simple thin-wall spherical approximation of the ventricles [40] was applied to study the ventricular contractility at different preload and calcium kinetics in myocytes. The model was further validated [41], being able to reproduce some experimental data including the effects of an anesthetic on hemodynamics through its influence on the Ca^{2+} sensitivity of myofilaments. A more sophisticated 3D electromechanical model of a DCM heart was used to estimate the efficiency of an LV-assist device [42]. Although these authors used more realistic anatomy of the ventricles than the one in our model and included electrophysiological processes, the description of the active tension was too simplified to reproduce the changes in active tension and its regulation caused by the mutations. Several models were suggested to describe heart remodeling (see reviews [43,44]). Some models of this kind describe electromechanics of failing heart including concentric and eccentric hypertrophy.

For example, model [45] described electrophysiology, ventricle anatomy, and passive myocardial mechanics in some detail, while the specification of active tension did not allow one to simulate the effects of the Tpm mutations. A 3D simulation of the diastolic function of DCM heart that considered its detailed remodeling including the changes in a number of sarcomeres in cardiomyocytes was presented [46]. Concentric hypertrophy was simulated by a detailed 3D electromechanical model [47], where cardiac muscle electrophysiology was described by a bidomain model and its mechanics were specified by a detailed cardiac cell model [48]. The authors examined the effects of heart remodeling (concentric hypertrophy) caused by aortic stenosis on heart performance and did not investigate any effects of mutations of sarcomere proteins.

4.5. Limitations

Our simulation was based on a simplified axisymmetric LV model. The 3D models [49] are too expensive computationally for a thorough investigation of the effects of the Tpm mutations. These effects are caused by rather uniform cell-level changes in cardiomyocytes, so that a 2D approximation appears to be sufficient for their simulation. Besides, no 3D data are available for the LV geometry of patients with the Tm mutations considered here. For these reasons, the use of a model with reduced dimensionally and detailed description of the cell-level mechanics and Ca^{2+} regulation seems to be a reasonable simplification.

In our simulations, we did not take into account any alteration of the distribution of the orientation of cardiac fibers in the LV wall as a long-time effect of HCM and DCM accompanying the remodeling of the LV geometry [50,51]. This can be the reason for the discrepancy between our estimation of the LV twist upon HCM and some clinical data. The absence of the right ventricle in our finite element model may be another reason for the discrepancy in the LV twist calculation. It should also be noticed that the LV remodeling was set in a straightforward manner without the consideration of any assumptions on the kinetics of growth and/or structural rearrangement [43,44,46]. We also were not able to reproduce possible alterations in myocardium electrical excitability and conductivity including the remodeling of the gap junctions accompanying the cardiomyopathies [52], because our model did not contain a description of myocardium electrophysiology.

Supplementary Materials: The following are available online at http://www.mdpi.com/2227-7390/8/7/1169/s1, Table S1: Parameters of the myocardium cell model, Table S2: LV shape approximation parameters, Table S3: Parameters of the hemodynamic model, Table S4: Values of the parameters for the regulation block of the myocardium cell-model for the simulation of HCM- and DCM-associated mutations, Table S5: Geometrical parameters of the normal LV and the LVs with cardiomyopathy. Video S1: Animation of the changes in the shape and sarcomere lengths during the simulation of a heartbeat. Left to right: the normal, HCM, and DCM LV models. The animation starts with the atrial systole. The color code is for sarcomere length in µm.

Author Contributions: Conceptualization, F.S. and A.T.; software, F.S. and A.K.; validation, F.S.; investigation, numerical simulations and the results analysis, F.S., A.K. and A.O.; visualization—A.K. and A.O., writing—original draft preparation, F.S.; writing—review and editing, A.K., A.O. and A.T.; supervision, A.T. All authors have read and agreed to the published version of the manuscript.

Funding: This research was funded by the Russian Foundation for Basic Research, grant numbers 18-31-00065 (for F.S.) and 17-00-00066 (a part of a complex grant 17-00-0071, for A.T.).

Conflicts of Interest: The authors declare no conflict of interest. The funders had no role in the design of the study; in the collection, analyses, or interpretation of data; in the writing of the manuscript, or in the decision to publish the results.

Appendix A

A.1. Cell Model of Myocardium [18]

Cauchy stress tensor **T** was specified by the following equation:

$$\mathbf{T} = \left(\frac{\partial \Phi(I_1, I_2)}{\partial I_1} + I_1 \frac{\partial \Phi(I_1, I_2)}{\partial I_2}\right)\mathbf{F} - \frac{\partial \Phi(I_1, I_2)}{\partial I_2}\mathbf{F}^2 - p\mathbf{E} + \mathbf{B}(T_{tit} + T_A) \tag{A1}$$

F is the Finger deformation tensor, I_1, I_2 are its first and second invariants;
E is the unit tensor, p is the Lagrange factor caused by incompressibility;
B is the anisotropy tensor equal to the tensor square of the unit vector $\vec{c_f}$ aligned with the direction of muscle fibers in deformed muscle $\mathbf{B} = \vec{c_f} \otimes \vec{c_f}$.

The isotropic elastic strain energy Φ was taken in the form similar to that used by Guccione et al. [53]:

$$\Phi = a_0 \exp\left(a_1\left((I_1 - 3)^2 - 2(I_2 - 2I_1 + 3)\right)\right), \tag{A2}$$

where a_0, a_1 are constant parameters. The difference from the work [53] is the absence of anisotropic part, which, in our model, was included only in the last term of the Equation (A1).

T_{tit} is the anisotropic passive tension of the intra-sarcomere cytoskeleton mainly caused by titin filaments; T_A is the active tension produced by the actin–myosin interaction. Titin tension was specified by the equation based on the worm-like chain model [26].

$$T_{tit} = \begin{cases} N_M \frac{6k_BT}{L_p}\left(\frac{1}{4}\left(1 - \frac{L_s - L_{s0}}{L_c}\right)^{-2} + \frac{L_s - L_{s0}}{L_c} - \frac{1}{4}\right), L_s \geq L_{s0}, \\ N_M \frac{9k_BT}{L_pL_c}(L_s - L_{s0}), L_s < L_{s0}. \end{cases} \tag{A3}$$

L_s and L_{s0} are the deformed and reference sarcomere lengths, N_M is the number of the myosin filaments per unit cross-section area of muscle in its initial reference state; L_c is the total, or 'contour', length of a titin molecule, L_p is so-called persistence length; k_B and T are the Boltzmann constant and absolute temperature. The deformed sarcomere length is described by the equation $L_s = L_{s0}\sqrt{\vec{c_{f_0}}\mathbf{G}\vec{c_{f_0}}}$, where **G** is the right Cauchy–Green deformation tensor, $\vec{c_{f_0}}$ is the unit vector aligned with fibers in unstrained muscle.

Cross-bridge kinetics was based on the Lymn–Taylor cycle, in which a specific part of a myosin molecule (myosin head) can be in a free state or attached to actin filament in two different ways: a weakly bound and force generating strongly bound state. Thus, the active tension was specified as $T_A = E_{cb}N_M N_{cb}W(L_s)n(\delta + \theta h)$. Here, N_{cb} is the total number of myosin heads per one-half of a myosin filament; $W(L_s)$ is the length of the overlap zone of the thick and thin filaments in a half-sarcomere normalized for its maximal value; E_{cb} is the constant cross-bridge stiffness, n is the probabilities of a myosin head being attached to the actin filament, θ is the fraction of strongly bound cross-bridges among n, and h is a cross-bridge distortion during transition from the weakly bound state to the strongly bound one. The kinetic equations for n and θ are as follows

$$\frac{\partial n}{\partial t} = f_+(\delta)(A_1 - n) - f_-(\delta)n, \tag{A4}$$

$$\frac{\partial n\theta}{\partial t} = H_+(\delta)n(1-\theta) - H_-(\delta)n\theta \Rightarrow (H_+, H_- \gg f_+, f_-) \Rightarrow \theta = \frac{H_+(\delta)}{H_+(\delta) + H_-(\delta)}. \tag{A5}$$

Here, f_+, f_-, H_+, H_- are kinetic rates that depend on ensemble-averaged cross-bridge distortion δ. The equation for the normalized cross-bridge distortion $\delta' = \delta/h$ was

$$\frac{\partial \delta'}{\partial t} = \frac{1}{2h}\frac{\partial L_s}{\partial t} - \frac{(A_1 - n)}{n}f_+(\delta')\delta', \tag{A6}$$

and the kinetic rates were set as

$$f_+(\delta') = f_+^0\begin{cases} 1, \delta' \leq 0, \\ \frac{\delta_0^2}{(\delta_0 - \delta')^2}, \delta' > 0, \end{cases} f_-(\delta') = f_+^0\begin{cases} b_{cb} + c_{cb}\delta'^2, \delta' \leq 0 \\ b_{cb} + \frac{\delta'}{\delta_0 - \delta'}, \delta' > 0, \end{cases} \frac{H_+(\delta')}{H_-(\delta')} = e^{-\Delta\delta'}. \tag{A7}$$

In Equation (4), A_1 is the probability that a binding site of actin in the overlap region of the actin and myosin filaments is in the open state for the myosin head. The similar probability for the region outside the overlap zone was denoted by A_2. The kinetics of these variables depended on c (Ca^{2+} concentration in cytoplasm normalized by 'normal' half-activation concentration $c^0{}_{50}$), cooperativity parameter m (the Hill coefficient), sarcomere length via parameter k_s, and the number of strongly bound cross-bridges via parameter k_n. W_a is the length of the overlap zone normalized by the actin length per half sarcomere.

$$\frac{\partial A_1}{\partial t} = \begin{cases} \alpha_+\left(c(1-A_1)^{\frac{1}{m}} - \frac{c_{50}A_1^{\frac{1}{m}}}{(1+k_s(L_s-L_{s0})/L_{s0})(1+k_n n\theta)}\right), & \frac{\partial W_a(L_s)}{\partial t} \leq 0, \\ \alpha_+\left(c(1-A_1)^{\frac{1}{m}} - \frac{c_{50}A_1^{\frac{1}{m}}}{(1+k_s(L_s-L_{s0})/L_{s0})(1+k_n n\theta)}\right) + \frac{\partial W_a(L_s)}{\partial t}\frac{(A_2-A_1)}{W_a(L_s)}, & \frac{\partial W_a(L_s)}{\partial t} > 0, \end{cases} \quad (A8)$$

$$\frac{\partial A_2}{\partial t} = \begin{cases} \alpha_+\left(c(1-A_2)^{\frac{1}{m}} - \frac{c_{50}A_2^{\frac{1}{m}}}{(1+k_s(L_s-L_{s0})/L_{s0})}\right) + \frac{\partial W_a(L_s)}{\partial t}\frac{(A_2-A_1)}{(1-W_a(L_s))}, & \frac{\partial W_a(L_s)}{\partial t} \leq 0, \\ \alpha_+\left(c(1-A_2)^{\frac{1}{m}} - \frac{c_{50}A_2^{\frac{1}{m}}}{(1+k_s(L_s-L_{s0})/L_{s0})}\right) + \frac{\partial W_a(L_s)}{\partial t}, & \frac{\partial W_a(L_s)}{\partial t} > 0, \end{cases} \quad (A9)$$

$$\left(1 + \frac{k_{BC}B_{Ca}}{(c \cdot c_{50}^0 + k_{BC})^2}\right)\frac{\partial c}{\partial t} = I_{Ca}(t) - \frac{(Y_{1Ca}/c_{50}^0)(c^2 - c_0^2)}{c^2 \cdot c_{50}^0 + Y_{2Ca}^2} - (C_{Tn}/c_{50}^0)\frac{\partial(A_1 W_a(L_s) + A_2(1 - W_a(L_s)))}{\partial t}. \quad (A10)$$

α_+ is a characteristic rate constant of the Ca-troponin; $I_{Ca}(t) = I_{Ca}^0(exp(-k_{1Ca}t) - exp(-k_{2Ca}t))$ is a given inflow of Ca^{2+} ions to the cell normalized by $c^0{}_{50}$. The parameters of this block of the model and their values are presented in Table S1.

A.2. Left Ventricle Approximation

LV approximation was set by the following expressions for Cartesian coordinates (r, z) through the curvilinear coordinates (γ, ψ) with parameters ε, r_{in}, r_{out}, h_{in}, and h_{out} [16,22].

$$\begin{aligned} r &= (r_{in} + \gamma(r_{out} - r_{in}))(\varepsilon\cos\psi + (1-\varepsilon)(1-\sin\psi)), \\ z &= (h_{in} + \gamma(h_{out} - h_{in}))(1-\sin\psi) + (1-\gamma)(h_{out} - h_{in}). \end{aligned} \quad (A11)$$

A.3. Hemodynamics Model [17]

Passive pressures of the right ventricle and the atria were described as follows.

$$\begin{aligned} P_{LA_{Pas}} &= E_{1LA} \cdot \left(e^{E_{2LA}V_{LA}(t)} - e^{E_{2LA}V_{0LA}} + \mu_{a1}\frac{\partial V_{LA}(t)}{\partial t}\right), \\ P_{RA\_Pas} &= E_{1RA} \cdot \left(e^{E_{2RA}V_{RA}(t)} - e^{E_{2RA}V_{0RA}} + \mu_{a1}\frac{\partial V_{RA}(t)}{\partial t}\right), \\ P_{RV\_Pas} &= E_{1RV} \cdot \left(e^{E_{2RV}V_{RV}(t)} - e^{E_{2RV}V_{0RV}}\right). \end{aligned} \quad (A12)$$

Here, P_{**_Pas} are passive elastic parts of chamber pressures, V_{**} are their volumes, and V_{0**}, E_1, E_2, and μ_{a1} are constant parameters. Subindices LA, RA, and RV stand for the left atrium, right atrium, and right ventricle, respectively. Active pressures of the right ventricle and the atria were found from ordinary differential equations. Due to the introduction of time delay with relaxation time τ and the analogue of force–velocity equation, those described the pressure time-course more accurately than the pressure–volume dependencies with time-dependent stiffness coefficients commonly used in other lumped parameter models.

$$\begin{aligned} \tau\frac{\partial P_{LA\_Act}(t)}{\partial t} + P_{LA\_Act}(t) &= F_{LA\_Act}(t) \cdot \left(\mu_{a2}\frac{\partial V_{LA}(t)}{\partial t} + E_{3LA}\right), \\ \tau\frac{\partial P_{RA\_Act}(t)}{\partial t} + P_{RA\_Act}(t) &= F_{RA\_Act}(t) \cdot \left(\mu_{a2}\frac{\partial V_{RA}(t)}{\partial t} + E_{3RA}\right), \\ \tau\frac{\partial P_{RV\_Act}(t)}{\partial t} + P_{RV\_Act}(t) &= F_{RV\_Act}(t) \cdot \left(\mu_v\frac{\partial V_{RV}(t)}{\partial t} + E_{3RV}\right). \end{aligned} \quad (A13)$$

Here, P_{**_Act} are active parts of chamber pressures, and $E3$, μ_v, and μ_{a2} are constant parameters. F_{**_Act} depended on activation time-functions $e(t)$ and on the volumes V providing the Starling's law for the atria and the right ventricle; k_V, V_{Min}, and V_{Max} are the parameters for the pressure–volume relation.

$$F_{LA\_Act} = e_a(t) \cdot \left(k_V + (1-k_V)\right) \cdot \left(\frac{V_{LA}-V_{LA\_Min}}{V_{LA\_Max}-V_{LA\_Min}}\right),$$
$$F_{RA\_Act} = e_a(t) \cdot \left(k_V + (1-k_V)\right) \cdot \left(\frac{V_{RA}-V_{RA\_Min}}{V_{RA\_Max}-V_{RA\_Min}}\right), \quad (A14)$$
$$F_{RV\_Act} = e_v(t) \cdot \left(k_V + (1-k_V)\right) \cdot \left(\frac{V_{RV}-V_{RV\_Min}}{V_{RV\_Max}-V_{RV\_Min}}\right).$$

Activation functions were set as follows

$$e_a = \begin{cases} 0.5\left(1 - \cos\left(2\frac{(t-t_a)\pi}{T_a}\right)\right), & t_a \leq t < t_a + T_a \\ 0, & \text{otherwise,} \end{cases}$$
$$e_v = \begin{cases} 0.5\left(1 - \cos\left(2\frac{(t-t_v)\pi}{T_v}\right)\right), & t_v \leq t < t_v + T_v \\ 0, & \text{otherwise,} \end{cases} \quad (A15)$$

where t_a and t_v are the times of contraction initiations for the atria and ventricles, respectively; and T_a, T_v are the systole durations.

The full system of ordinary differential equations for other hemodynamic variables is

$$\frac{dV_{RV}}{dt} = Q_{iRV} - Q_{oRV},$$
$$\frac{dV_{RA}}{dt} = Q_{iRA} - Q_{iRV},$$
$$\frac{dV_{LV}}{dt} = Q_{iLV} - Q_{oLV} + C_{LV}\frac{dP_{LV}}{dt},$$
$$\frac{dV_{LA}}{dt} = Q_{iLA} - Q_{iLV}, L_{iLV}\frac{dQ_{iLV}}{dt} + R_{iLV}Q_{iLV} = P_{LA} - P_{LV},$$
$$L_{oLV}\frac{dQ_{oLV}}{dt} + R_{oLV}Q_{oLV} = P_{LV} - P_{A1},$$
$$L_A\frac{dQ_A}{dt} + R_AQ_A = P_{A1} - P_{A2},$$
$$Q_{iRA} = \frac{P_V - P_{RA}}{R_{iRA}},$$
$$Q_{iRV} = \frac{P_{RA} - P_{RV}}{R_{iRV}}, \quad (A16)$$
$$L_{oRV}\frac{dQ_{oRV}}{dt} + R_{oRV}Q_{oRV} = P_{RV} - P_{APulm}, C_{A1}\frac{dP_{A1}}{dt} = Q_{oLV} - Q_A,$$
$$C_{A2}\frac{dP_{A2}}{dt} = Q_A - \frac{P_{A2}-P_V}{R_{per}} - \frac{P_{A2}-P_C}{R_C},$$
$$C_V\frac{dP_V}{dt} = \frac{P_{A2}-P_V}{R_{per}} - Q_{iRA},$$
$$C_C\frac{dP_C}{dt} = \frac{P_{A2}-P_C}{R_C},$$
$$C_{APulm}\frac{dP_{APulm}}{dt} = Q_{oRV} - \frac{P_{APulm}-P_{VPulm}}{R_{perPulm}},$$
$$C_{VPulm}\frac{dP_{VPulm}}{dt} = Q_{iLA} + \frac{P_{APulm}-P_{VPulm}}{R_{perPulm}}.$$

Q_{iRV}, Q_{oRV}, Q_{iLV}, Q_{oLV}, Q_{iRA}, and Q_{iLA} are blood flows through the tricuspid valve, pulmonary valve, mitral valve, aortic valve, and the flows through systemic and pulmonary veins, respectively; Q_A is the arterial flow. R and L are hydraulic and inertial resistances of the vessels and chamber entrances (subindex i) and exits (subindex o); R_{per} and $R_{perPulm}$ are the peripheral vascular resistances of systemic and pulmonary circulation systems; C represents the compliances of the vascular reservoirs presented in the model; R_C and C_C characterize the viscoelastic properties of the systemic arteries. Indexes $A1$, $A2$, V, $APulm$, and $VPulm$ corresponded to the aorta, large arteries, systemic veins, pulmonary arteries, and veins, respectively.

A.4. Local Hydraulic Valve Resistances in the Cases of the Valve Pathologies

The equations for the flows through the aortic and mitral valves in Equation (A16) were modified introducing the quadratic hydraulic valvular resistance and the dependencies of the resistances on orifice area.

$$L_{**}\frac{S_0}{S}\frac{dQ_{**}}{dt} + \left(R_{**} + R_{sq}|Q_{**}|\right)Q_{**} = \Delta p,$$
$$R_{sq} = C_{sq}\varepsilon_{Re}\zeta_{sq}S^{-2},$$
$$\varepsilon_{Re} = \sum_{i=0}^{5} b_i(lg(Re))^i, \quad (A17)$$
$$\zeta_{sq} = \left(C_\zeta\left(1 - \frac{S}{S_0}\right)^{0.375} + 1 - \frac{S}{S_0}\right)^2.$$

S is the orifice area, and S_0 is an area of a completely open valve in healthy conditions; Re is the Reynolds number, C_ζ and b_i are parameters.

References

1. Moraczewska, J. Thin filament dysfunctions caused by mutations in tropomyosin Tpm3.12 and Tpm1.1. *J. Muscle Res. Cell Motil.* **2020**, *41*, 39–53. [CrossRef]
2. Redwood, C.; Robinson, P. Alpha-tropomyosin mutations in inherited cardiomyopathies. *J. Muscle Res. Cell Motil.* **2013**, *34*, 285–294. [CrossRef]
3. Watkins, H.; Ashraan, H.; Redwood, C. Inherited cardiomyopathies. *N. Engl. J. Med.* **2011**, *364*, 1643–1656. [CrossRef]
4. Opie, L.H.; Commerford, P.J.; Gersh, B.J.; Pfeffer, M.A. Controversies in ventricular remodelling. *Lancet* **2006**, *367*, 356–367. [CrossRef]
5. Nagueh, S.F.; Shah, G.; Wu, Y.; Torre-Amione, G.; King, N.M.; Lahmers, S.; Witt, C.C.; Becker, K.; Labeit, S.; Granzier, H.L. Altered titin expression, myocardial stiffness, and left ventricular function in patients with dilated cardiomyopathy. *Circulation* **2004**, *110*, 155–162. [CrossRef]
6. Røe, Å.T.; Aronsen, J.M.; Skårdal, K.; Hamdani, N.; Linke, W.A.; Danielsen, H.E.; Sejersted, O.M.; Sjaastad, I.; Louch, W.E. Increased passive stiffness promotes diastolic dysfunction despite improved Ca^{2+} handling during left ventricular concentric hypertrophy. *Cardiovasc. Res.* **2017**, *113*, 1161–1172. [CrossRef]
7. Villemain, O.; Correia, M.; Mousseaux, E.; Baranger, J.; Zarka, S.; Podetti, I.; Soulat, G.; Damy, T.; Hagège, A.; Tanter, M.; et al. Myocardial stiffness evaluation using noninvasive shear wave imaging in healthy and hypertrophic cardiomyopathic adults. *JACC Cardiovasc. Imaging* **2019**, *12*, 1135–1145. [CrossRef]
8. Gordon, A.M.; Homsher, E.; Regnier, M. Regulation of contraction in striated muscle. *Physiol. Rev.* **2000**, *80*, 853–924. [CrossRef]
9. Lehman, W. Switching muscles on and off in steps: The McKillop-Geeves three-state model of muscle regulation. *Biophys. J.* **2017**, *112*, 2459–2466. [CrossRef]
10. Land, S.; Niederer, S.A. A spatially detailed model of isometric contraction based on competitive binding of troponin I explains cooperative interactions between tropomyosin and crossbridges. *PLoS Comput. Biol.* **2015**, *1*, e1004376. [CrossRef]
11. Metalnikova, N.A.; Tsaturyan, A.K. A mechanistic model of Ca regulation of thin filaments in cardiac muscle. *Biophys. J.* **2013**, *105*, 941–950. [CrossRef] [PubMed]
12. Smith, D.A.; Maytum, R.; Geeves, M.A. Cooperative regulation of myosin-actin interactions by a continuous flexible chain I: Actin-tropomyosin systems. *Biophys. J.* **2003**, *84*, 3155–3167. [CrossRef]
13. Lakdawala, N.K.; Dellefave, L.; Redwood, C.R.; Sparks, E.; Cirino, A.L.; Depalma, S.; Colan, S.D.; Funke, B.; Zimmerman, R.S.; Robinson, P.; et al. Familial dilated cardiomyopathy caused by an alpha-tropomyosin mutation: The distinctive natural history of sarcomeric dilated cardiomyopathy. *J. Am. Coll. Cardiol.* **2010**, *55*, 320–329. [CrossRef] [PubMed]
14. Lynn, M.L.; Tal Grinspan, L.; Holeman, T.A.; Jimenez, J.; Strom, J.; Tardiff, J.C. The structural basis of alpha-tropomyosin linked (Asp230Asn) familial dilated cardiomyopathy. *J. Mol. Cell Cardiol.* **2017**, *108*, 127–137. [CrossRef]

15. Sequeira, V.; Wijnker, P.J.; Nijenkamp, L.L.; Kuster, D.W.; Najafi, A.; Witjas-Paalberends, E.R.; Regan, J.A.; Boontje, N.; Ten Cate, F.J.; Germans, T.; et al. Perturbed length-dependent activation in human hypertrophic cardiomyopathy with missense sarcomeric gene mutations. *Circ. Res.* **2013**, *112*, 1491–1505. [CrossRef]
16. Syomin, F.A.; Tsaturyan, A.K. Mechanical model of the left ventricle of the heart approximated by axisymmetric geometry. *Russ. J. Numer. Anal. Math. Model.* **2017**, *32*, 327–337. [CrossRef]
17. Syomin, F.A.; Zberia, M.V.; Tsaturyan, A.K. Multiscale simulation of the effects of atrioventricular block and valve diseases on heart performance. *Int. J. Numer. Method. Biomed. Eng.* **2019**, *35*, e3216. [CrossRef]
18. Syomin, F.A.; Tsaturyan, A.K. A simple model of cardiac muscle for multiscale simulation: Passive mechanics, crossbridge kinetics and calcium regulation. *J. Theor. Biol.* **2017**, *420*, 105–116. [CrossRef]
19. De Tombe, P.P.; Mateja, R.D.; Tachampa, K.; Ait Mou, Y.; Farman, G.P.; Irving, T.C. Myofilament length dependent activation. *J. Mol. Cell Cardiol.* **2010**, *48*, 851–858. [CrossRef]
20. Kopylova, G.; Berg, V.; Bershitsky, S.; Matyushenko, A.; Shchepkin, D. The effects of cardiomyopathy-associated mutations in the head-to-tail overlap junction of a-tropomyosin on the calcium regulation of the actin-myosin interaction. *FEBS Open Bio* **2019**, *9*, 111.
21. Gupte, T.M.; Haque, F.; Gangadharan, B.; Sunitha, M.S.; Mukherjee, S.; Anandhan, S.; Rani, D.S.; Mukundan, N.; Jambekar, A.; Thangaraj, K.; et al. Mechanistic heterogeneity in contractile properties of α-Tropomyosin (TPM1) mutants associated with inherited cardiomyopathies. *J. Biol. Chem.* **2015**, *290*, 7003–7015. [CrossRef] [PubMed]
22. Pravdin, S.F.; Berdyshev, V.I.; Panfilov, A.V.; Katsnelson, L.B.; Solovyova, O.; Markhasin, V.S. Mathematical model of the anatomy and fibre orientation field of the left ventricle of the heart. *Biomed. Eng. Online* **2013**, *12*, 54. [CrossRef] [PubMed]
23. Le Ven, F.; Bibeau, K.; De Larochelliere, E.; Tizon-Marcos, H.; Deneault-Bissonnette, S.; Pibarot, P.; Deschepper, C.F.; Larose, E. Cardiac morphology and function reference values derived from a large subset of healthy young Caucasian adults by magnetic resonance imaging. *Eur. Heart J. Cardiovasc. Imaging* **2016**, *17*, 981–990. [CrossRef] [PubMed]
24. Haland, T.F.; Hasselberg, N.E.; Almaas, V.M.; Dejgaard, L.A.; Saberniak, J.; Leren, I.S.; Berge, K.E.; Haugaa, K.H.; Edvardsen, T. The systolic paradox in hypertrophic cardiomyopathy. *Open Heart* **2017**, *4*, e000571. [CrossRef] [PubMed]
25. Huang, X.; Yue, Y.; Wang, Y.; Deng, Y.; Liu, L.; Di, Y.; Sun, S.; Chen, D.; Fan, L.; Cao, J. Assessment of left ventricular systolic and diastolic abnormalities in patients with hypertrophic cardiomyopathy using real-time three-dimensional echocardiography and two-dimensional speckle tracking imaging. *Cardiovasc. Ultrasound* **2018**, *16*, 23. [CrossRef]
26. Marko, J.F.; Siggia, E.D. Statistical mechanics of supercoiled DNA. *Phys. Rev. E* **1995**, *52*, 2912–2938. [CrossRef]
27. Suga, H.; Sagawa, K. Instantaneous pressure-volume relationships and their ratio in the excised, supported canine left ventricle. *Circ. Res.* **1974**, *35*, 117–126. [CrossRef]
28. Sagawa, K. The end-systolic pressure-volume relation of the ventricle: Definition, modifications and clinical use. *Circulation* **1981**, *63*, 1223–1227. [CrossRef]
29. Kanzaki, H.; Nakatani, S.; Yamada, N.; Urayama, S.; Miyatake, K.; Kitakaze, M. Impaired systolic torsion in dilated cardiomyopathy: Reversal of apical rotation at mid-systole characterized with magnetic resonance tagging method. *Basic Res. Cardiol.* **2006**, *101*, 465–470. [CrossRef]
30. Omar, A.M.S.; Bansal, M.; Sengupta, P.P. Advances in echocardiographic imaging in heart failure with reduced and preserved ejection fraction. *Circ. Res.* **2016**, *119*, 357–374. [CrossRef]
31. Rady, M.; Ulbrich, S.; Heidrich, F.; Jellinghaus, S.; Ibrahim, K.; Linke, A.; Sveric, K.M. Left ventricular torsion—a new echocardiographic prognosticator in patients with non-ischemic dilated cardiomyopathy. *Circ. J.* **2019**, *83*, 595–603. [CrossRef]
32. van der Bijl, P.; Bootsma, M.; Hiemstra, Y.L.; Ajmone Marsan, N.; Bax, J.J.; Delgado, V. Left ventricular 2D speckle tracking echocardiography for detection of systolic dysfunction in genetic, dilated cardiomyopathies. *Eur. Heart J. Cardiovasc. Imaging* **2018**, *20*, 694–699. [CrossRef] [PubMed]
33. Urbano Moral, J.A.; Ariez Godinez, J.A.; Maron, M.S.; Malik, R.; Eagan, J.E.; Patel, A.R.; Pandian, N.G. Left ventricular twist mechanics in hypertrophic cardiomyopathy assessed by three-dimensional speckle tracking echocardiography. *Am. J. Cardiol.* **2011**, *108*, 1788–1795. [CrossRef]
34. Soullier, C.; Obert, P.; Doucende, G.; Nottin, S.; Cade, S.; Perez-Martin, A.; Messner-Pellenc, P.; Schuster, I. Exercise response in hypertrophic cardiomyopathy. *Circ. Cardiovasc. Imaging* **2012**, *5*, 324–332. [CrossRef]

35. Abozguia, K.; Nallur-Shivu, G.; Phan, T.T.; Ahmed, I.; Kalra, R.; Weaver, R.A.; McKenna, W.J.; Sanderson, J.E.; Elliott, P.; Frenneaux, M.P. Left ventricular strain and untwist in hypertrophic cardiomyopathy: Relation to exercise capacity. *Am. Heart J.* **2010**, *159*, 825–832. [CrossRef]
36. Carasso, S.; Yang, H.; Woo, A.; Vannan, M.A.; Jamorski, M.; Wigle, E.D.; Rakowski, H. Systolic myocardial mechanics in hypertrophic cardiomyopathy: Novel concepts and implications for clinical status. *J. Am. Soc. Echocardiogr.* **2008**, *21*, 675–683. [CrossRef]
37. Kawaguchi, M.; Hay, I.; Fetics, B.; Kass, D.A. Combined ventricular systolic and arterial stiffening in patients with heart failure and preserved ejection fraction. *Circulation* **2003**, *107*, 714–720. [CrossRef]
38. Westermann, D.; Kasner, M.; Steendijk, P.; Spillmann, F.; Riad, A.; Weitmann, K.; Homann, W.; Poller, W.; Pauschinger, M.; Schultheiss, H.P.; et al. Role of left ventricular stiffness in heart failure with normal ejection fraction. *Circulation* **2008**, *117*, 2051–2060. [CrossRef]
39. Bozkurt, S. Mathematical modeling of cardiac function to evaluate clinical cases in adults and children. *PLoS ONE* **2019**, *14*, 1–20. [CrossRef]
40. Kosta, S.; Negroni, J.; Lascano, E.; Dauby, P.C. Multiscale model of the human cardiovascular system: Description of heart failure and comparison of contractility indices. *Math. Biosci.* **2017**, *284*, 71–79. [CrossRef]
41. Lascano, E.C.; Felice, J.I.; Wray, S.; Kosta, S.; Dauby, P.C.; Cabrera-Fischer, E.I.; Negroni, J.A. Experimental assessment of a myocyte-based multiscale model of cardiac contractile dysfunction. *J. Theor. Biol.* **2018**, *456*, 16–28. [CrossRef]
42. Bakir, A.; Al Abed, A.; Stevens, M.C.; Lovell, N.H.; Dokos, S. A multiphysics biventricular cardiac model: Simulations with a left-ventricular assist device. *Front. Physiol.* **2018**, *9*, 1259. [CrossRef] [PubMed]
43. Genet, M.; Lee, L.C.; Baillargeon, B.; Guccione, J.M.; Kuhl, E. Modeling pathologies of diastolic and systolic heart failure. *Ann. Biomed. Eng.* **2016**, *44*, 112–127. [CrossRef] [PubMed]
44. Lee, L.; Kassab, G.; Guccione, J. Mathematical modeling of cardiac growth and remodeling. *Wires. Syst. Biol. Med.* **2016**, *8*, 211–226. [CrossRef] [PubMed]
45. Berberoglu, E.; Solmaz, H.O.; Goktepe, S. Computational modeling of coupled cardiac electromechanics incorporating cardiac dysfunctions. *Eur. J. Mech. A Solid.* **2014**, *48*, 60–73. [CrossRef]
46. Sahli Costabal, F.; Choy, J.S.; Sack, K.L.; Guccione, J.M.; Kassab, G.S.; Kuhl, E. Multiscale characterization of heart failure. *Acta Biomat.* **2019**, *86*, 66–76. [CrossRef]
47. Bianco, F.D.; Franzone, P.C.; Scacchi, S.; Fassina, L. Electromechanical effects of concentric hypertrophy on the left ventricle: A simulation study. *Comput. Biol. Med.* **2018**, *99*, 236–256. [CrossRef]
48. Land, S.; Niederer, S.A.; Aronsen, J.M.; Espe, E.K.S.; Zhang, L.; Louch, W.E.; Sjaastad, I.; Sejersted, O.M.; Smith, N.P. An analysis of deformation-dependent electromechanical coupling in the mouse heart. *J. Physiol.* **2012**, *590*, 4553–4569. [CrossRef]
49. Trayanova, N.A.; Constantino, J.; Gurev, V. Electromechanical models of the ventricles. *Comput. Methods Biomech. Biomed. Engin.* **2011**, *301*, H279–H286. [CrossRef]
50. von Deuster, C.; Sammut, E.; Asner, L.; Nordsletten, D.; Lamata, P.; Stoeck, C.T.; Kozerke, S.; Razavi, R. Studying dynamic myofiber aggregate reorientation in dilated cardiomyopathy using in vivo magnetic resonance diffusion tensor imaging. *Circ. Cardiovasc. Imaging* **2016**, *9*, e005018. [CrossRef]
51. Tseng, W.Y.; Dou, J.; Reese, T.G.; Wedeen, V.J. Imaging myocardial fiber disarray and intramural strain hypokinesis in hypertrophic cardiomyopathy with MRI. *J. Magn. Reson. Imaging* **2006**, *23*, 1–8. [CrossRef] [PubMed]
52. Kessler, E.L.; Boulaksil, M.; van Rijen, H.V.M.; Vos, M.A.; Veen, T.A.B. Passive ventricular remodeling in cardiac disease: Focus on heterogeneity. *Front. Physiol.* **2014**, *5*, 482. [CrossRef] [PubMed]
53. Guccione, J.M.; McCulloch, A.D.; Waldman, L.K. Passive material properties of intact ventricular myocardium determined from a cylindrical model. *J. Biomech. Eng.* **1991**, *113*, 42–55. [CrossRef]

© 2020 by the authors. Licensee MDPI, Basel, Switzerland. This article is an open access article distributed under the terms and conditions of the Creative Commons Attribution (CC BY) license (http://creativecommons.org/licenses/by/4.0/).

Article

Dynamics of Periodic Waves in a Neural Field Model

Nikolai Bessonov [1], Anne Beuter [2,3], Sergei Trofimchuk [4] and Vitaly Volpert [5,6,7,*]

1. Institute of Problems of Mechanical Engineering, Russian Academy of Sciences, 199178 Saint Petersburg, Russia; nickbessonov1@gmail.com
2. Bordeaux INP, Avenue des Facultes, 33400 Talence, France; anne.beuter@wanadoo.fr
3. CorStim SAS, 700 Avenue du Pic Saint Loup, 34090 Montpellier, France
4. Instituto de Matematica y Fisica, Universidad de Talca, Casilla 747, Talca, Chile; trofimch@inst-mat.utalca.cl
5. Institut Camille Jordan, UMR 5208 CNRS, University Lyon 1, 69622 Villeurbanne, France
6. INRIA Team Dracula, INRIA Lyon La Doua, 69603 Villeurbanne, France
7. Peoples' Friendship University of Russia (RUDN University), 6 Miklukho-Maklaya St, 117198 Moscow, Russia
* Correspondence: volpert@math.univ-lyon1.fr

Received: 12 March 2020; Accepted: 29 June 2020; Published: 2 July 2020

Abstract: Periodic traveling waves are observed in various brain activities, including visual, motor, language, sleep, and so on. There are several neural field models describing periodic waves assuming nonlocal interaction, and possibly, inhibition, time delay or some other properties. In this work we study the influences of asymmetric connectivity functions and of time delay for symmetric connectivity functions on the emergence of periodic waves and their properties. Nonlinear wave dynamics are studied, including modulated and aperiodic waves. Multiplicity of waves for the same values of parameters is observed. External stimulation in order to restore wave propagation in a damaged tissue is discussed.

Keywords: neural field model; integro-differential equation; waves; brain stimulation

---

1. Introduction

1.1. Brain Activity and Periodic Travelling Waves

The brain displays a variety of highly nonlinear, complex dynamics across multiple spatial and temporal scales [1]. About 86×10^9 neurons of the human brain entertain complex and fluctuating interactions. Understanding the dynamics of these interactions and the mechanisms underlying their control remains a technical and theoretical challenge. In other words, when healthy brain processes evolve toward abnormal and pathological states as a result of disease, degeneration or traumatic injury, how can a therapeutic intervention be used to reposition the control parameters and guide the dynamics back toward a healthy state? These brain processes are described today as interacting networks of nodes/hubs and edges which for the whole brain constitute the human connectome [2]. While the connectome is focused on anatomical connections, the dynamics of the networks are represented by functional connections (also called the dynome [1]). For example, brain functional connections described as a graph explore how signals are transmitted along neuroanatomical pathways and interact with local dynamics. Functional connections are often investigated via modeling [1]. One possibility is to use mathematical models to identify how an outside intervention such as neural electrical stimulation can modify local dynamics and how local dynamics will in turn affect other brain regions.

Cortical brain dynamics are investigated by means of periodic traveling waves (TW) characterized by their speed and frequency [3]. They describe the distribution of electric potential in the brain cortex. They are measured as a mean field potential (averaged macroscopic level). Propagating waves are observed during various types of brain activity. They provide subthreshold depolarization to individual neurons and increase their spiking probability. According to Muller et al. [3], TW "travel over spatial scales that range from the mesoscopic (single cortical areas and millimetres of cortex) to the macroscopic (global patterns of activity over several centimeters) and extend over temporal scales from tens to hundreds of milliseconds." It has been proposed that TW mediate information transfer in the cortex.

Propagating waves increase the probability of neuron firing due to depolarization of neuronal membrane [4]. In [3] it was suggested that TW can be "spontaneously generated by recurrent circuits or evoked by external stimuli and travel along brain networks at multiple scales, transiently modulating spiking and excitability as they pass." The phase relations between oscillations in different cortical regions produce the TW, and depending on the distance, axonal conduction delays can reach up to tens of milliseconds. These TW correlate with the subject's performance, propagate in specific directions and synchronize distributed cortical networks that are communicating [5].

Botella-Soler et al. [6] identified for each subject the set of intracranial contacts that showed a larger percentage of detected events during slow wave cortical activity. They called these contacts "hubs" because the slow wave events in their travel through the cortical networks seemed to have a great probability of passing through the region close to the contact. Using probabilities, they were able to reconstruct a preferential propagation network for each subject. Slow waves have been reported to propagate across cortical areas at about 1m/s with multiple propagation paths and several points of origin. It seems that the slow waves have a preference to start in the prefrontal cortex and to end in posterior and temporal regions of the cortex. These waves appear to shape and strengthen neuronal networks.

Let us also note that time delays are intrinsic to the dynamics of brain networks, nodes and edges. Propagation speed along axons depends on axonal length and diameter. Conduction times along neural circuits also depend on degree of fiber myelination. The introduction of time delays in models can lead to significant changes in brain dynamics. They can be averaged or treated as distributed delays. In this paper we address the question of the effects of these delays in the dynamics of periodic cortical waves.

Thus, according to the biological observations, TW propagate in the cortex, activating and coordinating different parts of the brain. In this work, we study some of their properties. In the next section, we introduce the model. Then we present stability analysis which determines the conditions of wave appearance. Their nonlinear dynamics will be discussed in Section 3.

1.2. Neural Field Model

Neural field models were first introduced in [7]. Periodic traveling waves are described by several models (see [8–12] and Appendix A). In this work we consider one equation model with delay, and we discuss two mechanisms of the emergence of such waves, which were not sufficiently investigated previously. The first mechanism is related to the asymmetric connectivity functions [13], and the second one is determined by the delay in the response function. It is known that the loss of stability of the homogeneous in the space solution in this model does not lead to the bifurcation of periodic waves [8]. We show that they still appear for some larger values of time delay. We consider the one-dimensional neural field equation for the electric potential in the brain cortex written in the form

$$\frac{\partial u}{\partial t} = D\frac{\partial^2 u}{\partial x^2} + W_a - W_i - \sigma u, \tag{1}$$

where D is the diffusion coefficient; W_a and W_i are given by the expressions

$$W_a(x,t) = \int_{-\infty}^{\infty} \phi_a(x-y) S_a\left(u\left(y, t - \frac{|x-y|}{q_a} - \tau_a\right)\right) dy, \tag{2}$$

$$W_i(x,t) = \int_{-\infty}^{\infty} \phi_i(x-y) S_i\left(u\left(y, t - \frac{|x-y|}{q_i} - \tau_i\right)\right) dy \tag{3}$$

and characterize neuron activation and inhibition. These expressions describe the intensity of signal coming from all points y to the point x; $S_a(u)$ and $S_i(u)$ are smooth functions; q_a and q_i are the excitation speeds; $|x - y|/q_{a,i}$ is the time delay due to the excitation propagation from the point y to the point x; ϕ_a and ϕ_i are the connectivity functions,

$$\phi_a(r) = \begin{cases} a_1 e^{-b_1 r}, & r > 0 \\ a_3 e^{b_3 r}, & r < 0 \end{cases}, \quad \phi_i(r) = \begin{cases} a_2 e^{-b_2 r}, & r > 0 \\ a_4 e^{b_4 r}, & r < 0 \end{cases}, \tag{4}$$

where a_i, b_i are some positive constants. Response functions $S_a(u)$ and $S_i(u)$ are non-negative, non-decreasing functions usually considered as sigmoid-type functions; τ_a and τ_i are time delays in neuron response to the activating and inhibitory signals. The last term on the right-hand side of Equation (1) describes signal decay with a decay rate $\sigma > 0$.

Neural field models are often considered without the diffusion term in the studies of both brain oscillations and traveling waves (see, e.g., [14–17]). In the case of small diffusion coefficients, its influence is not essential [18]. It can describe the ephaptic effect [19], ion diffusion and gap junction. The integral terms in Equation (1) characterize nonlocal neuron communication due to axons and dendrites. The speeds q_a and q_i of electric impulse propagation along axons are of the order 2–4 m/s, while the speed of wave propagation is one-two orders of magnitude less [20] (p. 213). Thus, we will consider a large speed limit $q_a = q_i \to \infty$:

$$\frac{\partial u}{\partial t} = D \frac{\partial^2 u}{\partial x^2} + \int_{-\infty}^{\infty} \left(\phi_a(x-y) S_a(u(y, t - \tau_a)) - \phi_i(x-y) S_i(u(y, t - \tau_i)) \right) dy - \sigma u. \tag{5}$$

Up to the diffusion term, which is not very essential for small diffusion coefficients, this model is a particular case of the two equation model considered in [8]. If the kernels in the two equations are the same, then the system can be reduced to the single equation.

Different neural field models describe the propagation of periodic traveling waves (see Appendix A). In this work we will consider two other mechanisms of their emergence. One of them is related to asymmetric kernels ϕ_a and ϕ_i, whose existence is confirmed by the experimental observations and used in theoretical considerations [13,21]. Another one is determined by the secondary bifurcation due to time delay for symmetric connectivity functions. Similarly to [8], we observe that the loss of stability of the homogeneous in space stationary solution leads to the appearance of periodic time oscillations independent of the space variable or of stationary periodic in space solutions. Periodic traveling waves bifurcate in the instability region, and they are unstable close to the bifurcation point. We will see that they can become stable under further change of parameters.

We studied the dynamics of the periodic waves, including their non-uniqueness for the same values of parameters. This property seems to us important, since the co-existence of waves with different frequencies is experimentally observed.

2. Stability

2.1. Linearization and Eigenvalues

In this section we will consider the Equation (5) on the interval $0 < x < L$ with periodic boundary conditions. We extend the function $u(x, t)$ by periodicity on the whole axis, $-\infty < x < \infty$, so that the integrals in Equation (5) are well defined. Let u_0 be a solution of the equation

$$\phi_a^* S_a(u) + \phi_i^* S_i(u) - \sigma u = 0,$$

where $\phi_a^* = \int_{-\infty}^{\infty} \phi_a(x)dx$, $\phi_i^* = \int_{-\infty}^{\infty} \phi_i(x)dx$. Then u_0 is a stationary solution of Equation (5). Linearizing this equation about u_0, we obtain the eigenvalue problem:

$$Dv'' + S_a'(u_0)e^{-\lambda \tau_a}\int_{-\infty}^{\infty} \phi_a(x-y)v(y)dy - S_i'(u_0)e^{-\lambda \tau_i}\int_{-\infty}^{\infty} \phi_i(x-y)v(y)dy - \sigma v = \lambda v. \quad (6)$$

Here, v is a small perturbation of the stationary solution u_0. Applying the Fourier transform, we get

$$S_a'(u_0)e^{-\lambda \tau_a}\tilde{\phi}_a(\xi) - S_i'(u_0)e^{-\lambda \tau_i}\tilde{\phi}_i(\xi) - D\xi^2 - \sigma = \lambda, \quad (7)$$

where

$$\tilde{\phi}_a(\xi) = \frac{a_1 b_1}{b_1^2 + \xi^2} + \frac{a_3 b_3}{b_3^2 + \xi^2} + i\xi\left(\frac{a_1}{b_1^2 + \xi^2} - \frac{a_3}{b_3^2 + \xi^2}\right),$$

$$\tilde{\phi}_i(\xi) = \frac{a_2 b_2}{b_2^2 + \xi^2} + \frac{a_4 b_4}{b_4^2 + \xi^2} + i\xi\left(\frac{a_2}{b_2^2 + \xi^2} - \frac{a_4}{b_4^2 + \xi^2}\right)$$

are Fourier transforms of the functions ϕ_a and ϕ_i, respectively. Let us note that Fourier transform of non-integrable functions is considered in the sense of generalized functions. Using Fourier series instead of Fourier transform allows one to use a classical function instead of generalized functions. However, the advantage of the Fourier transform is that the periodicity of solutions is not imposed.

Set $\lambda = iv$. Separating the real and imaginary parts in Equation (7), we obtain:

$$S_a'(u_0)\cos(v\tau_a)\left(\frac{a_1 b_1}{b_1^2 + \xi^2} + \frac{a_3 b_3}{b_3^2 + \xi^2}\right) + S_a'(u_0)\xi\sin(v\tau_a)\left(\frac{a_1}{b_1^2 + \xi^2} - \frac{a_3}{b_3^2 + \xi^2}\right) - \\ S_i'(u_0)\cos(v\tau_i)\left(\frac{a_2 b_2}{b_2^2 + \xi^2} + \frac{a_4 b_4}{b_4^2 + \xi^2}\right) - S_i'(u_0)\xi\sin(v\tau_i)\left(\frac{a_2}{b_2^2 + \xi^2} - \frac{a_4}{b_4^2 + \xi^2}\right) - D\xi^2 = \sigma, \quad (8)$$

$$-S_a'(u_0)\sin(v\tau_a)\left(\frac{a_1 b_1}{b_1^2 + \xi^2} + \frac{a_3 b_3}{b_3^2 + \xi^2}\right) + S_a'(u_0)\xi\cos(v\tau_a)\left(\frac{a_1}{b_1^2 + \xi^2} - \frac{a_3}{b_3^2 + \xi^2}\right) + \\ S_i'(u_0)\sin(v\tau_i)\left(\frac{a_2 b_2}{b_2^2 + \xi^2} + \frac{a_4 b_4}{b_4^2 + \xi^2}\right) - S_i'(u_0)\xi\cos(v\tau_i)\left(\frac{a_2}{b_2^2 + \xi^2} - \frac{a_4}{b_4^2 + \xi^2}\right) = v. \quad (9)$$

2.2. Symmetric Connectivity Functions with Time Delay

If the symmetry condition

$$a_1 = a_3, a_2 = a_4, b_1 = b_3, b_2 = b_4 \quad (10)$$

is satisfied, then Equations (8) and (9) are as follows:

$$S_a'(u_0)\cos(v\tau_a)\frac{a_1 b_1}{b_1^2 + \xi^2} - S_i'(u_0)\cos(v\tau_i)\frac{a_2 b_2}{b_2^2 + \xi^2} - D\xi^2/2 = \sigma/2, \quad (11)$$

$$-S_a'(u_0)\sin(v\tau_a)\frac{a_1 b_1}{b_1^2 + \xi^2} + S_i'(u_0)\sin(v\tau_i)\frac{a_2 b_2}{b_2^2 + \xi^2} = v/2. \quad (12)$$

We can express ξ^2 through v from the last equation and substitute them into Equation (11). The resulting equation with respect to v should be solved numerically or asymptotically. Since the calculations are sufficiently complex, we will consider here a simplified case where $\tau_a = 0$. Numerical simulations show that periodic waves can exist in this case (Section 3). Our aim here is to analyze their bifurcations from the homogeneous in space solution. Assuming that $v \geq 0$, we also get the conjugate solution $-v$.

If $\tau_a = 0$, then Equations (11) and (12) are as follows:

$$S'_a(u_0) \frac{a_1 b_1}{b_1^2 + \xi^2} - S'_i(u_0) \cos(\nu \tau_i) \frac{a_2 b_2}{b_2^2 + \xi^2} - D\xi^2/2 = \sigma/2, \qquad (13)$$

$$S'_i(u_0) \sin(\nu \tau_i) \frac{a_2 b_2}{b_2^2 + \xi^2} = \nu/2. \qquad (14)$$

By the stability boundary we will understand here the value $\sigma = \sigma_0$ such that system (13), (14) does not have solutions for $\sigma > \sigma_0$, and it has at least one solution for $\sigma < \sigma_0$. This stability boundary σ_0 is uniquely defined and it can be negative.

Proposition 1. *At the stability boundary, $\xi = 0$ or $\nu = 0$.*

Proof. Consider Equation (14) as an equation with respect to ν for a fixed ξ. It has a solution $\nu \neq 0$ if and only if

$$2\tau_i S'_i(u_0) \frac{a_2 b_2}{b_2^2 + \xi^2} > 1. \qquad (15)$$

Suppose that this condition is satisfied for some $\xi_0 > 0$, and denote by $\nu(\xi)$ solution of Equation (14) in the vicinity of $\xi = \xi_0 > 0$, and $\nu_0 = \nu(\xi_0) > 0$. Assume that the corresponding value σ_0 belongs to the stability boundary.

Taking into account Equation (14), we can write Equation (13) as follows:

$$2S'_a(u_0) \frac{a_1 b_1}{b_1^2 + \xi^2} - \nu \cot(\nu \tau_i) - D\xi^2 = \sigma. \qquad (16)$$

Let ξ_1 be sufficiently close to ξ_0 such that condition (15) remains satisfied, and $\xi_1 < \xi_0$. Considering $\nu(\xi_0) < \pi/\tau_i$, we get $\nu(\xi_1) > \nu(\xi_0)$. Therefore,

$$\nu(\xi_1) \cot(\nu(\xi_1) \tau_i) < \nu(\xi_0) \cot(\nu(\xi_0) \tau_i).$$

Set

$$F(\xi) = 2S'_a(u_0) \frac{a_1 b_1}{b_1^2 + \xi^2} - \nu(\xi) \cot(\nu(\xi) \tau_i) - D\xi^2.$$

Thus, $F(\xi_1) > F(\xi_0)$. Hence, $\sigma_0 = F(\xi_0)$ cannot belong to the stability boundary since $\sigma_1 = F(\xi_1) > \sigma_0$, and system (13), (14) has a solution for $\sigma = \sigma_1$. This contradiction shows that $\xi_0 = 0$ when $\nu_0 \neq 0$.

Let $2\pi/\tau_i < \nu(\xi_0) < 5\pi/(2\tau_i)$. Then there is another solution $\nu_1(\xi_0)$ of Equation (14) such that $\pi/(2\tau_i) < \nu_1(\xi_0) < \pi/\tau_i$. Therefore, $\cos(\nu_1(\xi_0) \tau_i) < 0$, $\cos(\nu(\xi_0) \tau_i) > 0$, and

$$\sigma_0 = F(\xi_0) < 2S'_a(u_0) \frac{a_1 b_1}{b_1^2 + \xi_0^2} - \nu_1(\xi_0) \cot(\nu_1(\xi_0) \tau_i) - D\xi_0^2.$$

Hence, σ_0 cannot belong to the stability boundary.

Finally, in all other cases with $\nu(\xi_0) > 5\pi/(2\tau_i)$ we obtain a contradiction similarly to the two cases considered before. This contradiction shows that at the stability boundary, either $\xi = 0$ or $\nu = 0$. □

Let us now determine the stability boundary with respect to time delay for a fixed σ. If $\nu = 0$, then from Equation (13) we get

$$\sigma = \frac{\alpha_1}{b_1^2 + \xi^2} - \frac{\alpha_2}{b_2^2 + \xi^2} - D\xi^2, \qquad (17)$$

where $\alpha_1 = 2S'_a(u_0)a_1b_1$, $\alpha_2 = 2S'_i(u_0)a_2b_2$. Assuming that $\nu \neq 0$, we set $\beta_i = \alpha_i/(b_i^2 + \xi^2)$, $i = 1, 2$. Then $\sin(\nu\tau_i) = \nu/\beta_2$,

$$\beta_1 - \sqrt{\beta_2^2 - \nu^2} = \sigma, \quad \nu^2 = \beta_2^2 - (\beta_1 - \sigma)^2.$$

Example 1. *Consider the values of parameters: $S'_a(u_0) = S'_i(u_0) = 20$, $a_1 = a_2 = 4$, $b_1 = 40$, $b_2 = 20$, $\alpha_1 = 6400$, $\alpha_2 = 3200$, $\sigma = 0.01$, Then we find*

$$\xi = 0, \beta_1 = 4, \beta_2 = 8, \nu^2 \approx 48, \nu \approx 6.93, \ \tau_i = \frac{1}{\nu}\arcsin\frac{\nu}{\beta_2} = 0.151,$$

and

$$\xi = \pi, \beta_1 = 3.975, \beta_2 = 7.805, \nu^2 \approx 45.123, \nu \approx 6.717, \ \tau_i = \frac{1}{\nu}\arcsin\frac{\nu}{\beta_2} = 0.154$$

Hence, the homogeneous in space oscillations appear for a smaller value of time delay than periodic traveling waves. Therefore, these waves are unstable in the vicinity of the bifurcation point. We will see in the next section that they become stable for larger values of time delay.

2.3. Asymmetric Connectivity Function without Time Delay

If conditions (10) do not not hold, then eigenvalue (7) has a nonzero imaginary part, $\lambda(\xi) = \alpha(\xi) + i\nu(\xi)$,

$$\alpha(\xi) = \frac{s_a a_1 b_1}{b_1^2 + \xi^2} + \frac{s_a a_3 b_3}{b_3^2 + \xi^2} - \frac{s_i a_2 b_2}{b_2^2 + \xi^2} - \frac{s_i a_4 b_4}{b_4^2 + \xi^2} - D\xi^2 - \sigma,$$

$$\nu(\xi) = s_a \xi \left(\frac{a_1}{b_1^2 + \xi^2} - \frac{a_3}{b_3^2 + \xi^2}\right) - s_i \xi \left(\frac{a_2}{b_2^2 + \xi^2} - \frac{a_4}{b_4^2 + \xi^2}\right), \qquad (18)$$

$s_a = S'_a(u_0)$, $s_i = S'_i(u_0)$. Condition $\alpha(\xi) = 0$ determines the stability boundary,

$$\sigma = \frac{s_a a_1 b_1}{b_1^2 + \xi^2} + \frac{s_a a_3 b_3}{b_3^2 + \xi^2} - \frac{s_i a_2 b_2}{b_2^2 + \xi^2} - \frac{s_i a_4 b_4}{b_4^2 + \xi^2} - D\xi^2 \equiv \Phi_2(\xi). \qquad (19)$$

If $\sigma < \Phi_2(\xi)$ for some values of ξ, then the solution loses its stability due to a pair of complex conjugate eigenvalues $\lambda(\xi) = \alpha(\xi) \pm i\nu(\xi)$. For $\alpha = 0$ corresponding to the stability boundary, the bounded solution of the linearized equation can be written as follows:

$$u(x,t) = e^{i\nu t}e^{i\xi x} + e^{-i\nu t}e^{-i\xi x} = \cos(\nu t + \xi x).$$

We find ξ from equality $\alpha(\xi) = 0$ and ν from (18). The frequency of the periodic wave equals ξ and its speed $c = -\nu/\xi$ can be determined from (18):

$$c = -s_a \left(\frac{a_1}{b_1^2 + \xi^2} - \frac{a_3}{b_3^2 + \xi^2}\right) + s_i \left(\frac{a_2}{b_2^2 + \xi^2} - \frac{a_4}{b_4^2 + \xi^2}\right). \qquad (20)$$

Different waves with the frequencies satisfying condition $\sigma < \Phi(\xi)$ can exist for the same values of parameters. Their speed can be increasing or decreasing functions of frequency according to (20).

3. Numerical Results

Numerical simulations of Equation (5) in a bounded interval $0 \leq x \leq L$ with periodic boundary conditions will be started in the case without time delay and continued with the case of time delay in the response functions $S_a(u) = S_b(u) = \arctan(hu)$, $h > 0$. We will finish this section with modeling of the stimulation of the damaged tissue in order to restore wave propagation.

The numerical method uses an implicit finite difference scheme with Thomas's algorithm for the inversion of the tridiagonal matrix. Initial conditions are considered either in the form of a piece-wise constant function or a sinus-like function with a given periodicity. The continuation method is used to follow the branches of solutions. In this case, the value of some parameter gradually changes. As an initial condition for the new value of parameter, we take the result of the simulation obtained for the previous value of parameter. The time step was usually taken as 0.05 and the space step 0.005. The accuracy of numerical simulations was controlled by decreasing the time and space steps. The typical values of parameters used in numerical simulations are presented in Table A1. Their specific values are give below.

3.1. Wave Propagation without Time Delay

In the case without time delay ($\tau_a = \tau_i = 0$) and with symmetric connectivity functions ϕ_a and ϕ_i, where $a_1 = a_3, a_2 = a_4, b_1 = b_3, b_2 = b_4$, periodic in space stationary solutions bifurcate from the constant solution. Linear stability analysis shows that traveling waves with nonzero speed do not bifurcate in this case.

If the connectivity functions are not symmetric, then traveling waves with nonzero speed are observed (see Figures 1 and 2). Linear stability analysis allows the determination of the wave speed and frequency near the bifurcation point. Let us consider an example with the following values of parameters: $a_1 = 3, a_2 = 4, a_3 = 3, a_4 = 1, b_1 = 40, b_2 = 20, b_3 = 40, b_4 = 20, D = 10^{-3}, h = 20, L = 2$. Analysis of the function $\alpha(\xi)$ (Section 2.3) shows that it has a maximum at $\xi = 20.53$. It is positive for $\sigma < \sigma_c = 0.4$ resulting in the bifurcation of the periodic wave with the speed given by Equation (20). For the chosen values of parameters, we get $c = -0.54$. Numerical simulations show the emergence of a periodic wave with 7 periods in the interval $[0, L]$ and with the corresponding frequency $\xi = 21.98$. Since the number of periods is an integer, the frequency of the wave observed in numerical simulations is not precisely equal to the analytical value. The wave with the closest frequency emerges since the corresponding eigenvalue has the maximal real part in comparison with the waves with other periods. The speed of this wave $c = -0.52$ is close to the analytical value.

For the values of parameters far from the stability boundary, two types of solutions are observed in numerical simulations: periodic waves with a constant speed (Figure 1) and aperiodic waves with oscillating speed (Figure 2). In the first case, the wave has conventional form $w(x - ct)$, where $w(x)$ is a periodic in space function and c is a constant. In the second case, the amplitude of spatial peaks and the wave speed oscillate.

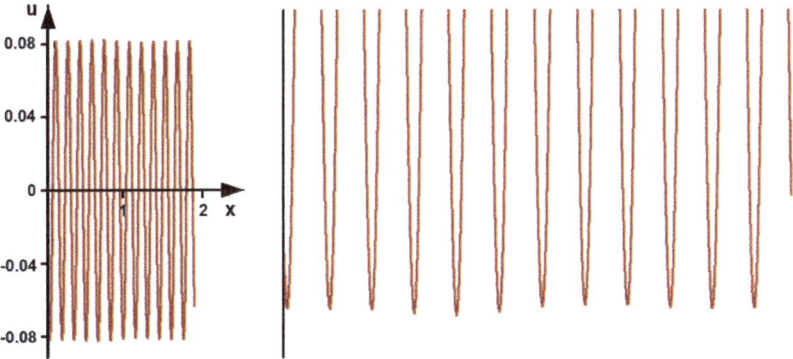

Figure 1. A snapshot of periodic wave described by Equation (5) for $\tau_a = \tau_i = 0$ (left) and a zoom-in on the lower part of the graph (right). The values of parameters are as follows: $D = 10^{-4}, \sigma = 0.01, L = 2$, $a_1 = a_2 = 0.6, a_3 = a_4 = 4, b_1 = b_3 = 40, b_2 = b_4 = 20$. Here $S_a(u) = S_i(u) = \arctan(hu)$, and $h = 20$. Here and in all examples below, the stationary solution $u_0 = 0$.

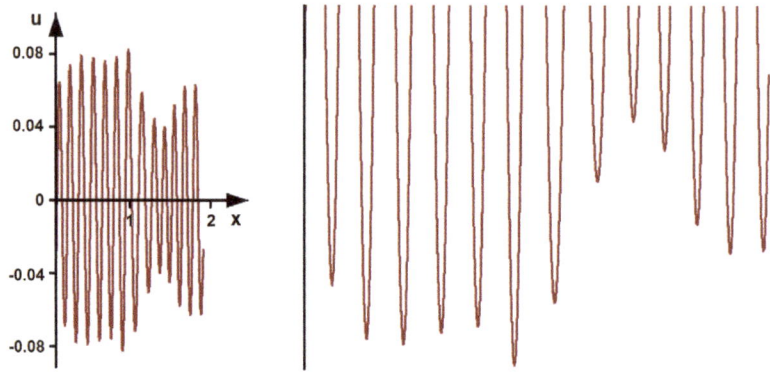

Figure 2. A snapshot of aperiodic wave described by Equation (5) for $\tau_a = \tau_i = 0$ (left) and a zoom-in on the lower part of the graph (right). Periodic (Figure 1) and aperiodic waves can exist for the same values of parameters. Convergence of solutions to one of them is determined by the initial conditions. The values of parameters are as follows: $D = 10^{-4}$, $\sigma = 0.01$, $L = 2$, $a_1 = a_2 = 0.6$, $a_3 = a_4 = 4$, $b_1 = b_3 = 40$, $b_2 = b_4 = 20$. Here $S_a(u) = S_i(u) = \arctan(hu)$, and $h = 20$.

Different periodic regimes can co-exist for the same values of parameters. Figure 3 shows the dependence of the wave speed on the values $a_1 (= a_2)$ for all other parameters fixed. There are three branches of solutions corresponding to different spatial frequencies. Thus, there are three different periodic waves for the same values of parameters with the speeds depending on their frequency.

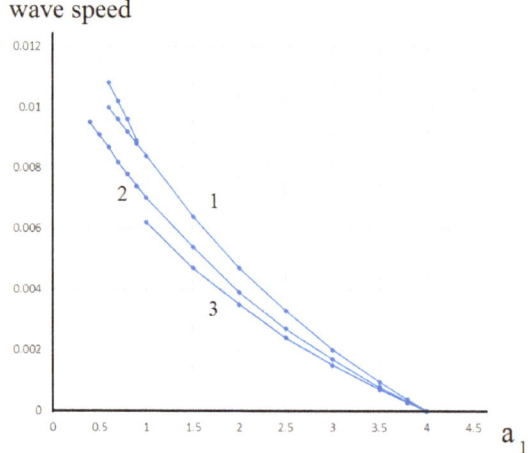

Figure 3. The speed of periodic waves for different values of the parameters $a_1 = a_2$. Three curves correspond to different values of the spatial frequency: (1) solution with 12 periods in the interval $[0, L]$, (2) 13 periods, (3) 14 periods. Branching in curve 1 shows the transition to modulated oscillations with the maximal and minimal values of the oscillating speed. The values of parameters are as follows: $D = 10^{-4}$, $\sigma = 0.01$, $L = 2$, $a_3 = a_4 = 4$, $b_1 = b_3 = 40$, $b_2 = b_4 = 20$. Here $S_a(u) = S_i(u) = \arctan(hu)$, and $h = 20$.

For a sufficiently small a_1, transition to modulated waves can occur (curve 1) with the amplitude and speed depending on time. Such solutions can be qualitatively approximated by the function $u(x,t) = (k_1 + \epsilon \sin(k_2 x + k_3 t)) \sin(k_4 x + k_5 t)$, where k_i are some constants. For even smaller values of

a_1, transition to aperiodic oscillations is observed for all three branches of solutions. These oscillations can coexist with periodic or modulated periodic waves for the same values of parameters.

3.2. Time Delay and Symmetric Connectivity Functions

3.2.1. Initial Conditions

As it was discussed in the previous section, the loss of stability of the homogeneous in space stationary solution leads either to the homogeneous in space time oscillations or to the stationary periodic in space solutions. Consider time delay as a bifurcation parameter. If it exceeds a critical value, then time oscillations emerge. Periodic traveling waves bifurcate for a larger value of time delay, and they are unstable in the vicinity of the bifurcation point. Numerical simulations show that they can become stable under a further increase of τ. Nevertheless, periodic time oscillations homogeneous in space remain stable. Therefore, we need to choose some particular initial conditions in order to get periodic traveling waves.

The simulations presented in this section were carried out with two types of initial conditions. In the first case, we consider the equation

$$\frac{\partial u}{\partial t} = D\frac{\partial^2 u}{\partial x^2} + I(x,t) \tag{21}$$

on the time interval $0 \leq t \leq T_0$. Here $I(x,t) = I_0 \cos(px+qt)$, and the initial condition $u(x,0) = 0$. The result of this simulation is considered as the initial condition for Equation (5). We set $I_0 = 0.5, q = 0.015, T_0 = 20$; the value of p is taken $3, 6, 9$ depending on the required space periodicity.

The second type of initial conditions is used in the continuation method. The results of the simulations of Equation (5) for some values of parameters are used as initial conditions for some other values of parameters.

3.2.2. Multiplicity of Waves and Parameter Dependence

An example of periodic traveling waves with different initial conditions and the same values of parameters are shown in Figure 4. We observe one, two and three-period waves generated by function $I(x,t)$ for $p = 3, 6$ and 9. These waves have different speeds and amplitudes. The waves with larger wavelengths have higher speeds and amplitudes. If time delay is sufficiently small, then the waves become unstable, and the transition to periodic time oscillations independent of the space variable is observed. If the value I_0 is small enough, then the solution converges to the homogeneous time oscillations.

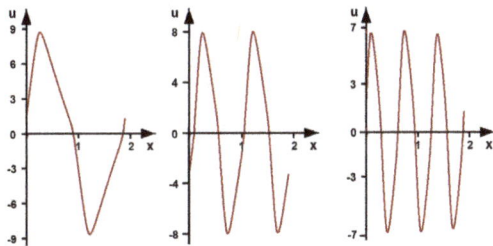

Figure 4. Three types of periodic waves observed for the same values of parameters and having different spatial frequency and speed: one-period wave with speed -0.027 (left); two-period wave with speed -0.012 (middle); three-period wave with speed -0.0094. The values of parameters are as follows: $D = 10^{-4}, \sigma = 0.01, L = 2, \tau_a = 0, \tau_i = 12, a_1 = a_2 = a_3 = a_4 = 4, b_1 = b_3 = 40, b_2 = b_4 = 20$. Here $S_a(u) = S_i(u) = \arctan(hu)$, and $h = 20$.

Increase of time delay leads to the increase of the wave amplitude and to the decrease of its speed (Figure 5). Wave propagation is determined by the transmission of activating signal between the neurons. Since time delay retards this transmission, the speed decreases. This result was previously obtained analytically for monotone waves [22]. Since the activating signal propagates slower with time delay, the local response and the wave amplitude increase.

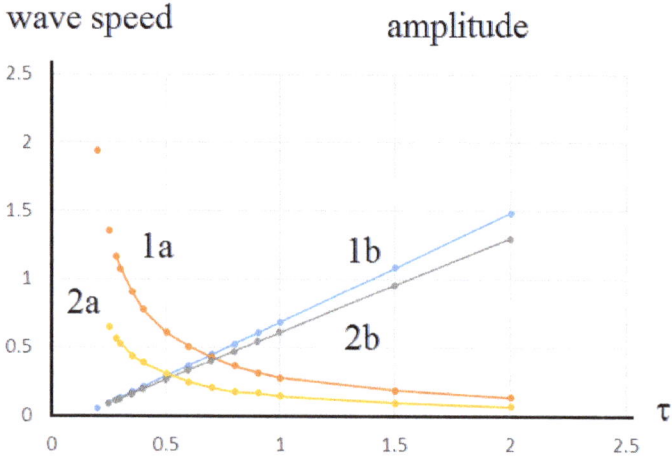

Figure 5. The absolute value of the speed (1a, 2a) and the amplitude (1b, 2b) for the one-period wave (1a, 1b) and two-period wave (2a, 2b) depending on τ_i. Connected points correspond to stable solutions; separate points to unstable solutions. The latter lead to the appearance of stationary solutions periodic in space. The values of parameters are as follows: $D = 10^{-4}$, $\sigma = 0.01$, $\tau_a = 0$, $a_1 = a_2 = a_3 = a_4 = 4$, $b_1 = b_3 = 40$, $b_2 = b_4 = 20$. Here $S_a(u) = S_i(u) = \arctan(hu)$, and $h = 20$.

Let us note that the critical value of time delay $\tau_i \approx 1.5$ found in numerical simulations and the value of the wave speed near the bifurcation point $c \approx -2$ correspond to the analytical values determined in the example in Section 2.2 (Figure 5, curves 1a, 1b). The analytical value of the speed $c = -\nu/\xi = -2.13$ at the bifurcation point is slightly different from the numerical value. Since the periodic wave is unstable near the bifurcation point, we can determine its speed in numerical simulations only at some distance from the critical value τ_i.

Figure 6 shows the dependence of periodic waves on the value b_2 (equal to b_4). The wave amplitude and speed decrease with the increase of b_2. There is a critical value $b_2 \approx 40$ for which the speed becomes zero, and a transition to another branch of solutions is observed. These are stationary, periodic in space solutions with growing amplitude as b_2 increases. It is interesting to note the existence of weakly oscillating time periodic solutions in a narrow interval between traveling waves and stationary solutions.

The results presented above were obtained for a single delay τ_i in the inhibition term while $\tau_a = 0$. If we fix $\tau_i = 1$ and increase τ_a beginning from $\tau_a = 0$, then the amplitude and the speed of traveling waves are not monotonous. The former first decreases, passes through the minimum and then increases, while the latter increases in the beginning and decreases for larger values of the delay (not shown).

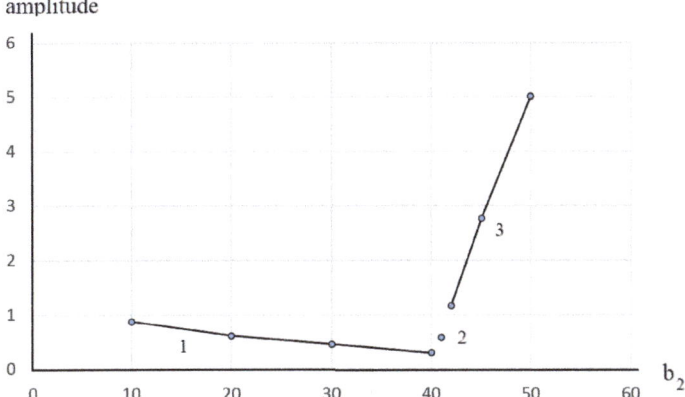

Figure 6. Dependence of the amplitude of the observed regimes on $b_2(=b_4)$. Periodic traveling waves are located on branch 1; stationary solutions on branch 3. A short branch 2 contains weakly oscillating solutions. The values of parameters are as follows: $D = 10^{-4}$, $\sigma = 0.01$, $L = 2$, $\tau_a = 0$, $\tau_i = 1$, $a_1 = a_2 = a_3 = a_4 = 4$, $b_1 = b_3 = 40$. Here $S_a(u) = S_i(u) = \arctan(hu)$, and $h = 20$.

3.3. Stimulation

Wave propagation can be different in a damaged tissue [23]. In this section we discuss its possible restoration by external stimulation. This question was considered in [24] for monotonous wave fronts.

3.3.1. Exact Solution of the Stimulation Problem

Let us write Equation (5) for the normal tissue

$$\frac{\partial u}{\partial t} = D \frac{\partial^2 u}{\partial x^2} + J(u) - \sigma u, \tag{22}$$

where

$$J(u) = \int_{-\infty}^{\infty} \phi_a(x-y) S_a\left(u\left(y, t - \tau_a\right)\right) dy - \int_{-\infty}^{\infty} \phi_i(x-y) S_i\left(u\left(y, t - \tau_i\right)\right) dy,$$

and a similar equation for the damaged tissue

$$\frac{\partial v}{\partial t} = D \frac{\partial^2 v}{\partial x^2} + J^*(v) - \sigma v, \tag{23}$$

where

$$J^*(v) = \int_{-\infty}^{\infty} \phi_a^*(x-y) S_a^*\left(v\left(y, t - \tau_a\right)\right) dy - \int_{-\infty}^{\infty} \phi_i^*(x-y) S_i^*\left(v\left(y, t - \tau_i\right)\right) dy,$$

and * denotes the functions for the damaged tissue. In the presence of stimulation $I(x, t)$ this equation is as follows:

$$\frac{\partial z}{\partial t} = D \frac{\partial^2 z}{\partial x^2} + J^*(z) - \sigma z + I(x, t). \tag{24}$$

We will choose this function such that solution $z(x, t)$ of Equation (24) becomes equal solution $u(x, t)$ of Equation (22). We set $z(x, t) = u(x, t)$, and substitute $u(x, t)$ in Equation (24). Then

$$I(x, t) = \frac{\partial u}{\partial t} - D \frac{\partial^2 u}{\partial x^2} - J^*(u) + \sigma u = J(u) - J^*(u).$$

Thus, the stimulation function

$$I(x, t) = J(u) - J^*(u)$$

gives solution of the complete reconstruction problem.

3.3.2. Approximate Solution of the Stimulation Problem

The stimulation function suggested above uses the solution $u(x,t)$ which may not be known for the patient with the damaged brain tissue. In this case, some approximate solutions of the stimulation problem can be used. Consider the integral $J^*(v)$ in the following form:

$$J^*(v) = \int_{-\infty}^{\infty} W(x)W(y)\left(\phi_a(x-y)S_a\left(v(y,t-\tau_a)\right) - \phi_i(x-y)S_i\left(v(y,t-\tau_i)\right)\right) dy,$$

where $W(x) = w_0 < 1$ for $x_1 \leq x \leq x_2$, $x_1, x_2 \in (0, L)$, and $W(x) = 1$ outside the interval $[x_1, x_2]$. Hence, the connectivity function decreases if x or y belongs to the damaged area $[x_1, x_2]$. If we set $w_0 = 0$, then the connectivity function vanishes if one of the two points x or y (neurons) belongs to the damaged area.

Without damage, the periodic traveling wave solution of Equation (22) and the integral $J(u)$ can be approximated by a cosine function (Figure 7). Hence, we will look for the stimulation function in the form approximating the periodic wave: $I(x,t) = I_0(x)\cos(px+qt)$. We set $p = 6, q = 1$ to approximate the frequency and the speed of the wave in the normal tissue. We set $I_0(x) = i_0$ for $x_1 \leq x \leq x_2$, and $I_0(x) = i_1$ outside the interval $[x_1, x_2]$. The results of numerical simulations are presented in Figure 7 with the comparison of the normal tissue, damaged tissue without stimulation and damaged tissue with stimulation. We observe a periodic wave in the normal tissue. The behavior of the solution is completely different in the damaged tissue. The solution is close to 0 at the damaged interval (green in the middle figure), and it oscillates periodically in time from both sides of this interval. Stimulation restores the wave propagation with the same frequency and speed. The choice of the stimulation amplitude $i_0 = 0.6$ and $i_1 = 0.1$ (for the example in the figure) is important. If we set $i_0 = 0.6, i_1 = 0$, the stimulation is not successful; there is no wave propagation. Thus, stimulation should be done not only inside the damaged interval but also around it.

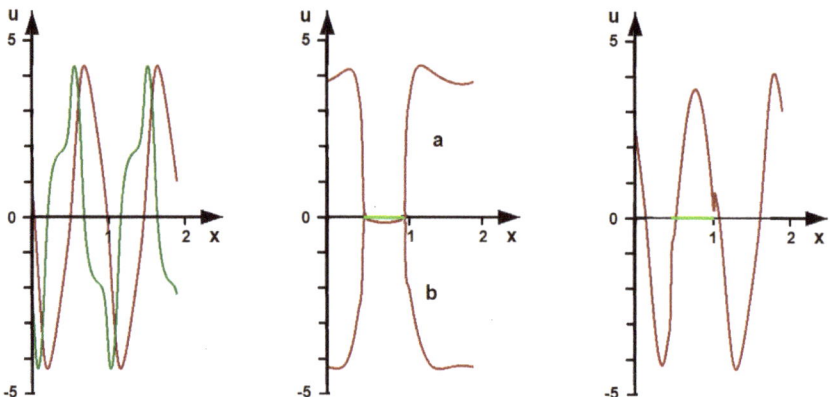

Figure 7. Periodic traveling wave in the normal tissue and the integral $J(u)$ (**left**). Snapshots of solutions in the damaged tissue for two different moments of time (**a**) and (**b**) (**middle**). Solution with stimulation becomes close to the periodic wave (**right**). The values of parameters are as follows: $D = 10^{-4}$, $\sigma = 0.01$, $L = 2$, $\tau_a = 0, \tau_i = 1$, $a_1 = a_2 = a_3 = a_4 = 4$, $b_1 = b_3 = 40, b_2 = b_4 = 20$, $S_a(u) = S_i(u) = \arctan(hu)$, $h = 20$, $p = 6, q = 1$, $i_0 = 0.6$, $i_1 = 0.1$. Green interval $[0.5, 1.07]$ shows the damaged tissue, $w_0 = 0$.

4. Discussion

Brain functioning is determined by large scale networks of epicenters (hubs) located in the cortex and connected by white matter fiber tracks. The structure of these networks depends on the particular type of the brain activity (motor, language and so on) and on the inter-individual variation. The epicenters exchange information by means of signaling along the cortex or along the white matter fibers. Apparently, this signaling occurs in the form of traveling waves. Periodic traveling waves are observed in thalamus, visual cortex, hippocampus and other parts of the brain. There is more and more evidence that they play the key roles in brain functioning. However, their exact roles and organization are not known. Their characteristics, such as speed, frequency and amplitude, can vary in wide limits (Appendix B), and it is not clear how they are initiated and stop, and how they are related to some particular types of activities. For example, periodic waves are observed in the beginning of produced speech, and they inverse their direction at the end [25]. The corresponding mechanisms governing these activities are not yet understood. These investigations are at the stage of accumulation of biological data and of elaboration of different models whose role and utility will become clearer with time.

Neural field models are widely used to study brain patterns, including stationary structures, pulses, wave fronts, and periodic waves. There are a number of models which describe periodic waves (Appendix A) since it is relatively easy to find these waves on the basis of linear stability analysis of the homogeneous in space stationary solution. The mechanisms of this instability can be related to a combination of inhibition, time delay, refractoriness (also a variant of time delay) and nonlocal interaction. In this work we study the influence of asymmetry of connectivity functions and of time delay in neuron response.

Neuron connectivity is provided by numerous axons whose density decrease approximately exponentially as a function of distance [26,27]. In modeling, the connectivity functions are usually considered to be symmetric; that is, axon connections between points x and y have the same density as between y and x. There is some evidence that connectivity can be asymmetric [13].

In this work we study periodic waves emerging due to the asymmetry of the connectivity function and due to time delay in the response function. Both of them correspond to the existing biological mechanisms. They allow us to study periodic waves in the minimal model which consists of a single integro-differential equation. Other models contain two equations (Appendix A). In the case of asymmetric connectivity function, periodic waves bifurcate due to the loss of stability of the homogeneous in space stationary solution. They become unstable, while bifurcating periodic waves are stable. This is different in the case of time delay. Periodic waves bifurcate from the unstable stationary solution because of space-independent oscillations. Increasing time delay leads to their stabilization.

Though the model contains quite many parameters, using some combination of parameters reduces their number. Next, there are some additional relations between the parameters which determine the stability boundary. Furthermore, there are some physical estimates, such as the decay rate of neuron connectivity function or the value of time delay. Finally, the wave speed and frequency, which depend on parameters, should be in some experimentally observed range. Altogether, these constraints determine some limited intervals of parameter variation (Appendix B).

Cortex Damage and Stimulation

The network of hubs related to some brain function (connectome) can be damaged because of stroke or other factors leading to partial or complete loss of the corresponding function. Post-stroke patient recovery is often incomplete, and usually limited to six months after the accident. Various stimulation techniques are discussed in the literature, but their results are controversial [28].

Post-stroke brain damage can influence propagation of brain waves between the hubs of the connectome. Traveling waves in the cortex can have several functions, including activation of some of its parts. This activation facilitates firing of individual neurons [4]. TW reflect information originating from right and left hemispheres, traveling short and long distances, and containing various time delays. In other words, TW coordinate multiple faster and slower speech events by preparing the arrival of signals traveling along white matter tracts to specific hubs. In particular, direct brain recording (ECoG) showed the presence of TW during consonant-vowel syllables' pronunciation [25]. Summarizing the results by Gross et al. [29] examining how brain waves help us make sense of speech in healthy subjects, Weaver [30] indicated the presence of different time scales in speech production from tens of milliseconds (phoneme) to hundreds of milliseconds (intonation). Today, the language networks have been identified with precision, including phonological processing, speech planning, language semantics, spatial cognition and other functions [31].

We address in this work the question about possible restoration of cortex waves by external stimulation. We show that appropriate choice of injected current allows the recovery of wave speed and frequency, possibly leading to a better communication between the hubs of the connectome. This proof of concept is the first step on the long road to a possible application of this approach to patient rehabilitation.

Author Contributions: Conceptualization, V.V.; Data curation, A.B.; Formal analysis, S.T.; Methodology, A.B. and V.V.; Software, N.B.; Writing original draft, V.V. All authors have read and agreed to the published version of the manuscript.

Funding: A. Beuter was funded by CorStim SAS. S. Trofimchuk was partially supported by FONDECYT (Chile), project 1190712. This research was also funded by "RUDN University Program 5-100" and the French-Russian project PRC2307.

Conflicts of Interest: The authors declare no conflict of interest.

Appendix A. Periodic Waves in Different Models

Two equations with time delay.

In the work [8] a single neuron population model (excitation)

$$\tau \frac{\partial u}{\partial t} = \int_{-\infty}^{\infty} P(x-y)\psi(u(y,t-d))du - u \tag{A1}$$

and a two population model (excitation and inhibition)

$$\tau \frac{\partial u}{\partial t} = \int_{-\infty}^{\infty} \left(P_{11}(x-y)\psi_1(u(y,t-d)) - P_{12}(x-y)\psi_2(v(y,t-d))\right)dy - u \tag{A2}$$

$$\tau \frac{\partial v}{\partial t} = \int_{-\infty}^{\infty} \left(P_{21}(x-y)\psi_1(u(y,t-d)) - P_{22}(x-y)\psi_2(v(y,t-d))\right)dy - v \tag{A3}$$

are considered. Here $P(x)$ and $P_{ij}(x)$ are symmetric positive functions. A particular example of step-wise constant functions is studied. Periodic traveling waves cannot exist for the first model. They are observed for the second model. The existence of pulse solutions in a similar model without delay was studied in [32].

One equation with distributed speed and delay.

Equation with a distributed propagation speed and time delay is considered in [9]:

$$L\left(\frac{\partial u}{\partial t}\right) = \alpha \int_0^{\infty} g(v) \int_{-\infty}^{\infty} K(z) S(u(x+z, t-|z|/v)) dz dv +$$

$$\beta \int_0^{\infty} f(\tau) \int_{-\infty}^{\infty} F(z) S(u(x+z, t-\tau)) dz d\tau, \tag{A4}$$

where L is a second-order differential operator; the functions $K(z)$ and $F(z)$ include both activatory and inhibitory kernels. Different regimes are observed: periodic in time and independent of space, stationary periodic in space, periodic traveling waves.

The model with distributed time delay is considered in [10]

$$u(x,t) = \int_{-\infty}^{\infty} w(x-y) dy \int_{-\infty}^{t} \eta(t-s) f(u(y, s - |x-y|/v)) ds.$$

The same regimes as above and oscillating Turing structures are observed.

Neural field model with linear adaptation.

It is a two equation model

$$\tau \frac{\partial u}{\partial t} = -u - \beta v + \int_{D} w(x-y) F(u(y,t)) dy, \qquad \text{(A5)}$$

$$\frac{1}{\alpha} \frac{\partial v}{\partial t} = u - v, \qquad \text{(A6)}$$

where the second variable represents a linear adaptation (see [11] and the references therein). A large variety of waves and patterns are observed, including stationary periodic in space solutions, traveling waves, modulated traveling waves and stationary and oscillating bumps.

Neural field model with refractoriness.

Periodic traveling waves are also found in one-equation model without inhibition term but with neuron refractoriness (time delay after firing) [12]:

$$\frac{1}{r} \frac{\partial u}{\partial t} = -u + \left(1 - \int_{t-1}^{t} u(x,s) ds\right) f(w \otimes u).$$

Here $\otimes$ denotes spatial convolution.

Appendix B. The Values of Parameters

Periodic traveling waves behave as $\cos(\beta t + \xi x)$ with the time frequency β, space frequency ξ and speed $c = -\beta/\xi$. With the interval length $L = 2$ cm, we get $\xi = 2\pi/L \approx 3$ cm^{-1}.

Let us consider the example of the simulations in Figure 5 with $\tau = 1$ and wave speed 0.3. If this value of τ corresponds to the characteristic time delay 10 ms ([9,33] Chapter 2.3.1), that is, the time unit in the simulation corresponds to 0.01 s, then $c = 0.3 \times 100 = 30$ cm/s.

The time frequency $\beta = c\xi = 90$ s^{-1} belongs to the upper limit of the observed range. The value of the wave speed, and respectively, the time frequency linearly dependent on it can be decreased by the variation of parameters a_i and b_i.

Connectivity functions can be estimated from the data in [26,27]. It exponentially decreases with the rate of decrease in the interval 3–10 times at the at the distance 0.03 cm. This corresponds to the exponential $\exp(-\mu x)$ with μ in the range $30 \div 40$ cm^{-1}.

Propagation speed measures vary depending of the methodology used. When macroscopic waves are recorded from EEG or from ECoG which have low spatial and high temporal resolutions, the propagation speeds varies between 1 and 10 m/s. As indicated by Muller et al. [3], these results are compatible with the range of axonal conduction speeds of myelinated white matter fibers in the cortex. However, when measuring mesoscopic waves' propagation speed using local field potential (LFP) from multielectrode arrays (MEAs) or from optical imaging signals recorded with voltage-sensitive dyes (VSDs) having high spatial and temporal resolution, the propagation speeds varies from 0.1 to 0.8 m per second, consistent with the axonal conduction speed of the unmyelinated long-range horizontal fibers within the superficial layers of the cortex, as indicated by Muller et al. [3].

Summary of parameters.

Parameters of the model are summarized in the following table. Let us note that all parameters are dimensionless. They can be recalculated to dimensional values as in the example above.

Table A1. Typical values of parameters of the model.

Parameter	Name	Unit	Typical Value
D	diffusion coefficient	length^2/time	10^{-4}
L	length of the interval	length	2
a_1, a_2, a_3, a_4	factors in connectivity functions	1/length	$1 \div 5$
b_1, b_2, b_3, b_4	exponents in connectivity functions	1/length	$20 \div 40$
$S'_a(0), S'_i(0)$	growth rate of response functions	1/time	20
τ_a, τ_i	time delay in response functions	time	$1 \div 10$
σ	potential decay rate	1/time	0.01

References

1. Kopell, N.; Gritton, H.J.; Whittington, M.A.; Kramer, M.A. Beyond The Connectome: The Dynome. *Neuron* **2014**, *83*, 1319–1328. [CrossRef] [PubMed]
2. Sporns, O. The human connectome: Origins and challenges. *NeuroImage* **2013**, *80*, 53–61. [CrossRef] [PubMed]
3. Muller, L.; Chavane, F.; Reynolds, J.; Sejnowski, T.J. Cortical travelling waves: Mechanisms and computational principles. *Nat. Rev. Neurosci.* **2018**, *19*, 255–268. [CrossRef] [PubMed]
4. Wu, J.Y.; Huang, X.; Zhang, C. Propagating Waves of Activity in the Neocortex: What They Are, What They Do. *Neuroscientist* **2008**, *14*, 487–502. [CrossRef]
5. Zhang, H.; Watrous, A.J.; Patel, A.; Jacobs, J. Theta and Alpha Oscillations Are Travelling Waves in the Human Neocortex. *Neuron* **2018**, *98*, 1269–1281. [CrossRef] [PubMed]
6. Botella-Soler, V.; Valderrama, M.; Crépon, B.; Navarro, V.; Quyen, M.L.V. Large-Scale Cortical Dynamics of Sleep Slow Waves. *PLoS ONE* **2012**, *7*, e30757. [CrossRef]
7. Wilson, H.R.; Cowan, J.D. A Mathematical Theory of the Functional Dynamics of Cortical and Thalamic Nervous Tissue. *Kybernetik* **1973**, *13*, 55–80. [CrossRef]
8. Senk, J.; Korvasova, K.; Schuecker, J.; Hagen, E.; Tetzlaff, T.; Diesmann, M.; Helias, M. Conditions for travelling waves in spiking neural networks. *arXiv* **2018**, arXiv:1801.06046v1.
9. Atay, F.M.; Hutt, A. Neural Fields with Distributed Transmission Speeds and Long-Range Feedback Delays. *SIAM J. Appl. Dyn. Syst.* **2006**, *5*, 670–698. [CrossRef]
10. Venkov, N.A.; Coombes, S.; Matthews, P.C. Dynamic instabilities in scalar neural field equations with space-dependent delays. *Physica D* **2007**, *232*, 1–15. [CrossRef]
11. Ermentrout, G.B.; Folias, S.E.; Kilpatrick, Z.P. Spatiotemporal Pattern Formation in Neural Fields with Linear Adaptation. In *Neural Fields*; Coombes, S., beim Graben, P., Potthast, R., Wright, J., Eds.; Springer: Berlin/Heidelberg, Germany, 2014; pp. 119–151.
12. Meijer, H.G.E.; Coombes, S. Travelling waves in a neural field model with refractoriness. *J. Math. Biol.* **2014**, *68*, 1249–1268. [CrossRef] [PubMed]
13. Pinotsis, D.A.; Hansen, E.; Friston, K.J.; Jirsa, V.K. Anatomical connectivity and the resting state activity of large cortical networks. *Neuroimage* **2013**, *65*, 127–138. [CrossRef] [PubMed]
14. Modolo, J.; Bhattacharya, B.; Edwards, R.; Campagnaud, J.; Legros, A.; Beuter, A. Using a virtual cortical module implementing a neural field model to modulate brain rhythms in Parkinson's disease. *Front. Neurosci.* **2010**, *4*, 45. [CrossRef] [PubMed]
15. Ermentrout, B.; McLeod, J.B. Existence and uniqueness of travelling waves for a neural network. *Proc. R. Soc. Edinburgh* **1994**, *134A*, 1013–1022. [CrossRef]
16. Chen, Z.; Ermentrout, B.; Wang, X.J. Wave Propagation Mediated by GABAB Synapse and Rebound Excitation in an Inhibitory Network: A Reduced Model Approach. *J. Comput. Neurosci.* **1998**, *5*, 53–69. [CrossRef]

17. Amari, S. Dynamics of pattern formation in lateral-inhibition type neural fields. *Biol. Cybern.* **1977**, *27*, 77–87. [CrossRef]
18. Bessonov, N.; Beuter, A.; Trofimchuk, S.; Volpert, V. Estimate of the travelling wave speed for an integro-differential equation. *Appl. Math. Lett.* **2019**, *88*, 103–110. [CrossRef]
19. Buzsaki, G.; Anastassiou, C.A.; Koch, C. The origin of extracellular fields and currents—EEG, ECoG, LFP and spikes. *Nat. Rev. Neurosci.* **2016**, *13*, 407–420. [CrossRef]
20. Pinto, D.J.; Ermentrout, G.B. Spatially structured activity in synapticalaly coupled neuronal networks: I. Travelling fronts and pulses. *SIAM J. Appl. Math.* **2001**, *62*, 206–225. [CrossRef]
21. Kilpatrick, Z.P. Coupling layers regularizes wave propagation in laminar stochastic neural fields. *Phys. Rev. E* **2014**, *89*, 022706. [CrossRef]
22. Moussaoui, A.; Volpert, V. Speed of wave propagation for a nonlocal reaction-diffusion equation. *Appl. Anal.* **2018**, 1–15. [CrossRef]
23. Rabiller, G.; He, J.-W.; Nishijima, Y.; Wong, A.; Liu, J. Perturbation of brain oscillations after ischemic stroke: a potential biomarker for post-stroke function and therapy. *Int. J. Mol. Sci.* **2015**, *16*, 25605–25640. [CrossRef]
24. Beuter, A.; Balossier, A.; Trofimchuk, S.; Volpert, V. Modelling of post-stroke stimulation of cortical tissue. *Math. Biosci.* **2018**, *305*, 146–159. [CrossRef] [PubMed]
25. Rapela, J. Travelling waves appear and disappear in unison with produced speech. *arXiv* **2018**, arXiv:1806.09559v1.
26. Pelt, J.; van Ooyen, A. Estimating neuronal connectivity from axonal and dendritic density fields. *Front. Comput. Neurosci.* **2013**, *7*, 160.
27. van Ooyen, A.; Carnell, A.; de Ridder, S.; Tarigan, B.; Mansvelder, H.D.; Bijma, F.; Gunst, M.D.; Pelt, J.V. Independently Outgrowing Neurons and Geometry- Based Synapse Formation Produce Networks with Realistic Synaptic Connectivity. *PLoS ONE* **2014**, *9*, e85858. [CrossRef]
28. Carter, A.R.; Connor, L.T.; Dromerick, A.W. Rehabilitation after stroke: Current state of the science. *Curr. Neurol. Neurosci. Rep.* **2010**, *10*, 158–166. [CrossRef]
29. Gross, J.; Hoogenboom, N.; Thut, G.; Schyns, P.; Panzeri, S.; Belin, P.; Garrod, S. Speech Rhythms and Multiplexed Oscillatory Sensory Coding in the Human Brain. *PLoS Biol.* **2013**, *11*, e1001752. [CrossRef]
30. Weaver, J. How Brain Waves Help Us Make Sense of Speech. *PLoS Biol.* **2013**, *11*, e1001753. doi:10.1371/journal.pbio.1001753. [CrossRef]
31. Sarubbo, S.; Benedictis, A.D.; Merler, S.; Mandonnet, E.; Balbi, S.; Granieri, E.; Duffau, H. Towards a Functional Atlas of Human White Matter. *Hum. Brain Mapp.* **2015**. [CrossRef]
32. Pinto, D.J.; Ermentrout, G.B. Spatially structured activity in synapticalaly coupled neuronal networks: II. Lateral inhibition and standing pulses. *SIAM J. Appl. Math.* **2001**, *62*, 226–243. [CrossRef]
33. Gerstner, W.; Kistler, W.M.; Naud, R.; Paninski, L. *Neuronal Dynamics: From Single Neurons to Networks and Models of Cognition*; Cambridge University Press: Cambridge, UK, 2014.

 © 2020 by the authors. Licensee MDPI, Basel, Switzerland. This article is an open access article distributed under the terms and conditions of the Creative Commons Attribution (CC BY) license (http://creativecommons.org/licenses/by/4.0/).

Article

Mathematical Modeling Shows That the Response of a Solid Tumor to Antiangiogenic Therapy Depends on the Type of Growth

Maxim Kuznetsov [1,2]

1. P.N. Lebedev Physical Institute of the Russian Academy of Sciences, 53 Leninskiy Prospekt, Moscow 119991, Russia; kuznetsovmb@mail.ru
2. Peoples Friendship University of Russia (RUDN University), 6 Miklukho-Maklaya Street, Moscow 117198, Russia

Received: 5 April 2020; Accepted: 6 May 2020; Published: 11 May 2020

Abstract: It has been hypothesized that solid tumors with invasive type of growth should possess intrinsic resistance to antiangiogenic therapy, which is aimed at cessation of the formation of new blood vessels and subsequent shortage of nutrient inflow to the tumor. In order to investigate this effect, a continuous mathematical model of tumor growth is developed, which considers variables of tumor cells, necrotic tissue, capillaries, and glucose as the crucial nutrient. The model accounts for the intrinsic motility of tumor cells and for the convective motion, arising due to their proliferation, thus allowing considering two types of tumor growth—invasive and compact—as well as their combination. Analytical estimations of tumor growth speed are obtained for compact and invasive tumors. They suggest that antiangiogenic therapy may provide a several times decrease of compact tumor growth speed, but the decrease of growth speed for invasive tumors should be only modest. These estimations are confirmed by numerical simulations, which further allow evaluating the effect of antiangiogenic therapy on tumors with mixed growth type and highlight the non-additive character of the two types of growth.

Keywords: mathematical oncology; spatially distributed modeling; reaction-diffusion-convection equations; computer experiment

MSC: 35K57; 35Q92; 92C05

---

1. Introduction

The use of mathematical methods has currently become a necessity in oncology. The identification of tumor cells [1], the design of nanomedical systems [2], and real-time adaptation of radiotherapy [3] are just a few examples of problems for which the solution already benefits from the use of simulation studies. One specific mathematical approach is the modeling of tumor growth and treatment, wherein a whole tumor and its microenvironment are considered as a single complex system. Its main goals are gaining new insights into various aspects of cancerous tumors [4,5] and the suggestions for the optimization of treatment protocols [6,7].

Clearly, a mathematical model of tumor growth must capture at least some of the most essential features of cancer cells' behavior and their interaction with the microenvironment. The most notable property of cancer cells is their ability for unlimited growth under favorable conditions [8]. However, tumor growth in tissue is restrained, first of all by the limited availability of nutrients. These aspects were taken into consideration already in the first non-spatially distributed phenomenological models of tumor growth [9,10]. Accounting for another hallmark of cancer—tissue invasion and metastasis—is possible in models that explicitly consider the spatial distribution of cancer cells. In continuous models, the intergrowth of tumor in normal tissue is usually realized via the introduction of parabolic terms, which include the intrinsic motility of tumor cells as a parameter [11,12]. As a rule, these models neglect the convective motion, which arises due to the proliferation of tumor cells. This process by itself can provide an increase in tumor volume even under zero cell motility. Accounting for it can be realized via hyperbolic equations and by itself results in a compact type of tumor growth [13,14]. Such a mechanism is often crucial for low-grade tumors, while the invasive type of growth usually begins to play an increasing role during their progression.

One more hallmark of cancer is sustained angiogenesis, i.e., the formation of new blood vessels that leads to an increase in tumor nutrient supply and thus promotes its growth. This process was incorporated in various ways in mathematical models of different complexity, from the ones governed by ordinary differential equations [15,16], to the complicated multiscale hybrid models [17,18].

The first antiangiogenic drug, bevacizumab, was approved for medical use in 2004 and is utilized currently along with about a dozen other angiogenesis inhibitors. Today, the majority of approved administration schemes, which include antiangiogenic drugs, combine them with various chemotherapeutic agents, which are aimed at direct cell killing [19]. Early experiments on mouse tumor models showed promising results regarding the use of antiangiogenic drugs in monotherapy regimes as well, since their administration as single agents allowed achieving significant delays in tumor growth. However, in the majority of the clinical trials, the administration of mono-AAT did not lead to any notable increase in survival [20]. Such a discrepancy is supposed to be linked to one obvious qualitative difference between preclinical and clinical tests: while the former were mostly conducted on localized primary tumors, the latter were mainly focused on the late-stage diseases [21]. Only relatively recently have clinical trials for early-stage localized tumors been initiated. Their preliminary data suggest the efficiency of mono-AAT for such tumors, at least in terms of tumor mass reduction [22].

Various mechanisms of resistance to treatment have been proposed in order to explain this effect [23]. One such mechanism is the accentuated invasiveness of tumor cells, which allows them to move away from nutrient-deprived zones. There exists a significant amount of experimental evidence that AAT accelerates tumor progression towards increasingly invasive phenotypes [21,24]. Based on this observation, it has been hypothesized that the tumors, which initially have an invasive phenotype, should possess intrinsic resistance to AAT, while compactly growing tumors should be more susceptible to it [23].

In this work, a continuous model of a monoclonal tumor growth in tissue is presented, which is able to reproduce this effect. The model allows accounting for both types of tumor growth, as well as their combination. Analytical estimations of tumor growth speed are obtained for tumors with both of the pure types of growth, which allow assessing the effect of antiangiogenic therapy on them. Numerical simulations are presented, which show good agreement with analytical estimations. They further allow evaluating the effect of antiangiogenic therapy on tumors with a mixed type of growth and provide insights into the mechanism of the interaction between the two types of growth, highlighting their non-additive character.

2. Model

2.1. Equations

The mathematical model of tumor growth, considered herein, represented a simplification of the model, previously developed by our research group. Its different versions were used for the investigation of several aspects of tumor growth and treatment [25–27]. There were four variables in this version of the model, which were a function of space and time coordinates, x and t: the density of tumor cells $n(x,t)$, the fraction of necrotic tissue $m(x,t)$, the concentration of glucose $g(x,t)$, and the density of capillaries surface area $c(x,t)$. All the variables, as well as all the used parameters should be strictly non-negative due to their physical meaning. The one-dimensional planar case was considered in this work, which was suitable for the consideration of a large spherically-symmetrical tumor. The following set of equations governs the dynamics of the model variables:

$$\begin{aligned}
\text{tumor cells:} \quad & \frac{\partial n}{\partial t} = \overbrace{Bn \cdot [1 - \sigma(g)]}^{\text{proliferation}} - \overbrace{Mn \cdot \sigma(g)}^{\text{death}} + \overbrace{D_n \frac{\partial^2 n}{\partial x^2}}^{\text{migration}} - \overbrace{\frac{\partial (In)}{\partial x}}^{\text{convection}}; \\
\text{necrotic tissue:} \quad & \frac{\partial m}{\partial t} = \overbrace{Mn \cdot \sigma(g)}^{\text{death}} - \overbrace{\frac{\partial (Im)}{\partial x}}^{\text{convection}}; \\
\text{glucose:} \quad & \frac{\partial g}{\partial t} = \overbrace{Pc[1 - g]}^{\text{inflow}} - \overbrace{Qn \cdot [1 - \sigma(g)]}^{\text{consumption}} + \overbrace{D_g \frac{\partial^2 g}{\partial x^2}}^{\text{diffusion}}; \\
\text{capillaries:} \quad & \frac{\partial c}{\partial t} = \overbrace{-R[n + m]c}^{\text{degradation}};
\end{aligned} \quad (1)$$

$$\text{where} \quad \sigma(g) = \frac{1}{2}[1 - \tanh(\epsilon\{g - g_{cr}\})];$$

$$\nabla I = Bn \cdot [1 - \sigma(g)] + D_n \frac{\partial^2 n}{\partial x^2}.$$

2.1.1. Tumor Cells

The term tumor cell proliferation implies that it happens ceaselessly under a sufficient level of glucose, which was chosen to be the key nutrient, since it is indispensable for cell division [28]. With the fall of glucose level, the rate of proliferation slows down, and tumor cells die by necrosis. The drainage of necrotic tissue was neglected. The exact rates of the processes of cell proliferation and death were governed by a sigmoid function $\sigma(g)$. Tumor cells were able to migrate throughout the tissue, which was governed by a diffusion-like term. The convective terms described the bulk motion of tissue elements, the velocity field **I** being determined by the dynamics of tumor cells. The expression for it was derived under the assumption of the constancy of the total density of tumor cells, necrotic tissue, and normal cells, which was normalized to unity. The normal cells were not considered in the model explicitly; however, it was assumed that the passive motion due to the arising convective flow was the only part of their dynamics.

2.1.2. Glucose and Capillaries

The dynamics of glucose is comprised of its inflow from the capillaries into the tissue, consumption by tumor cells, and diffusion throughout the tissue. The inflow of glucose is governed primarily by the process of passive diffusion through the walls of capillaries [29]. Therefore, the rate of glucose inflow is proportional to the density of the capillaries' surface area and to the difference in glucose concentrations in blood and in tissue. Glucose concentration in blood was considered to be constant and was normalized to unity.

The capillary network degrades inside the tumor. This process has various reasons of a mechanical [30] and chemical nature [31], the details of which are difficult to account for in a reliable manner. Therefore, the degradation of capillaries was described by a rather phenomenological term. The volume of capillaries was considered to be negligible compared to the volume of cells, and therefore, their dynamics did not affect the convective velocity field. The convective motion of capillaries was also neglected. The normal density of capillaries' surface area was normalized to unity.

2.1.3. Angiogenesis and Antiangiogenic Therapy

A more or less straightforward consideration of angiogenesis would require the introduction of an additional variable for the concentration of the main pro-angiogenic factor VEGF, which would affect the dynamics of capillaries. Such an approach has been utilized previously in different continuous models of tumor growth [12,32,33], including our models [27,34]. Herein, a more concise method was used for the sake of analytical study. It was assumed that during the growth of an untreated tumor, the concentration of VEGF was so high that it provided the maximum possible angiogenic effect throughout all the considered part of the tissue (the rate of this effect was limited by the number of receptors on endothelial cells). In fact, there are two separate effects, induced by VEGF, that affect the inflow of glucose from the capillaries and therefore needed to be considered in this model. One of them is the increase of the capillary density due to the formation of new capillaries. Another effect is the increase of the vascular permeability [35,36]. Due to the above-described assumption of the uniform action of VEGF, accounting for the latter effect came down to the increase in the value of parameter P, since it corresponded to the uniform identical increase in the permeability of all capillaries in the tissue. Assuming further that the local increase of the capillaries' surface area density, c, was proportional to its own local value throughout the tissue, one may conclude that for the consideration of the glucose inflow in tissue, an increase of c was equivalent to an analogous increase in the value of parameter P. Thus, the total effect of angiogenesis was accounted for in this model by the increase in the value of a single parameter P. Since AAT leads to the normalization of the structure of capillaries and further normalization of their density [37,38], its action as reflected by a decrease of P to a value, which was assumed to correspond to normal tissue.

2.2. Parameters

The parameters of the model were estimated according to the data of various experimental works. The basic set of parameters is listed in Table 1. The dimensionless model values of the parameters were the approximations of their normalized values, which were obtained with the use of the following normalization parameters: $t_n = 1$ h for time, $x_n = 10^{-2}$ cm for length, $g_n = 1$ mg/mL for glucose concentration, $c_n = 100$ cm^2/cm^3 for normal capillary surface area density, which was close to its average value for human muscle [29], and $n_n = 3 \times 10^8$ cells/mL for the maximum density of tumor cells. The latter value was taken from the experimental work on the in vitro growth of multicellular tumor spheroids [39]. The values for the proliferation rate of tumor cells and their glucose consumption rate were also estimated according to the data of this work; however, it was assumed that these values should be proportionally diminished during the growth of a relevant tumor in tissue. The basic coefficient of tumor cells' motility corresponded to highly motile glioma cells [40]. This parameter is set to zero in Section 3.1 in order to focus on the compact type of tumor growth and is varied in Section 3.3 in order to consider tumors with a mixed type of growth. The value for capillaries' degradation rate was chosen in order to correspond to experimental observations, which showed that capillaries with adequate blood filling were very scarce inside the core of the tumors with radii of several millimeters [41]. The rate of death of tumor cells was taken to be significantly higher that that of their proliferation. The values of the parameters of function $\sigma(g)$ could not be assessed straightforwardly, since the form of this function was chosen for phenomenological reasons. Therefore, it was merely taken to be rather close to a stepwise function, with the transition of tumor cells from proliferation to death happening at the concentrations of glucose an order of magnitude lower than that in blood. Importantly, the use of these values of variables M, g_{cr}, and ϵ allowed keeping the concentrations of glucose positive within the necrotic zone.

Table 1. Basic set of model parameters.

Parameter	Description	Estimated Value	Model Value	Based on
B	tumor cells' proliferation rate	0.01 h^{-1}	0.01	[39]
Q	tumor cells' glucose consumption rate	6.2×10^{-17} mol/(cells·s)	12	[39]
D_g	glucose diffusion coefficient	2.8×10^{-6} cm^2/s	100	[42]
P	angiogenesis parameter	1.1×10^{-5} cm/s	4	[29]
g_{cr}	critical level of glucose	0.56 mM	0.1	see the text
D_n	tumor cells' motility	2.4×10^{-4} cm^2/day	0.1	[40]
M	tumor cells' death rate	0.05 h^{-1}	0.05	see the text
R	capillaries' degradation rate	1.7×10^{-10} mL/(cells·s)	0.2	[41]
ϵ	tumor cells' sensitivity to glucose level	–	100	see the text

In order to consider the effect of angiogenesis on tumor growth, the value of angiogenesis parameter P was varied up to a value ten times greater than the basic one. Such a limit was selected as a very approximate product of two estimated factors. Firstly, experiments in various mouse tumor models showed that the local density of microvessels near a tumor could increase three to six times [43]. Secondly, in the work [34], it was shown that a 2.5-fold increase in the permeability of tumor capillaries to glucose due to the action of VEGF was a physiologically reasonable estimation.

2.3. Numerical Solving

During numerical simulations, the set of equations Equation (1) was solved in a region with a size of several centimeters. The exact size X was adjusted for each case in order to be sufficiently small to spare computational time without imposing noticeable edge effects, while for all the variables, the zero-flux boundary condition was used at both edges. The convective flow speed was set to zero at the left boundary, where $x = 0$, resulting in the following equation for it:

$$I(x,t) = \int_0^x Bn(r,t) \cdot [1 - \sigma(g(r,t))]dr + D_n \frac{\partial n(x,t)}{\partial x}.$$

The following initial conditions were used, which represented a normal tissue with a small, 0.01 mm in width, colony of tumor cells, located near the left boundary, where capillaries were absent:

$$\begin{cases} n = 1, \\ m = 0, \\ g = 1, \\ c = 0 \end{cases} \text{for } x <= 0.1; \quad \begin{cases} n = 0, \\ m = 0, \\ g = 1, \\ c = 1 \end{cases} \text{for } x > 0.1. \tag{2}$$

The method of splitting into physical processes was used for all variables, i.e., kinetic equations, diffusion equations, and convective equations were solved successively during each time step. The implicit Crank–Nicholson scheme was used for the glucose diffusion equation. Since the glucose diffusion term provided maximum rate of local change among all the variables and thus required a sufficiently small time step even for implicit solving, it was decided to solve the cell migration equation by a simpler explicit forward Euler scheme, and all kinetic equations by the explicit Euler method. Convective equations were solved using the flux-corrected transport algorithm with explicit anti-diffusion stage. The last method was introduced in the work [44], while other classical methods were described in many books (see, e.g., [45]). The choice of time and space steps is justified in the following sections. For optimization purposes, the function $\sigma(g)$ was not recalculated every time; instead, it was precalculated for about ten thousand values of g, evenly distributed on the segment $[0, 1]$, and only thusly obtained values were used during the calculations as approximations of the actual values of $\sigma(g)$.

The tumor growth speed $V_{gr}(t)$ was calculated as the rate of change of tumor radius, which was evaluated as the maximum space coordinate, at which $n \geq 0.1$. In all the simulations, after an initial transitional period, the speed of tumor growth seemed to tend asymptotically to some constant value. In order to ensure that this value could be estimated with suitable precision, for each simulation, the part of the function $V(t)$, obtained after the manually designated initial transitional period, was fitted via the least-squares method by a function of the form:

$$V_{gr}(t) = V + \sum_{i=1}^{k} a_i e^{-b_i t},$$

wherein the value $k = 3$ was selected manually as the one providing sufficiently fine approximation. A simulation was stopped manually if the current tumor growth speed and V, expressed in mm/week, were equal up to three decimal places. If it did not happen until the proximity of the tumor to the right boundary notably affected its growth speed, the simulation was rerun in a larger region. Further throughout the text, this limiting value V is meant for tumor growth speed.

For some simulations, the number of tumor cells was estimated analogically as:

$$N = \lim_{t \to \infty} \int_0^X n(x) dx,$$

the integral being calculated numerically by the elementary rectangle rule.

The computational codes were implemented in C++ and can be found in the Supplementary Materials.

3. Results

3.1. Compact Type of Growth

At first, let us obtain an analytical estimation for the growth speed of a solid tumor with zero cell motility, i.e., $D_n = 0$, which thus has a purely compact type of growth. For this purpose, let us seek the solution of Equation (1) in the form of a wave, traveling with constant shape and speed V in an infinite region. In such a case, the governing equations can be reduced to a system of ordinary differential equations by introducing the traveling coordinate frame $z = x - Vt$:

$$\begin{aligned}
\text{tumor cells:} \quad & Bn \cdot [1 - \sigma(g)] - Mn \cdot \sigma(g) - \frac{\partial(In)}{\partial z} + V\frac{\partial n}{\partial z} = 0; \\
\text{necrotic tissue:} \quad & Mn \cdot \sigma(g) - \frac{\partial(Im)}{\partial z} + V\frac{\partial m}{\partial z} = 0; \\
\text{glucose:} \quad & Pc[1 - g] - Qn \cdot [1 - \sigma(g)] + D_g \frac{\partial^2 g}{\partial z^2} + V\frac{\partial g}{\partial z} = 0; \\
\text{capillaries:} \quad & -R[n + m]c + V\frac{\partial c}{\partial z} = 0;
\end{aligned} \quad (3)$$

$$\begin{aligned}
\text{where} \quad & \sigma(g) = \frac{1}{2}[1 - \tanh(\epsilon\{g - g_{cr}\})]; \\
& I(z) = \int_{-\infty}^{z} Bn(r) \cdot [1 - \sigma(g(r))] dr; \\
& V = \lim_{z \to +\infty} I(z).
\end{aligned}$$

The living part of the tumor was considered as a planar front, which propagated towards the right boundary, which represented the normal tissue, and left the necrotic zone behind it. That is formalized by the following boundary conditions:

$$\begin{cases} n = 0, \\ m = 1, \\ \frac{\partial g}{\partial z} = 0, \\ c = 0, \\ I = 0 \end{cases} \text{for } z \to -\infty; \quad \begin{cases} n = 0, \\ m = 0, \\ g = 1, \\ c = 1, \\ \frac{\partial I}{\partial z} = 0, \end{cases} \text{for } z \to +\infty. \quad (4)$$

Further, in order to obtain an analytically tractable case, let us consider the following limitations:

- $\epsilon \to \infty$, i.e., all the tumor cells either proliferate or die at a given position at a given moment;
- $M \to \infty$, i.e., tumor cells die instantaneously;
- $R \to \infty$, i.e., there are no capillaries inside the tumor.

The last condition is sensible only if the tumor has a clear boundary, i.e.,

$$\exists x_b : \forall x > x_b, \; n(x) + m(x) = 0,$$

These limitations result in the following system:

$$\begin{aligned}
\text{tumor cells:} & \quad \begin{cases} n = 0 \; \text{if} \; g \leq g_{cr}, \\ Vn' - (In)' + Bn = 0 \; \text{if} \; g > g_{cr}; \end{cases} \\
\text{necrotic tissue:} & \quad m = \begin{cases} 1 \; \text{if} \; g \leq g_{cr}, \\ 0 \; \text{if} \; g > g_{cr}; \end{cases} \\
\text{glucose:} & \quad D_g g'' + V g' + Pc[1-g] - Qn = 0; \\
\text{capillaries:} & \quad c = \begin{cases} 1 \; \text{if} \; n+m = 0, \\ 0 \; \text{if} \; n+m > 0; \end{cases}
\end{aligned} \qquad (5)$$

$$\text{where} \quad I(z) = \int_{-\infty}^{z} Bn(r)\,dr; \; V = \lim_{z \to +\infty} I(z),$$

where primes denote differentiation with respect to z.

Finally, let us specify the form of the tumor cells' distribution in the sought solution as a limiting case of a piecewise function, which can be equal only to zero and one, like the functions for necrotic tissue and capillaries' distribution. Thus, the sought solution has the form depicted in Figure 1, where the origin of the z-axis is placed at the front of the tumor for convenience. This solution can be split up into three regions: (1) necrotic core, (2) proliferating rim of yet unknown width L, in which $n = 1$ and $g > g_{cr}$, and (3) normal tissue. The expression for the tumor growth speed is now simplified to:

$$V = I(0) = BL.$$

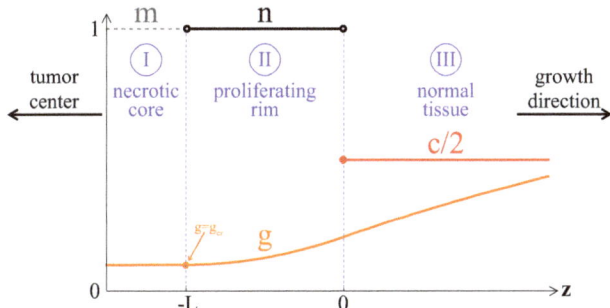

Figure 1. The form of the solution of Equation (5), which is searched for analytically.

The distribution of glucose has to be found in order to estimate tumor growth speed. This procedure, described in Appendix A, yields the following implicit expression for L:

$$1 - \frac{Q}{B} + \frac{QD_g}{B^2 L^2}[1 - e^{-BL^2/D_g}][1 - \frac{2}{1 + \sqrt{1 + 4D_g P/(B^2 L^2)}}] = g_{cr}. \qquad (6)$$

Its numerical solving resulted in $L \approx 1.325$ under the values of the parameters used, taken from the basic parameter set (see Table 1), which led to the tumor growth speed $V \approx 0.223$ mm/week. There was under a ten-fold greater value for angiogenesis parameter, i.e., $P = 40$, $L \approx 2.602$ and $V \approx 0.437$ mm/week, which meant almost a two-fold increase in tumor growth speed. These values were in good correspondence with the speeds of the in vitro growth of multicellular tumor spheroids, experimentally obtained in the work [39] (≈ 0.5 mm/week), based on which several values of the model parameters were estimated.

The estimation of V can be significantly simplified due to the presence of a small parameter:

$$\delta = \frac{BL^2}{D_g} = \frac{V}{D_g/L},$$

the smallness of which is due to the fact that the tumor growth speed should be significantly lower than the characteristic speed of glucose diffusion within the proliferating rim. For example, for the values of parameters from their basic set, $\delta \approx 1.7 \times 10^{-4}$. Expansion of the expression (6) in a Taylor series up to $o(\delta)$ yields:

$$g_{cr} = 1 - \frac{Q}{B} + \frac{Q}{B}[1 - \sqrt{\frac{B}{P}}\sqrt{\delta} - \frac{\delta}{2}(1 - \frac{B}{P}) + o(\delta)].$$

Moreover, it is convenient to neglect B/P as a small parameter compared to one, e.g., $B/P \approx 2.5 \times 10^{-3}$ for the basic values of the model parameters. This allows obtaining the equation for an approximate value of proliferating rim width, $\tilde{L}$:

$$g_{cr} = 1 - \frac{Q}{\sqrt{PD_g}}\tilde{L} - \frac{Q}{2D_g}\tilde{L}^2,$$

which can be solved, only its positive root having physical meaning. That leads to the following formula for approximate tumor growth speed:

$$\tilde{V} = B\tilde{L} = B\sqrt{D_g}\left\{\sqrt{\frac{1}{P} + 2\frac{1 - g_{cr}}{Q}} - \sqrt{\frac{1}{P}}\right\}, \tag{7}$$

which under the considered parameter values provides the values of tumor growth speed, equal up to three decimal places to the ones derived numerically from Equation (6). Note that this expression has a limit under $P \to \infty$:

$$\tilde{V}_{lim} = B\sqrt{\frac{2D_g}{Q}(1 - g_{cr})},$$

that is ≈ 0.651 mm/week under the basic set of parameters. Thus, the speed of growth of a considered tumor cannot increase more than three-fold due to angiogenesis. Under $P = 40$, it is about two-thirds of its limit value, while under $P = 80$, it would be around three-quarters of it.

These estimations allow suggesting that for compactly growing tumors, AAT may provide a several times decrease of their growth speed. Of note, there is no account for the drainage of necrotic tissue in the model, while its consideration should enhance the result. Moreover, in this case, AAT might be able to lead to a complete tumor growth stop, as well as its shrinkage, which is sometimes observed experimentally [46].

The estimations of tumor growth speed under finite values of ϵ, M, and R are quite difficult if at all impossible to perform analytically, and they were performed via numerical simulations. However, at first, it was examined how well the results, obtained in a numerical experiment, could correspond to the already obtained analytical estimations. For this purpose, the sets of four simulations with different time and space steps were conducted for each of the six values of the angiogenesis parameter within the considered range $P \in [4, 40]$. In each of the simulations, the time step τ and space step h

related as $\tau = h^2$. The values of R and M were chosen to be close to the maximum values allowed by numerical calculations, i.e., $R = M = 0.1/\tau$. The value of ϵ was chosen to be 10^5, which practically resulted in a stepwise function $\sigma(g)$ due to the method of its implementation in the code.

The dots in Figure 2a denote the values of tumor growth speed, obtained numerically in these simulations. The analytically obtained values are designated by crosses of the corresponding colors on the vertical axis. For every used value of P, the best fit quadratic function was built in order to approximate the numerical value for the tumor growth speed under $h \to 0$, $\tau \to 0$, $M \to \infty$, and $R \to \infty$. The thusly obtained values were in a good correspondence with the analytical ones, being only slightly smaller than them. The discrepancy between the values increased with the increase of P, constituting less than 1% under $P = 4$ and less than 5% under $P = 40$. Figure 2b–e shows the distributions of the model variables on the 20th day of simulations under the values of parameters, designated by the corresponding letters in Figure 2a. They showed that, expectedly, the numerically obtained profiles of tumor cells differed from stepwise functions; however, their front edges were getting steeper under finer discretization. Of note, the tumors that were obtained under coarser discretization, shown in Figure 2c,e, had greater radii on the 20th day than the corresponding tumors obtained under finer discretization, shown in Figure 2b,d. This did not reflect the relation between the growth speeds of the tumors in these simulations, since these profiles were obtained during the transitional period of tumor growth. As Figure 2a shows, tumor growth speed changed differently with the refinement of the discretization under different values of P.

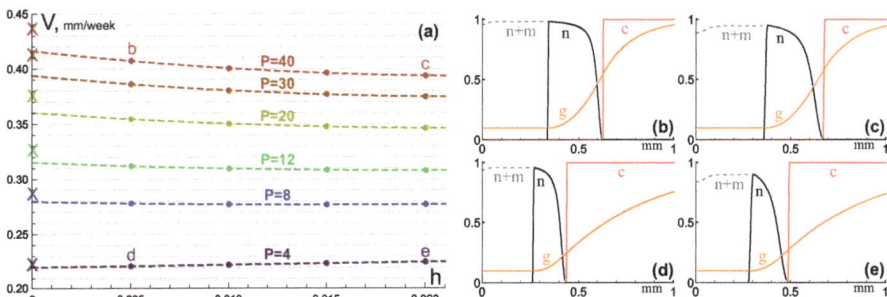

Figure 2. (a) The dots denote the values of tumor growth speed V, obtained numerically in the simulations of Equation (1) under designated values of angiogenesis parameter P and space step h, with time step $\tau = h^2$, $D_n = 0$, $R = M = 0.1/\tau$, $\epsilon = 10^5$ and the values of other parameters taken from the basic set. The dashed lines represent the best fit quadratic functions for every value of P. The crosses on the vertical axis mark the analytically derived values of tumor growth speed for the values of P used. (b–e) The distributions of the model variables on the 20th day of simulations under the values of the parameters, designated by the corresponding letters in (a).

The analogical results, obtained in simulations with the values of parameters ϵ, R, and M, taken from the basic set, are shown in Figure 3. With the use of the basic values for these parameter, the approximated numerical values for tumor growth speed increased modestly for each considered value of P, becoming $\approx$1.8–2.4% higher than the corresponding analytically estimated values. Certainly, a further decrease of either ϵ or R, as well as the increase of either proliferation rate B, or glucose consumption rate Q, or glucose diffusion coefficient D_g, would lead to a further increase in tumor growth speed. However, too large alterations of the values of these parameters would lead to physically meaningless results involving negative glucose concentrations and/or explosive tumor growth, thus indicating the limitations of the presented model. Of note, the decrease of M by itself had only a small effect on the tumor growth speed, as well as on the profile of glucose concentration, but its change largely influenced the border between the profiles of necrotic tissue and tumor cells, which were considered to be yet alive.

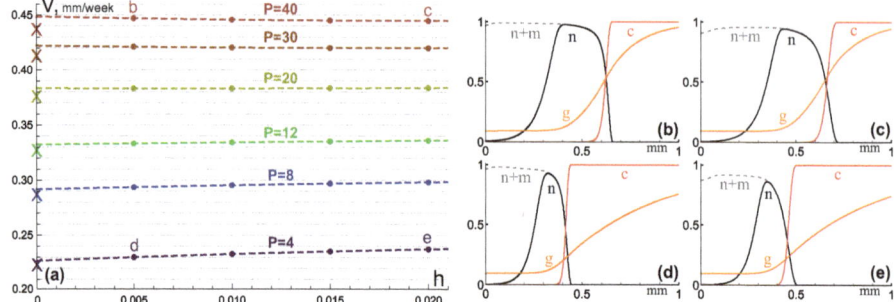

Figure 3. (a) The dots denote the values of tumor growth speed V, obtained numerically in the simulations of Equation (1) under designated values of angiogenesis parameter P and space step h, with time step $\tau = h^2$, $D_n = 0$, and the values of other parameters taken from the basic set. The dashed lines represent the best fit quadratic functions for every value of P. The crosses on the vertical axis mark the analytically derived values of tumor growth speed for the values of P used. (**b–e**) The distributions of the model variables on the 20th day of simulations under the values of the parameters, designated by the corresponding letters in (**a**).

3.2. Invasive Type of Growth

Now, let us obtain an analytical estimation for the growth speed of a solid tumor with non-zero cell motility D_n, but neglecting the bulk motion of tissue elements, expressed in Equation (1) by hyperbolic terms. Once again, the solution was sought in the form of a wave, traveling with a constant shape and speed V in an infinite region. The boundary conditions are now modified as:

$$\begin{cases} n = 0, \\ m = m^*, \\ g = g^*, \\ c = 0 \end{cases} \text{for } z \to -\infty; \qquad \begin{cases} n = 0, \\ m = 0, \\ g = 1, \\ c = 1 \end{cases} \text{for } z \to +\infty, \qquad (8)$$

where g^* and m^* are introduced as the limiting constant values of glucose concentration and necrotic tissue fraction at $z \to -\infty$. For the estimation of V, it is necessary to consider the asymptotic behavior of the solutions at $z = \pm\infty$. For this purpose, let us linearize the system (1) at the boundary values (8), neglecting the convective terms, and thus obtain two systems of ODEs with constant coefficients:

$$\begin{cases} D_n n''_- + V n'_- + B n_- \cdot [1 - \sigma(g^*)] - M n_- \cdot \sigma(g^*) = 0; \\ V m'_- + M n_- \cdot \sigma(g^*) = 0; \\ D_g g''_- + V g'_- + P c_- [1 - g^*] - Q n_- \cdot [1 - \sigma(g^*)] = 0; \\ V c'_- - R m^* c_- = 0. \end{cases}$$

and:

$$\begin{cases} D_n n''_+ + V n'_+ + B n_+ \cdot [1 - \sigma(1)] - M n_+ \cdot \sigma(1) = 0; \\ V m'_+ + M n_+ \cdot \sigma(1) = 0; \\ D_g g''_+ + V g'_+ - P g_+ - Q n_+ \cdot [1 - \sigma(1)] = 0; \\ V c'_+ - R[n_+ + m_+] = 0, \end{cases}$$

where primes denote differentiation with respect to $z = x - Vt$ and subscripts $-$ and $+$ refer to the linearized problems at $z \to -\infty$ and $z \to +\infty$, respectively. The solutions to these problems should be sought in the form $(n'_\pm, n_\pm, m_\pm, g'_\pm, g_\pm, c_\pm)^T \sim \mathbf{k}^\pm \exp(\mu^\pm z)$, which reduces these systems of linear differential equations to eigenvalue problems.

For $z \to -\infty$, the eigenvalues are:

$$\mu_{1,2}^- = 0, \quad \mu_3^- = Rm^-/V, \quad \mu_4^- = -V/D_g,$$
$$\mu_{5,6}^- = [-V/2 \pm \sqrt{V^2/4 + D_n\{M\sigma(g^-) - B[1-\sigma(g^-)]\}}]/D_n,$$

where μ_3^- is positive, μ_4^- is negative, and $\mu_{5,6}^-$ have opposite signs, since $M\sigma(g^-) - B[1-\sigma(g^-)] > 0$, i.e., the death rate exceeds the proliferation rate at sufficiently low values of z, which follows from the physical meaning of the model.

For $z \to +\infty$, the eigenvalues are:

$$\mu_{1,2}^+ = 0, \quad \mu_{3,4}^+ = [-V/2 \pm \sqrt{V^2/4 + D_g P}]/D_g,$$
$$\mu_{5,6}^+ = [-V/2 \pm \sqrt{V^2/4 - D_n\{B[1-\sigma(1)] - M\sigma(1)\}}]/D_n,$$

where $\mu_{3,4}^+$ have opposite signs and, depending on the value of V, $\mu_{5,6}^+$ are either complex numbers with negative real parts, or real numbers with opposite signs, or both are equal to zero. Therefore, for sufficiently low values of V, the solution may oscillate for large values of z, yielding regions with negative values of n and m, which is physically unrealistic. Thus, the following restriction on tumor growth speed is sufficient for physically reasonable solutions:

$$V \geq V_{min} = 2\sqrt{D_n\{B[1-\sigma(1)] - M\sigma(1)\}} \approx 2\sqrt{BD_n}, \tag{9}$$

the last two values being extremely close under the basic set of parameters, differing by less than $10^{-13}\%$ even under a decrease of ϵ up to the value of 20. For the basic set of parameters, $V_{min} \approx 1.063$ mm/week, which was in a good correspondence with the speed of growth of highly invasive tumors [40]. Formula (9) corresponds well to the well-known result regarding Fisher's equation, which can be written out in the notation used herein as:

$$\frac{\partial n}{\partial t} = Bn(1-n) + D_n \frac{\partial^2 n}{\partial x^2}. \tag{10}$$

For this equation, the range of the speed of monotone waves satisfies:

$$V \geq V_{min}^f = 2\sqrt{BD_n},$$

this result not being affected by the alteration of the proliferation term as long as the proliferation rate tends to B for $n \to 0$. For Fisher's equation, it is known that if its initial conditions have a compact support, i.e.,

$$n(x,0) = n_0(x) \geq 0, \quad n_0(x) = \begin{cases} 1 & \text{if } x \leq x_1, \\ 0 & \text{if } x \geq x_2, \end{cases} \tag{11}$$

where $x_1 < x_2$ and $n_0(x)$ is continuous in $x_1 < x < x_2$, then the solution $n(x,t)$ of Equation (10) evolves to a traveling wavefront solution with the speed $V_{min}^f = 2\sqrt{BD_n}$ [47].

The numerical simulations of the system (1) under the neglect of convective terms with the initial conditions (2), which represent a function with a compact support, show that they evolve to the wavefronts, traveling with speeds very close to $2\sqrt{BD_n}$. For the basic set of parameters, numerical simulations gave $V \approx 1.061$ mm/week, yielding the same tumor growth speed up to three decimal places (as well as the same number of cells $N \approx 1.383$) under space steps $h = \{0.2, 0.1, 0.05, 0.025\}$ and time steps $\tau = h^2$. Thus, much coarser discretization could provide more accurate results for the considered system under the neglect of convective terms. A dozen simulations under other values of parameters, chosen randomly from the ranges $B \in [0.05, 0.3]$ and $D_n \in [0.001, 0.4]$, as well produced the values of tumor growth speeds that differed from $2\sqrt{BD_n}$ by no more than 0.5%.

Since the considered system could be viewed as an augmentation of Fisher's model, the obtained results suggested that the introduced modifications should not affect the wave speed of its solution, which evolved from a function having a compact support, or at least they should lead to only a very small correction of it within the physiologically justified range of parameters. Most importantly, the parameter P is not present in Formula (9), which implies that angiogenesis should not influence the growth speed of a considered invasive tumor at all. Numerical modeling speaks in favor of this result. The simulations under six values of the angiogenesis parameter within the considered range $P \in [4, 40]$ yielded tumor growth speeds equal up to three decimal places, which is illustrated by Figure 4a. The simulations were run with space step $h = 0.1$ and time step $\tau = h^2 = 0.01$, which nevertheless did not allow saving a sufficient amount of computational time, compared to the simulations of Section 3.1, since in this case, the tumor growth speed tended sufficiently slower to a constant value, thus requiring longer simulation on larger domains.

Of note, the number of tumor cells notably grew with the increase of P, as Figure 4a shows, resulting in an ≈72% increase under 10-fold magnification of the basic value of $P = 4$. This effect was well noticeable when comparing the profiles of tumor cells under $P = 4$ and $P = 40$, which are depicted in Figure 4b,c on the 20th day of tumors growth.

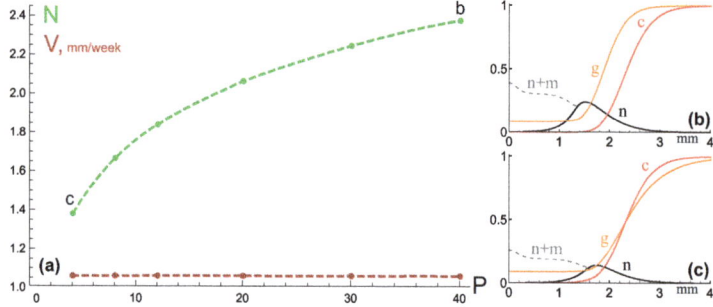

Figure 4. (a) The red dots denote the values of tumor growth speed V; the green dots denote the values of number of tumor cells N, which were obtained numerically in the simulations of Equation (1) with the neglect of hyperbolic terms (i.e., $\mathbf{I} = 0$) under designated values of angiogenesis parameter P, values of other parameters, taken from the basic set, space step $h = 0.1$, and time step $\tau = 0.01$. The dashed lines are interpolations of data points. (b,c) The distributions of the model variables on the 20th day of simulations under $P = 40$ and $P = 4$ correspondingly.

3.3. Mixed Type of Growth

The conclusion of the indifference of invasive tumor growth speed to angiogenesis was valid under the full neglect of the convective component of tumor growth. However, it was reasonable to assume that simultaneously accounting for it should nevertheless lead to a non-zero effect of AAT on the growth speed of an invasive tumor, since the convective component of tumor growth should be affected by the number of its cells. In order to evaluate the effect of antiangiogenic therapy on tumors with a mixed type of growth, the sets of six simulations under different values of angiogenesis parameter P were conducted for eight values of tumor cell motility D_n. As a compromise between computing resources and accuracy, the values of space step $h = 0.01$ and time step $\tau = 10^{-4}$ were used. The obtained tumor growth speeds are designated in Figure 5a, where the data for the case with zero cell motility, already shown in Figure 3a, is added.

Let introduce the parameter of "maximum antiangiogenic effect" as:

$$(V|_{P=40} - V|_{P=4})/V|_{P=40},$$

which indicates how the tumor growth speed would slow down upon a decrease of angiogenesis parameter P from 40 to four, the latter corresponding to the normal microvasculature level. Its dependence on tumor cell motility is shown in Figure 5b, while under $D_n = 0$, it was ≈48%. The graph suggests that the effect of mono-AAT should inversely correlate with the invasiveness of the tumor; however, it should be non-zero even for highly invasive tumors, since the maximum antiangiogenic effect, obtained under high values of tumor cell motility, $D_n = 0.1$ and $D_n = 0.33$, was ≈13% and ≈11%, correspondingly.

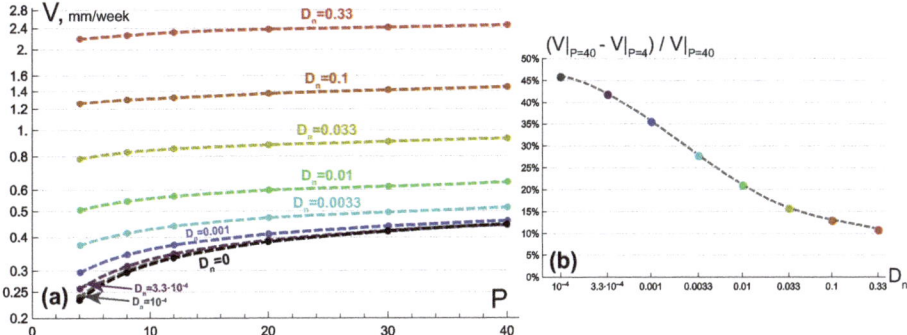

Figure 5. (a) The dots denote the values of tumor growth speeds V, obtained numerically in the simulations of Equation (1) with space step $h = 0.01$ and time step $\tau = 10^{-4}$, under designated values of angiogenesis parameter P and tumor cell motility D_n and the values of other parameters taken from the basic set. The dashed lines are interpolations of data points. (b) The graph of "maximum angiogenic effect", based on the data, shown in (a), which shows how the tumor growth speed would slow down upon the decrease of angiogenesis parameter P from 40 to four under designated values of D_n.

An interesting property of the considered tumors with a mixed type of growth was the fact that their growth speed was not equal to the sum of two speeds: V_{comp} that was obtained in the absence of cell motility—i.e., during purely compact growth—and $V_{inv} \approx 2\sqrt{BD_n}$ that was obtained in absence of convective flows—i.e., during purely invasive growth. This property is illustrated in Figure 6, which shows the dependence of the parameter $(V - V_{comp})/V_{inv}$ on P for different D_n. This parameter may be treated as the ratio of the increase in tumor growth speed, caused by the mobilization of initially immotile cells, to the speed of pure invasive growth itself. The reason for the non-additivity of the growth speeds was the fact that the redistribution of the tumor cells, caused by their migration, led to the change in the number of cells in the proliferating state and altered the convective velocity field. This effect had an ambiguous character. Firstly, the protrusion of tumor cells' profile towards the region with a normal density of microvasculature led to the increase in the effective rate of capillaries' degradation, which resulted in a decrease of glucose inflow to the tumor and a subsequent decrease of the pool of proliferating cells. Secondly, this protrusion itself enlarged the pool of tumor cells, located in the region with a concentration of glucose sufficient for their proliferation. The first aspect dominated under small values of D_n, leading to a much smaller increase in tumor growth speed, than $2\sqrt{BD_n}$. This effect was more pronounced under high values of P, since in this case, tumors had more cells, and therefore enlarged sizes of such protrusions that led to accelerated degradation of capillaries. According to the simulations, a tumor with motile cells may even grow slower than the same tumor with immobilized cells. The second aspect dominated under high values of D_n, resulting in the fact that the tumor growth speed was higher than $V_{conv} + V_{dif}$ under $D_n = 0.33$ for all the considered values of P.

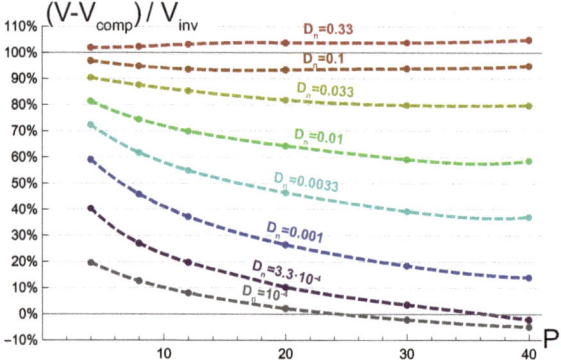

Figure 6. The dots denote the ratios of the differences of tumor growth speed V and V_{comp}, which is the growth speed of the same tumor under zero cell motility, to V_{inv}, which is the growth speed of the same tumor under the neglect of convective flows. All values were obtained numerically in the simulations of Equation (1) with space step $h = 0.01$ and time step $\tau = 10^{-4}$ under designated values of angiogenesis parameter P and tumor cell motility D_n and the values of other parameters taken from the basic set. The dashed lines are interpolations of data points.

4. Discussion

The goal of this paper was to study, by means of mathematical modeling, the effect of mono-antiangiogenic therapy (AAT) on monoclonal tumors, which have different types of growth. The analytical estimations performed allowed suggesting that AAT may provide a several times decrease of growth speed for compact tumors, but the decrease of growth speed for invasive tumors should be much less significant. Thus, the tumors, which initially had an invasive phenotype, should possess intrinsic resistance to AAT, while compactly growing tumors should be more susceptible to it. This conclusion corresponds well to the preclinical and clinical data [21,23]. To the best of my knowledge, this effect was not reproduced before via a continuous model of tumor growth, while this ability of the model should be crucial for the investigation of AAT.

The presented model was built on the fundamental principle that the diffusion and consumption of nutrients limit the growth of a solid tumor. This principle underlines a multitude of existing continuous models that account for spatio-temporal interactions between tumor cells and nutrients (see, e.g., [48] for a review). The main distinguishing feature of the presented model was the simultaneous accounting of the intrinsic motility of malignant cells and the convective motion, arising due to their proliferation, which allowed considering two types of tumor growth, as well as their combination. Despite the relative simplicity that this approach has under the assumption of the spherical symmetry of a tumor, it is relatively rarely used, and usually one of the two types of tumor growth is ignored in continuous models. Some of the few exceptions were the work [49], which provided a theoretical justification for the phenomenon of the dominance of a metastatically active population in an avascular heterogeneous tumor, and the work [50], which explained from a theoretical point of view the effect of tumor cells' migration within a multicellular tumor spheroid. The works [51,52], which also utilized such an approach, were devoted to the investigation of the fingering instability of the avascular tumor surface in the two-dimensional case. This instability could be regarded as a way for a tumor to counteract the diffusional limitations of nutrient inflow. While this phenomenon may affect the tumor growth speed quantitatively, the main qualitative results of the presented work, obtained under the assumption of radial symmetry, could hardly be affected in higher dimensions.

The presented model accounts for glucose as the main nutrient, since it is an indispensable substrate for cell division [28], while most of the relevant models consider tumor cell proliferation to be dependent on oxygen [48]. While such a difference may not be of major importance to the model presented herein, as well as to the models of avascular tumor growth, it should be noted that the approach for modeling the dynamics of oxygen with implicit accounting of tumor vasculature should differ from that for glucose due to the specific features of their blood transport, transvascular transport, and tumor metabolism, which were discussed in the previous paper by our research group [34].

The account of angiogenesis, utilized herein, is rather schematic; however, it allows considering the two major effects of AAT on a qualitative level, allowing performing analytical estimations that suit the purposes of this work. Under more straightforward consideration of angiogenesis, i.e., the introduction of a separate variable for VEGF, which would affect the dynamics of capillaries, one would obtain a distribution of capillaries with non-uniform alterations in their density and permeability [27,34]. However, for every simulation of such an extended model, there will exist a corresponding value of P, which would provide the same increase in tumor growth speed in the simplified model presented herein. Interestingly, the analytical expression for the speed of growth of a compact tumor with no capillaries inside it allowed obtaining a limit of its growth speed under the infinite increase in the number of capillaries and/or in their permeability. However, the estimations with physiologically based values of parameters suggested that achieving more than $\approx 80\%$ of this limit value of speed was unlikely for real tumors. The analytical expression for the speed of growth of invasive tumor indicated that angiogenesis did not influence it at all under the neglect of the convective component of tumor growth, which was confirmed by numerical simulations. However, the simulations, which accounted for both reasons of growth, suggested that in terms of tumor speed reduction, the maximum possible angiogenic effect for highly invasive monoclonal tumors should be around 10–15%.

Importantly, the neglect of convective motion would have led to the misleading conclusion of the indifference of tumor growth speed to angiogenesis for low-invasive tumors as well, since it was not affected by the exact value of tumor cell motility. It should be noted that the accounting of the convective component of tumor growth should be also crucial for modeling of other types of antitumor therapy, since the decrease in the number of tumor cells, caused by any treatment, in reaction-diffusion models should lead to an underestimated decrease of tumor growth speed.

The numerical simulations performed highlighted the non-additive character of the two types of tumor growth. Namely, the addition of a small enough motility to initially immobilized tumor cells should lead to a notably smaller increase in tumor growth speed that could be expected from the analytical estimations of the speed of tumor invasion. Interestingly, the simulations suggested that a low-invasive tumor may even grow slower than the same tumor with immobilized cells. Under sufficiently high cell motility, on the contrary, the two types of tumor growth produced a synergistic effect.

Supplementary Materials: The following are available online at http://www.mdpi.com/2227-7390/8/5/760/s1.

Funding: The publication has been prepared with the support of the "RUDN University Program 5-100".

Conflicts of Interest: The author declares no conflict of interest.

Abbreviations

The following abbreviations are used in this manuscript:

AAT   antiangiogenic therapy
VEGF   vascular endothelial growth factor

Appendix A. Analytical Estimation of Compact Tumor Growth Speed

The distribution of glucose in Figure 1 is searched as a piecewise function that must be continuously differentiable for the continuity of glucose concentration and its flow:

$$g(z) = g_I(z), \ z < -L;$$
$$g(z) = g_{II}(z), \ -L < z < 0;$$
$$g(z) = g_{III}(z), \ 0 < z;$$

$$g(-L) = g_I(-L) = g_{II}(-L) = g_{cr}, \ g'_I(-L) = g'_{II}(-L);$$
$$g(0) = g_{II}(0) = g_{III}(0), \ g'_{II}(0) = g'_{III}(0).$$

Within the necrotic core, the equation for glucose turns into:

$$D_g g''_I + BL g'_I = 0,$$

the general solution of which is:

$$g_I = C^I_1 + C^I_2 e^{-BLz/D_g},$$

for which the boundary conditions allow only for:

$$g_I = g_{cr}.$$

Within the proliferating rim, the equation for glucose is converted into:

$$-Q + D_g g''_{II} + BL g'_{II} = 0,$$

the general solution of which is:

$$g_{II} = C^{II}_1 + C^{II}_2 e^{-BLz/D_g} + \frac{Qz}{BL}.$$

Stitching of g_I and g_{II} at $z = -L$ yields:

$$g_I(-L) = g_{II}(-L): \ g_{cr} = C^{II}_1 + C^{II}_2 e^{BL^2/D_g} - \frac{Q}{B};$$
$$g'_I(-L) = g'_{II}(-L): \ 0 = -\frac{BL}{D_g} C^{II}_2 e^{BL^2/D_g} + \frac{Q}{BL}.$$

Therefore:

$$C^{II}_1 = g_{cr} + \frac{Q}{B}[1 - \frac{D_g}{BL^2}]; \ C^{II}_2 = \frac{QD_g}{B^2L^2} e^{-BL^2/D_g}.$$

For normal tissue, the glucose equation transforms into:

$$P[1 - g_{III}] + D_g g''_{III} + BL g'_{III} = 0,$$

the general solution of which is:

$$g_{III} = 1 + C^{III}_1 \exp(\frac{-BL - \sqrt{B^2L^2 + 4D_g P}}{2D_g} z) + C^{III}_2 \exp(\frac{-BL + \sqrt{B^2L^2 + 4D_g P}}{2D_g} z).$$

Since the glucose concentration is limited at $z \to +\infty$, then $C_2^{III} = 0$. Stitching of g_{II} and g_{III} at $z = 0$ results in:

$$g_{II}(0) = g_{III}(0): \quad C_1^{II} + C_2^{II} = 1 + C_1^{III};$$

$$g'_{II}(0) = g'_{III}(0): \quad -\frac{BL}{D_g}C_2^{II} + \frac{Q}{BL} = \frac{-BL - \sqrt{B^2L^2 + 4D_g P}}{2D_g}C_1^{III}.$$

Substituting the values of C_1^{II} and C_2^{II} into these equations allows obtaining the following implicit expression for L:

$$1 - \frac{Q}{B} + \frac{QD_g}{B^2L^2}[1 - e^{-BL^2/D_g}][1 - \frac{2}{1 + \sqrt{1 + 4D_g P/(B^2L^2)}}] = g_{cr}. \tag{A1}$$

References

1. Sarker, M.S.R.; Pokojovy, M.; Kim, S. On the Performance of Variable Selection and Classification via Rank-Based Classifier. *Mathematics* **2019**, *7*, 457. [CrossRef]
2. Geng, Y.; Dalhaimer, P.; Cai, S.; Tsai, R.; Tewari, M.; Minko, T.; Discher, D.E. Shape effects of filaments versus spherical particles in flow and drug delivery. *Nat. Nanotechnol.* **2007**, *2*, 249. [CrossRef] [PubMed]
3. Antico, M.; Prinsen, P.; Cellini, F.; Fracassi, A.; Isola, A.A.; Cobben, D.; Fontanarosa, D. Real-time adaptive planning method for radiotherapy treatment delivery for prostate cancer patients, based on a library of plans accounting for possible anatomy configuration changes. *PLoS ONE* **2019**, *14*, e0213002. [CrossRef] [PubMed]
4. Boucher, Y.; Baxter, L.; Jain, R. Interstitial pressure gradients in tissue-isolated and subcutaneous tumors: Implications for therapy. *Cancer Res.* **1990**, *50*, 4478–4484. [PubMed]
5. Gatenby, R.; Gawlinski, E.; Gmitro, A.; Kaylor, B.; Gillies, R. Acid-mediated tumor invasion: A multidisciplinary study. *Cancer Res.* **2006**, *66*, 5216–5223. [CrossRef]
6. Citron, M.; Berry, D.; Cirrincione, C.; Hudis, C.; Winer, E.; Gradishar, W.; Davidson, N.; Martino, S.; Livingston, R.; Ingle, J.; et al. Randomized trial of dose-dense versus conventionally scheduled and sequential versus concurrent combination chemotherapy as postoperative adjuvant treatment of node-positive primary breast cancer: first report of Intergroup Trial C9741/Cancer and Leukemia Group B Trial 9741. *J. Clin. Oncol.* **2003**, *21*, 1431–1439.
7. Chmielecki, J.; Foo, J.; Oxnard, G.; Hutchinson, K.; Ohashi, K.; Somwar, R.; Wang, L.; Amato, K.; Arcila, M.; Sos, M.; et al. Optimization of dosing for EGFR-mutant non-small cell lung cancer with evolutionary cancer modeling. *Sci. Transl. Med.* **2011**, *3*, 90ra59. [CrossRef]
8. Hanahan, D.; Weinberg, R.A. The hallmarks of cancer. *Cell* **2000**, *100*, 57–70. [CrossRef]
9. Laird, A.K. Dynamics of tumour growth. *Br. J. Cancer* **1964**, *18*, 490. [CrossRef]
10. Burton, A.C. Rate of growth of solid tumours as a problem of diffusion. *Growth* **1966**, *30*, 157–176.
11. Rockne, R.; Alvord, E.; Rockhill, J.; Swanson, K. A mathematical model for brain tumor response to radiation therapy. *J. Math. Biol.* **2009**, *58*, 561. [CrossRef] [PubMed]
12. Alfonso, J.; Köhn-Luque, A.; Stylianopoulos, T.; Feuerhake, F.; Deutsch, A.; Hatzikirou, H. Why one-size-fits-all vaso-modulatory interventions fail to control glioma invasion: In silico insights. *Sci. Rep.* **2016**, *6*, 37283. [CrossRef] [PubMed]
13. Ward, J.P.; King, J. Mathematical modeling of avascular-tumour growth. *Math. Med. Biol. J. IMA* **1997**, *14*, 39–69. [CrossRef]
14. Byrne, H.; Drasdo, D. Individual-based and continuum models of growing cell populations: A comparison. *J. Math. Biol.* **2009**, *58*, 657. [CrossRef]
15. Hahnfeldt, P.; Panigrahy, D.; Folkman, J.; Hlatky, L. Tumor development under angiogenic signaling: A dynamical theory of tumor growth, treatment response, and postvascular dormancy. *Cancer Res.* **1999**, *59*, 4770–4775.
16. Glick, A.; Mastroberardino, A. An Optimal Control Approach for the Treatment of Solid Tumors with Angiogenesis Inhibitors. *Mathematics* **2017**, *5*, 49. [CrossRef]

17. Macklin, P.; McDougall, S.; Anderson, A.; Chaplain, M.; Cristini, V.; Lowengrub, J. Multiscale modeling and nonlinear simulation of vascular tumour growth. *J. Math. Biol.* **2009**, *58*, 765–798. [CrossRef]
18. Kuznetsov, M.; Gorodnova, N.; Simakov, S.; Kolobov, A. Multiscale modeling of angiogenic tumor growth, progression, and therapy. *Biophysics* **2016**, *61*, 1042–1051. [CrossRef]
19. Vasudev, N.S.; Reynolds, A.R. Anti-angiogenic therapy for cancer: Current progress, unresolved questions and future directions. *Angiogenesis* **2014**, *17*, 471–494. [CrossRef]
20. Goel, S.; Wong, A.H.K.; Jain, R.K. Vascular normalization as a therapeutic strategy for malignant and nonmalignant disease. *Cold Spring Harb. Perspect. Med.* **2012**, *2*, a006486. [CrossRef]
21. Ebos, J.M.; Kerbel, R.S. Antiangiogenic therapy: Impact on invasion, disease progression, and metastasis. *Nat. Rev. Clin. Oncol.* **2011**, *8*, 210. [CrossRef] [PubMed]
22. Ebos, J.M.; Mastri, M.; Lee, C.R.; Tracz, A.; Hudson, J.M.; Attwood, K.; Cruz-Munoz, W.R.; Jedeszko, C.; Burns, P.; Kerbel, R.S. Neoadjuvant antiangiogenic therapy reveals contrasts in primary and metastatic tumor efficacy. *EMBO Mol. Med.* **2014**, *6*, 1561–1576. [CrossRef] [PubMed]
23. Bergers, G.; Hanahan, D. Modes of resistance to anti-angiogenic therapy. *Nat. Rev. Cancer* **2008**, *8*, 592. [CrossRef] [PubMed]
24. Pàez-Ribes, M.; Allen, E.; Hudock, J.; Takeda, T.; Okuyama, H.; Viñals, F.; Inoue, M.; Bergers, G.; Hanahan, D.; Casanovas, O. Antiangiogenic therapy elicits malignant progression of tumors to increased local invasion and distant metastasis. *Cancer Cell* **2009**, *15*, 220–231. [CrossRef]
25. Kolobov, A.; Kuznetsov, M. Investigation of the effects of angiogenesis on tumor growth using a mathematical model. *Biophysics* **2015**, *60*, 449–456. [CrossRef]
26. Kuznetsov, M.B.; Kolobov, A.V. Mathematical investigation of antiangiogenic monotherapy effect on heterogeneous tumor progression. *Comput. Res. Model.* **2017**, *9*, 487–501. [CrossRef]
27. Kuznetsov, M.B.; Gubernov, V.V.; Kolobov, A.V. Analysis of anticancer efficiency of combined fractionated radiotherapy and antiangiogenic therapy via mathematical modeling. *Russ. J. Numer. Anal. Math. Model.* **2018**, *33*, 225–242. [CrossRef]
28. Patra, K.C.; Hay, N. The pentose phosphate pathway and cancer. *Trends Biochem. Sci.* **2014**, *39*, 347–354. [CrossRef]
29. Levick, J.R. *An Introduction to Cardiovascular Physiology*; Butterworth-Heinemann: Oxford, UK, 2013.
30. Araujo, R.; McElwain, D. New insights into vascular collapse and growth dynamics in solid tumors. *J. Theor. Biol.* **2004**, *228*, 335–346. [CrossRef]
31. Holash, J.; Maisonpierre, P.; Compton, D.; Boland, P.; Alexander, C.; Zagzag, D.; Yancopoulos, G.; Wiegand, S. Vessel cooption, regression, and growth in tumors mediated by angiopoietins and VEGF. *Science* **1999**, *284*, 1994–1998. [CrossRef]
32. Swanson, K.R.; Rockne, R.C.; Claridge, J.; Chaplain, M.A.; Alvord, E.C.; Anderson, A.R. Quantifying the role of angiogenesis in malignant progression of gliomas: In silico modeling integrates imaging and histology. *Cancer Res.* **2011**, *71*, 7366–7375. [CrossRef] [PubMed]
33. Szomolay, B.; Eubank, T.D.; Roberts, R.D.; Marsh, C.B.; Friedman, A. Modeling the inhibition of breast cancer growth by GM-CSF. *J. Theor. Biol.* **2012**, *303*, 141–151. [CrossRef] [PubMed]
34. Kuznetsov, M.; Kolobov, A. Transient alleviation of tumor hypoxia during first days of antiangiogenic therapy as a result of therapy-induced alterations in nutrient supply and tumor metabolism—Analysis by mathematical modeling. *J. Theor. Biol.* **2018**, *451*, 86–100. [CrossRef] [PubMed]
35. Maeda, H.; Wu, J.; Sawa, T.; Matsumura, Y.; Hori, K. Tumor vascular permeability and the EPR effect in macromolecular therapeutics: A review. *J. Control. Release* **2000**, *65*, 271–284. [CrossRef]
36. Fu, B.M.; Shen, S. Structural mechanisms of acute VEGF effect on microvessel permeability. *Am. J. Physiol.-Heart Circ. Physiol.* **2003**, *284*, H2124–H2135. [CrossRef] [PubMed]
37. Abdollahi, A.; Lipson, K.E.; Sckell, A.; Zieher, H.; Klenke, F.; Poerschke, D.; Roth, A.; Han, X.; Krix, M.; Bischof, M.; et al. Combined therapy with direct and indirect angiogenesis inhibition results in enhanced antiangiogenic and antitumor effects. *Cancer Res.* **2003**, *63*, 8890–8898. [PubMed]
38. Dings, R.P.; Loren, M.; Heun, H.; McNiel, E.; Griffioen, A.W.; Mayo, K.H.; Griffin, R.J. Scheduling of radiation with angiogenesis inhibitors anginex and Avastin improves therapeutic outcome via vessel normalization. *Clin. Cancer Res.* **2007**, *13*, 3395–3402. [CrossRef]
39. Freyer, J.; Sutherland, R. A reduction in the in situ rates of oxygen and glucose consumption of cells in EMT6/Ro spheroids during growth. *J. Cell. Physiol.* **1985**, *124*, 516–524. [CrossRef]

40. Swanson, K.; Alvord, E., Jr.; Murray, J. A quantitative model for differential motility of gliomas in grey and white matter. *Cell Proliferat.* **2000**, *33*, 317–329. [CrossRef]
41. Stamatelos, S.; Kim, E.; Pathak, A.; Popel, A. A bioimage informatics based reconstruction of breast tumor microvasculature with computational blood flow predictions. *Microvasc. Res.* **2014**, *91*, 8–21. [CrossRef]
42. Tuchin, V.; Bashkatov, A.; Genina, E.; Sinichkin, Y.; Lakodina, N. In vivo investigation of the immersion-liquid-induced human skin clearing dynamics. *Tech. Phys. Lett.* **2001**, *27*, 489–490. [CrossRef]
43. Dickson, P.V.; Hamner, J.B.; Sims, T.L.; Fraga, C.H.; Ng, C.Y.; Rajasekeran, S.; Hagedorn, N.L.; McCarville, M.B.; Stewart, C.F.; Davidoff, A.M. Bevacizumab-induced transient remodeling of the vasculature in neuroblastoma xenografts results in improved delivery and efficacy of systemically administered chemotherapy. *Clin. Cancer Res.* **2007**, *13*, 3942–3950. [CrossRef] [PubMed]
44. Boris, J.P.; Book, D.L. Flux-corrected transport. I. SHASTA, a fluid transport algorithm that works. *J. Comput. Phys.* **1973**, *11*, 38–69. [CrossRef]
45. Press, W.H.; Teukolsky, S.A.; Vetterling, W.T.; Flannery, B.P. *Numerical Recipes 3rd Edition: The Art of Scientific Computing*; Cambridge University Press: Cambridge, UK, 2007.
46. Tamaskar, I.; Garcia, J.A.; Elson, P.; Wood, L.; Mekhail, T.; Dreicer, R.; Rini, B.I.; Bukowski, R.M. Antitumor effects of sunitinib or sorafenib in patients with metastatic renal cell carcinoma who received prior antiangiogenic therapy. *J. Urol.* **2008**, *179*, 81–86. [CrossRef]
47. Kolmogorov, A.N. Étude de l'équation de la diffusion avec croissance de la quantité de matière et son application à un problème biologique. *Bull. Univ. Moskow Ser. Intern. Sect. A* **1937**, *1*, 1–25.
48. Roose, T.; Chapman, S.J.; Maini, P.K. Mathematical models of avascular tumor growth. *SIAM Rev.* **2007**, *49*, 179–208. [CrossRef]
49. Kolobov, A.; Polezhaev, A.; Solyanik, G. The role of cell motility in metastatic cell dominance phenomenon: Analysis by a mathematical model. *Comput. Math. Methods Med.* **2000**, *3*, 63–77. [CrossRef]
50. Thompson, K.; Byrne, H. Modelling the internalization of labelled cells in tumour spheroids. *Bull. Math. Biol.* **1999**, *61*, 601–623. [CrossRef]
51. Franks, S.; King, J. Interactions between a uniformly proliferating tumour and its surroundings: Uniform material properties. *Math. Med. Biol.* **2003**, *20*, 47–89. [CrossRef]
52. Franks, S.; King, J. Interactions between a uniformly proliferating tumour and its surroundings: Stability analysis for variable material properties. *Int. J. Eng. Sci.* **2009**, *47*, 1182–1192. [CrossRef]

© 2020 by the author. Licensee MDPI, Basel, Switzerland. This article is an open access article distributed under the terms and conditions of the Creative Commons Attribution (CC BY) license (http://creativecommons.org/licenses/by/4.0/).

Article

Optimization of Dose Fractionation for Radiotherapy of a Solid Tumor with Account of Oxygen Effect and Proliferative Heterogeneity

Maxim Kuznetsov [1,2,*] and Andrey Kolobov [1]

[1] Division of Theoretical Physics, P.N. Lebedev Physical Institute of the Russian Academy of Sciences, 53 Leninskiy Prospekt, 119991 Moscow, Russia; scilpi@mail.ru
[2] Nikolsky Mathematical Institute, Peoples' Friendship University of Russia (RUDN University), 6 Miklukho-Maklaya Street, 117198 Moscow, Russia
* Correspondence: kuznetsovmb@mail.ru

Received: 26 June 2020; Accepted: 18 July 2020; Published: 22 July 2020

Abstract: A spatially-distributed continuous mathematical model of solid tumor growth and treatment by fractionated radiotherapy is presented. The model explicitly accounts for three time and space-dependent factors that influence the efficiency of radiotherapy fractionation schemes—tumor cell repopulation, reoxygenation and redistribution of proliferative states. A special algorithm is developed, aimed at finding the fractionation schemes that provide increased tumor cure probability under the constraints of maximum normal tissue damage and maximum fractional dose. The optimization procedure is performed for varied radiosensitivity of tumor cells under the values of model parameters, corresponding to different degrees of tumor malignancy. The resulting optimized schemes consist of two stages. The first stages are aimed to increase the radiosensitivity of the tumor cells, remaining after their end, sparing the caused normal tissue damage. This allows to increase the doses during the second stages and thus take advantage of the obtained increased radiosensitivity. Such method leads to significant expansions in the curative ranges of the values of tumor radiosensitivity parameters. Overall, the results of this study represent the theoretical proof of concept that non-uniform radiotherapy fractionation schemes may be considerably more effective that uniform ones, due to the time and space-dependent effects.

Keywords: mathematical oncology; spatially-distributed modeling; reaction-diffusion-convection equations; computer experiment; gradient descent

MSC: 35K57; 35Q92; 92C05

---

1. Introduction

Approximately half of the patients, diagnosed with cancer, undergo radiotherapy (RT) [1]. The effect of irradiation on cancer and normal cells can be mathematically expressed via classical linear-quadratic model, which is known to fit experimental data well in a wide range of clinical parameters [2]. According to this model, the fraction of cells, which survive after a single radiation dose D, can be estimated as

$$S(D) = e^{-\alpha D - \beta D^2}, \qquad (1)$$

where α and β are radiosensitivity parameters of cells. Usually cancer cells have higher values of linear radiosensitivity parameter α that corresponding normal cells. However, normal tissues as a rule have greater α/β ratio, which restricts the use of high radiation doses [3]. One option to reduce normal tissue damage is to concentrate the radiation dose within the tumor mass. However, such option carries significant risks even for tumors with clear boundaries, due to the leakage radiation [4].

Other option to spare normal tissues is to fractionate the total dose, that is, to divide it into much smaller fractions, administered over a period of several weeks. The efficiency of a fractionation scheme depends on the effects that are widely referred to as the four "R"s of radiotherapy [5]. Two of the effects—reoxygenation and redistribution of cell cycle—indicate that the radiosensitivity of a cell depends on the surrounding concentration of oxygen and on the current stage of its cell cycle. In particular, hypoxic and non-proliferating cells are more radioresistant [6]. The third effect is the repopulation of tumor cells that takes place between the irradiations. The fourth effect is the repair of sublethal damage, which can be neglected, unless the time interval between irradiations is as short as several hours [7].

The most typical RT fractionation schemes consist of fractions of 1.8 to 2.0 Gy, delivered once a day on weekdays within the period of several weeks [3]. However, different fractionation protocols were shown to lead to improvement in tumor cure and patient survival for most of the patients for some of the tumor types [8]. Importantly, in clinical practice optimization of RT fractionation is significantly complicated by a number of factors, including the great variability of tumors, belonging to the same type, and related ethical problems, associated with the fact that alternative protocols may worsen outcome for some of the patients. Of note, in the vast majority of the tested schemes the irradiation doses are distributed equally between the fractions. At that, the varied parameters of the schemes are the number of the fractions, the interval between them and the fractional dose, which are related through the constraint on total normal tissue damage.

Given the practical difficulties, mathematical modeling can be a powerful tool, providing insights into the problem of optimization of RT fractionation. Different approaches exist in this field that have their pros and cons. The use of non-spatially distributed phenomenological models of tumor growth, expressed in ordinary differential equations, often allows to obtain the globally optimal solutions for the considered problems via analytical methods [9–11]. However, these methods become complicated and even unsolvable under introduction of complex non-linear terms, aimed to account for time- and space-dependent effects [12] or for tumor-specific features [13]. In such cases, various heuristic approaches are used, ranging from direct comparison of the schemes [14] to more complex techniques like simulated annealing [13], which can not guarantee the global optimality of the solution. However, such methods can yield significant results, one of which has already been verified in a preclinical study [13].

Crucially, the use of ordinary differential equations can allow for only phenomenological consideration of the reoxygenation and redistribution of the cell cycle—the effects that result are the spatiotemporal variability of tumor cells' radiosensitivity [12,14]. Their explicit consideration is possible in spatially-distributed models, which can be divided in two types—a continuous model, expressed in partial differential equations (PDEs), and discrete models that usually treat every tumor cell as a separate agent but use PDEs for the consideration of dynamics of nutrients and other substances. However, to the best of our knowledge, the existing works on RT fractionation optimization that use continuous spatially-distributed models, only account for homogeneous and constant radiosensivity of tumor cells. That leads to the conclusion of optimality of standard radiation fractionation schemes [15,16]. Reoxygenation and redistribution of cell cycle can be straightforwardly incorporated into agent-based models. However, the complexity of such models and numerical costs of their simulations lead to practical impossibility of solving optimization tasks via them, at least under current level of computer technology. As a rule, in the corresponding works several fractionation schemes are compared directly, moreover, the considered numbers of tumor cells are several orders of magnitude less that the relevant numbers for the human tumors [17–19]. These factors significantly limit the usefulness of such models.

In this work, we present a spatially-distributed continuous mathematical model of solid tumor growth and treatment by fractionated RT that explicitly accounts for tumor cell repopulation, reoxygenation and redistribution of proliferative states. With the use of a specially-developed algorithm, we find the optimized fractionation schemes for varied radiosensitivity of tumor cells under the values of model parameters that correspond to different degrees of tumor malignancy. The resulting schemes lead to significant expansions in the curative ranges of the values of tumor radiosensitivity parameters.

2. Model

2.1. Equations for Tumor Growth

The mathematical model of tumor growth, considered herein, was based on our models, previously used for the investigation of various aspects of tumor growth and treatment [20–23]. There were five variables in this version of the model, which were the functions of space and time coordinates, r and t: the density of tumor cells $n(r,t)$, the density of normal cells $h(r,t)$, the fraction of necrotic tissue $m(r,t)$, the concentration of glucose $g(r,t)$ and the concentration of oxygen $w(r,t)$. The main simplification of this version of the model was the absence of an explicit variable for the capillaries, from which the nutrients flow into the tissue. Instead it was assumed that their local density was proportional to the local fraction of the normal cells in the tissue.

The following set of equations governed the dynamics of the model variables under the absence of radiotherapy:

$$\text{tumor cells:} \quad \frac{\partial n}{\partial t} = \overbrace{Bn\frac{g}{g+g^*}}^{\text{proliferation}} - \overbrace{\epsilon M_h(w)n}^{\text{death}} + \overbrace{D_n \Delta n}^{\text{migration}} - \overbrace{\nabla(In)}^{\text{convection}};$$

$$\text{normal cells:} \quad \frac{\partial h}{\partial t} = -\overbrace{[M_h(w) + Mn]h}^{\text{death}} - \overbrace{\nabla(Ih)}^{\text{convection}};$$

$$\text{necrotic tissue:} \quad \frac{\partial m}{\partial t} = \overbrace{\epsilon M_h(w)n + [M_h(w) + Mn]h}^{\text{cell death}} - \overbrace{\nabla(Im)}^{\text{convection}};$$

$$\text{glucose:} \quad \frac{\partial g}{\partial t} = \overbrace{P_g h[1-g]}^{\text{inflow}} - \overbrace{[Q_n^g n + Q_h^g h]\frac{g}{g+g^*}}^{\text{consumption}} + \overbrace{D_g \Delta g}^{\text{diffusion}}; \qquad (2)$$

$$\text{oxygen:} \quad \frac{\partial w}{\partial t} = \overbrace{P_w h[S(w_A) - S(w)]}^{\text{inflow}}$$

$$- \overbrace{[\{Q_n^\omega \frac{g}{g+g^*} + Q_h^\omega \frac{g^*}{g+g^*}\}n + Q_h^\omega h]\frac{\omega}{\omega+\omega^*}}^{\text{consumption}} + \overbrace{D_w \Delta w}^{\text{diffusion}};$$

where $n + m + h = 1;$

$$M_h(w) = \begin{cases} 0 & \text{if } w \geq w^*; \\ M[\{w/w^*\}^2 - 2\{w/w^*\} + 1] & \text{if } w < w^*; \end{cases}$$

$$S(w) = 1/[1 + \{w_{0.5}/w\}^x].$$

2.1.1. Dynamics of Cells and Necrotic Tissue

The term of tumor cells proliferation implied that the rate of this process was proportional to the rate of glucose consumption by tumor cells. This assumption was made on the basis that glucose is indispensable nutrient for biosynthesis [24]. Glucose is also a crucial energetic nutrient for tumor cells [25]; however, they are known to obtain energy under its depletion via multiple ways in order to increase their survival [26–28]. Therefore, the limiting nutrient in the model for the cell survival was oxygen, tumor cells being more or at least equally resistant to its depletion that the normal cells, that is, $\epsilon \leq 1$. The function of cell death rate $M_h(w)$ was chosen to be smooth, tending to its maximum value under the full absence of oxygen, and equal to zero for the levels of oxygen which exceed the critical value w^*. Normal cells also died in the presence of tumor cells, which was introduced in the model to coarsely reflect two processes: the inability of normal cells to remain viable in acidic tumor microenvironment [29] and the degradation of capillary network inside the tumor [30].

Tumor cells were able to migrate throughout the tissue, which was governed by a diffusion-like term. Directed migration of tumor cells was neglected [31]. The convective terms described the bulk motion of the tissue elements, the velocity field **I** being determined by the dynamics of tumor cells. The drainage of necrotic tissue was neglected. Due to the assumption of the constancy of the total density of tumor cells, necrotic tissue, and normal cells, which was normalized to unity, the following expression for the gradient of **I** was obtained:

$$\nabla \mathbf{I} = Bn \frac{g}{g + g^*} + D_n \Delta n. \tag{3}$$

2.1.2. Dynamics of Nutrients

Model dynamics of both nutrients accounted for the same physiological processes—inflow of nutrients from the capillary network into the tissue, their consumption by tumor and normal cells and their diffusion within the tissue, the latter being much faster for oxygen. The form of the terms for the inflow of oxygen and glucose differed due to significant distinctions in the mechanisms of their blood and transvascular transport. The inflow of glucose is governed primarily by the process of passive diffusion through the pores in the walls of capillaries [32]. Therefore, the rate of the glucose inflow was taken to be proportional to the difference in glucose concentrations in blood and in tissue, and to the capillaries density, which, as it was mentioned above, was assumed to be in linear relationship with the density of normal cells. Glucose concentration in blood was considered to be constant and was normalized to unity.

Oxygen, as lipid-soluble substance with low molecular weight, passes directly through the capillary walls and flows into the tissue at much greater rate that glucose. Oxygen levels in arterial and venous blood differ more than twice even under normal conditions [33], which implies that its blood concentration should not be treated as constant. Moreover, the inflow of oxygen into the tissue is not proportional to the difference of its concentrations in capillary blood and tissue due to the complicated blood transport of oxygen, which molecules are carried in blood in two forms—bound to hemoglobin and unbound from it. Overall, the used term for the oxygen inflow assumed that the rate of this process is proportional to the difference between the fraction of oxygen-saturated hemoglobin under two values of unbound oxygen concentration—the one in arterial blood, which enters the capillaries, and the one in tissue. The function $S(\omega)$ represented oxygen-hemoglobin dissociation curve, the form of which is well-known in physiology [34]. For more detailed explanation of the assumptions, which underpinned the term of oxygen inflow, we refer the readers to our previous work [35].

The nutrient consumption was described via the terms of the classical Michaelis-Menten type. Tumor cells are known to consume nutrients much faster that normal cells, in order to support their proliferative activity, therefore, $Q_n^g > Q_h^g$, $Q_n^\omega > Q_h^\omega$. The rate of oxygen consumption by tumor cells fell down to the rate of oxygen consumption by normal cells under the decrease in tumor cells proliferation rate, caused by the glucose shortage [36].

2.1.3. Numerical Solving of Tumor Growth Model

The set of Equations (2) was solved numerically with assumption of the spherical symmetry of the tumor. The size of the computational region L was adjusted in order to be sufficiently small to spare computational time without imposing noticeable edge effects. The convective flow speed was set to zero at the left boundary, which represented the center of the tumor, where initially $r = 0$. For all the variables, the zero-derivative boundary conditions were used at both edges. The following initial conditions were used for tumor cells, normal cells and necrotic tissue:

$$\begin{cases} n = 0.1, \\ m = 0, \quad for\ r <= 0.1; \\ h = 0.9 \end{cases} \quad \begin{cases} n = 0, \\ m = 0, \quad for\ r > 0.1. \\ h = 1 \end{cases} \tag{4}$$

Equations for glucose and oxygen were considered in the quasi-stationary approximation, due to the fast dynamics of these variables, and were solved using the tridiagonal matrix algorithm. The equation for normal cells was not solved explicitly, the relation $h = 1 - n - m$ being used for determining their density. For tumor cells and necrotic tissue the method of splitting into physical processes was used, that is, kinetic equations, migration equation and convective equations were solved successively during each time step. The implicit Crank-Nicholson scheme was used for the tumor cells migration equation. Convective equations were solved using the flux-corrected transport algorithm with the implicit anti-diffusion stage. The kinetic equations were solved by the simple explicit Euler method, which was justified by the relative smallness of the used time steps, adjusted for the solution of the transport equations. The flux-corrected transport algorithm was introduced in Reference [37], while other classical methods were described in many books (see, for example, Reference [38]). The choice of the time and space steps for different simulations is justified in Appendix A.

The computational code was implemented in C++ and can be found in the Supplementary Materials.

2.2. Equations for Radiotherapy

For the description of radiotherapy (RT), we relied on the classical linear-quadratic model, which was discussed in Section 1. We accounted for two effects that lead to spatiotemporal heterogeneity of the radiosensitivity of tumor cells. The first one is oxygen enhancement effect, which was introduced herein in the form presented in Reference [39], where the corresponding terms were deduced from the experimental data. The second effect is the decrease in radiosensitivity of quiescent cells, which was considered herein with the assumption that it fell along with the decrease in the cells' proliferation rate. We neglected the duration of each irradiation and assumed that the number of cells and the density of necrotic tissue changed in result of it instantaneously, which was realized in a code in a straightforward manner. We did not consider explicitly the death of normal cells due to RT, however, the total damage of the normal tissue was the crucial parameter for the optimization of RT fractionation, which is discussed in Section 2.3. Overall, the equations that expressed the densities of tumor cells and necrotic tissue after a single irradiation with the dose D through their values before it were as follows:

$$n_1|_{postRT} = n_1|_{preRT} \cdot exp(\{-\alpha [OER_\alpha(\omega) \cdot \gamma(g) \cdot D] - \beta [OER_\beta(\omega) \cdot \gamma(g) \cdot D]^2\}),$$
$$m|_{postRT} = m|_{preRT} + [n_1|_{preRT} - n_1|_{postRT}]; \qquad (5)$$

$$\text{where } OER_i(\omega) = \frac{\omega * OER_{i,m} + K_m}{\omega + K_m}, \; i = \alpha, \beta; \; \gamma(g) = \frac{g + kg^*}{g + g^*}.$$

2.3. Optimization of Radiotherapy Fractionation

The task of finding the optimized fractionation of RT was formalized the following way. During all the simulations, the first irradiation was performed at the moment $t = t_0$, when tumor radius, evaluated as the maximum space coordinate, at which $n + m \geq 0.1$, reached 1 cm. We considered the RT schemes, which consisted of 42 doses of radiation, some of which could be zero, which were administered successively at 24 h interval. Therefore, each scheme **D** could be expressed as a vector of non-negative numbers, representing the values of doses, expressed in grays: $\mathbf{D} = (D_i)$, $i \in [1, 42]$. As the standard reference scheme we used the following vector that corresponded to one of the typical courses in clinical practice, consisting of 30 doses of 2 Gy, delivered every weekday over six weeks [3]:

$$\mathbf{D}^{st} = (D_i^{st}), \; D_i^{st} = \begin{cases} 0 \; if \; i = 6 + 7[k-1] \lor i = 7k, \; k \in \mathbb{N}; \\ 2 \; otherwise; \end{cases} \; i \in [1, 42]. \qquad (6)$$

However, we did not impose the condition of the obligatory presence of days-off in the tested fractionation schemes. All the considered schemes had to satisfy two constraints, related to the normal tissue damage:

$$NTD_h(\mathbf{D}) \equiv \sum_{i=1}^{42} [(\alpha/\beta)_h \cdot D_i + D_i^2] \leq NTD_{max} \equiv NTD_h(\mathbf{D}^{st}); \qquad (7)$$

$$D_i < D_{max} \quad \forall i. \qquad (8)$$

The first inequality was analogical to the condition that the biologically effective dose, delivered to the normal tissue, could not exceed its value for the standard fractionation scheme. The second inequality corresponded to the acute reactions and indicated that each dose could not exceed a certain threshold. The aim of the search was to find the scheme $\mathbf{D}$, which would decrease the value of the following objective function $F(\mathbf{D})$ as much as possible:

$$F(\mathbf{D}) = \min_t(lgN(\mathbf{D}, t)), \text{ where } N(\mathbf{D}, t) \equiv \hat{n}\hat{r}^3 \cdot 4\pi \int_0^X n(\mathbf{D}, r, t) r^2 dr, \qquad (9)$$

where $\hat{n}$ and $\hat{r}$ are the normalization parameters of the model for the number of tumor cells and length. Thus, the aim of the optimization procedure was to find the fractionation scheme, leading to the most efficient eradication of tumor cells, which should correspond to the increase in the tumor cure probability (TCP). The following formula was used for the estimation of TCP:

$$TCP(\mathbf{D}) = e^{-\min_t(N(\mathbf{D},t))}, \qquad (10)$$

which can be interpreted as the fraction of eradicated tumors among the identical tumors that have undergone the same treatment, under the assumption that the number of surviving cells throughout these tumors at the end of the treatment follows a Poisson distribution with the average of N.

For the search of the optimized RT fractionation schemes and the optimized values of the objective functions, Algorithm 1 was developed and implemented in the program code. Its repeating steps 2 and 3 represented an adaptation of the classical gradient descent method for the considered problem. Like the classical method, these steps could find only local optimum, and the aim of step 4 was to try to further optimize this result. The meaning of the actions, performed during step 4, is explained in Section 3.3. By themselves, steps 2–4 could yield different results depending on the initial scheme. By testing different initial schemes under various model parameters, we found out that these steps most often produced the best results with the use of the most optimal uniform fractionation scheme as the initial. The search for such scheme was performed during the step 1. The procedures of normalization of the schemes, with the aim of their compliance with the above-mentioned restrictions, were performed iteratively.

Algorithm 1: Optimization of dose fractionation for radiotherapy.

Data: Distribution of variables of tumor growth model, governed by Equations (2), at $t = t_0$.
Run simulation of radiotherapy (RT), governed by Equations (5), with fractional doses $\mathbf{D^{st}}$;
Remember the value of the objective function F^{st};

Step 1. Search for the optimal uniform fractionation scheme: $j = 42$;

while $j = 42 \vee (DUF_1^j \leq D_{max} \wedge j \geq 1)$ **do**

$\quad DUF_i^j = \{0.5[\sqrt{(\alpha/\beta)_h^2 + 4 \cdot NTD_{max}/j} - (\alpha/\beta)_h] \text{ for } i \in [1,j];\ 0 \text{ for } i \in [j+1, 42]\}$;

$\quad$ **if** $DUF_1^j \leq D_{max}$ **then** Simulate RT with the scheme $\mathbf{DUF^j}$, remember FUF^j, $j = j-1$;

Choose $j_{opt}^{UF} : FUF^{j_{opt}^{UF}} = \min\limits_{i=j}^{42} FUF^i$;

if $FUF^{j_{opt}^{UF}} < F^{st}$ **then** $\mathbf{D^{opt}} = \mathbf{DUF^{j_{opt}^{UF}}}$, $F^{opt} = FUF^{j_{opt}^{UF}}$, $j_{max} = j_{opt}^{UF}$;

else $\mathbf{D^{opt}} = \mathbf{D^{st}}$, $F^{opt} = F^{st}$, $j_{max} = 42$;

$Stop = 0$;

while $Stop = 0$ **do**

$\quad k_n = 1$;

$\quad$ **while** $k_n > k_n^{min}$ **do**

$\quad\quad$ **Step 2.** Search for the "gradient":

$\quad\quad$ **for** $j = 1, 2, ..., j_{max}$ **do**

$\quad\quad\quad \mathbf{D^j} = \mathbf{D^{opt}}$; $D_j^j = D_j^{opt} + \delta_S$; Normalize $\mathbf{D^j}$ according to Equation (7), not altering
$\quad\quad\quad$ the doses, equal to D_{max};

$\quad\quad\quad$ Simulate RT with the scheme $\mathbf{D^j}$; Remember F^j;

$\quad\quad$ **Step 3.** Going down the "gradient": $n = 1$; $k_n = 1$; $F_D^0 = F^{opt}$; $Stop_{S3} = 0$;

$\quad\quad$ **while** $(F_D^n < F_D^{n-1} \vee (n = 1 \wedge k_n > k_n^{min})) \wedge Stop_{S3} = 0$ **do**

$\quad\quad\quad D_j^n = \{D_j^{opt} + k_n n \delta_D \cdot [F^{opt} - F^j] \text{ for } j \in [1, j_{max}];\ D_j^n = 0 \text{ for } j \in [j_{max}+1, 42]\}$;

$\quad\quad\quad$ Normalize $\mathbf{D^n}$ according to Equations (7) and (8);

$\quad\quad\quad$ Simulate RT with the scheme $\mathbf{D^n}$; Remember F_D^n;

$\quad\quad\quad$ **if** $F_D^n < F_D^{n-1}$ **then**

$\quad\quad\quad\quad$ **if** $k_n = 1$ **then** $n = n+1$;

$\quad\quad\quad\quad$ **else** $\mathbf{D^{opt}} = \mathbf{D^n}$, $F^{opt} = F_D^n$, $Stop_{S3} = 1$;

$\quad\quad\quad$ **else if** $n > 1$ **then** $\mathbf{D^{opt}} = \mathbf{D^{n-1}}$, $F^{opt} = F_D^{n-1}$;

$\quad\quad\quad$ **else** $k_n = k_n/2$. ;

$\quad$ **Step 4.** Trying to improve the final part of the scheme:

$\quad$ **if** $D_{j_{max}}^{opt} < D_{max}$ **then**

$\quad\quad j_{fin} = \min\limits_{j=1}^{j_{max}} j : D_j^{opt} > k_{fin} \cdot D_{j_{max}}^{opt}$; $NTD_{beg} \equiv \sum\limits_{i=1}^{j_{fin}-1}[(\alpha/\beta)_h \cdot D_i^{opt} + \{D_i^{opt}\}^2]$; $j = j_{max}$;

$\quad\quad$ **while** $j = j_{max} \vee (DSF_j^j \leq D_{max} \wedge j \geq j_{fin})$ **do**

$\quad\quad\quad DSF_i^j = \{D_i^{opt} \text{ for } i \in [1, j_{fin}-1];\ 0 \text{ for } i \in [j+1, 42]\}$;

$\quad\quad\quad 0.5[\sqrt{(\alpha/\beta)_h^2 + 4 \cdot (NTD_{max} - NTD_{beg})/(j - j_{fin} + 1)} - (\alpha/\beta)_h] \text{ for } i \in [j_{fin}, j]\}$;

$\quad\quad\quad$ **if** $DSF_j^j \leq D_{max}$ **then** Simulate RT with $\mathbf{DSF^j}$, remember FSF^j, $j = j-1$;

$\quad\quad$ Choose $j_{opt}^{SF} : FSF^{j_{opt}^{SF}} = \min\limits_{i=j}^{j_{max}} FSF^j$;

$\quad\quad$ **if** $FSF^{j_{opt}^{SF}} < F^{opt}$ **then** $\mathbf{D^{opt}} = \mathbf{DSF^{j_{opt}^{SF}}}$, $F^{opt} = FSF^{j_{opt}^{SF}}$, $j_{max} = j_{opt}^{SF}$;

$\quad\quad$ **else** $Stop = 1$;

$\quad$ **else** $Stop = 1$;

Result: $\mathbf{D^{opt}}$, F^{opt}.

2.4. Parameters

The values of the model parameters, with which the simulations were performed, are listed in Table 1. For some parameters, three values are designated, which belonged to the parameter sets that were assumed to correspond to different levels of tumor malignancy—high, intermediate and low. The way the values of these parameters changed with the increase in tumor malignancy reflected the stronger manifestation of some of the hallmarks of cancer: self-sufficiency in growth signals and insensitivity to anti-growth signals, evading apoptosis, stimulating growth of new vessels and invasion into normal tissue [40]. The dimensionless model values of the parameters were the approximations of their normalized values, which were obtained with the use of the following normalization parameters—$\hat{t} = 1$ h for time, $\hat{r} = 10^{-2}$ cm for length, $\hat{D} = 1$ Gy for radiation dose, $\hat{g} = 1$ mg/mL for glucose concentration, $\hat{\omega} = 1$ mM for oxygen concentration, $\hat{n} = 3 \times 10^8$ cells/mL for maximum density of cells. The latter value was taken from the experimental work on the in vitro growth of multicellular tumor spheroids [36]. The values of the proliferation rate of tumor cells B and their nutrients consumption rates Q_n^g and Q_n^ω were also estimated according to the data of this work with the assumption that these values should be proportionally diminished during the growth of a relevant tumor in tissue. Furthermore, it was assumed that B, Q_n^g and Q_n^ω should proportionally increase with tumor malignancy.

Table 1. Model parameters. Different values are designated for: HM—high malignant tumor, IM—intermediate malignant tumor, LM—low malignant tumor.

Parameter	Description	Model Value	Based on
Cells:			
B	tumor cells' proliferation rate	HM: 0.01 IM: 0.005 LM: 0.0025	[36] + see the text
M	normal cells' death rate parameter	0.01	[41]
ϵ	ratio of death rates of tumor and normal cells due to the lack of oxygen	HM: 0.3 IM: 0.7 LM: 1	[41] + see the text
D_n	tumor cells' motility	HM: 0.01 IM: 0.001 LM: 0	[42] + see the text
Nutrients:			
P_g	glucose inflow parameter	HM: 20 IM: 10 LM: 4	[32]
Q_n^g	tumor cells' glucose consumption rate	HM: 12 IM: 6 LM: 3	[36] + see the text
Q_h^g	normal cells' glucose consumption rate	0.3	[43]
g^*	Michaelis constant for glucose consumption rate	0.007	[44]
D_g	glucose diffusion coefficient	100	[45]
P_ω	oxygen inflow parameter	HM: 50.8 IM: 35.8 LM: 25.4	[46] + see the text
ω_A	oxygen concentration in artery	5.87	[47]
$\omega_{0.5}$	oxygen concentration, at which hemoglobin saturation is 50%	1.56	[48]

Table 1. Cont.

Parameter	Description	Model Value	Based on
χ	Hill coefficient for oxygen-hemoglobin dissociation curve	2.55	[48]
Q_n^ω	tumor cells' oxygen consumption rate	HM: 63 IM: 31.5 LM: 15.75	[36] + see the text
Q_h^ω	normal cells' oxygen consumption rate	8	[43]
ω^*	Michaelis constant for oxygen consumption rate	0.005	[44]
D_ω	oxygen diffusion coefficient	720	[49]
Radiotherapy:			
α	tumor cells' linear radiosensitivity parameter	0.07–0.21	see the text
β	tumor cells' quadratic radiosensitivity parameter	$\alpha/10$	see the text
$OER_{\alpha,m}$	maximum OER_α under aerobic conditions	2.5	[39]
$OER_{\beta,m}$	maximum OER_β under aerobic conditions	3	[39]
K_m	Michaelis constant for oxygen enhancement effect	0.193	[39]
k	ratio of radiosensitivity of quiescent and proliferating tumor cells	HM: 1 IM: 0.5 LM: 0.2	see the text
Optimization procedure:			
$(\alpha/\beta)_h$	alpha-beta ratio for normal tissue	3	[3]
D_{max}	maximum fractional dose	5	[11]
δ_S	the amount of radiation dose added to each fraction during the search for the "gradient"	0.2	see the text
δ_D	the coefficient of fractions alteration during the "descent"	4	see the text
k_n^{min}	minimum parameter of fractions alteration during the "descent"	0.001	see the text
k_{fin}	the threshold coefficient for determining the second stage of the scheme	0.98	see the text

The death rates parameters M and ϵ were assessed based on experimental data on cell behavior under extreme nutrient deprivation [41]. The death rate of tumor cells fell with the increase in tumor malignancy, reflecting the increased tolerance of malignant cells to nutrient deprivation. The coefficient of high malignant tumor cells' motility D_n was an order of magnitude lower than the value which corresponds to high malignant glioma, one of the most invasive types of cancer [42]. The parameter of glucose inflow P_g was estimated as the product of the experimental values of permeability of capillaries to glucose and normal capillary surface area density for human muscle [32]. The values of normal cells' rates of nutrients consumption for human muscle at rest were also used. The oxygen inflow parameter P_ω was adjusted so that the initial oxygen concentration lied within its normal range for human muscle at rest [46]. The values of P_g and P_ω for low malignant tumor were obtained under the assumption of absence of tumor-induced angiogenesis, that is, the formation of new blood vessels. Their increase with tumor malignancy reflected the stimulation of angiogenesis by tumor, and followed different trends due to the following reasoning (see our previous work [35] for details). As was mentioned in Section 2.1.2, the inflow of glucose should be proportional to the density of capillaries and to their permeability, both of these parameters increasing due to the tumor-induced angiogenesis. On the contrary, oxygen inflow in tissue is not affected by the alterations in the number and sizes of capillaries' pores. Moreover, it should be at first approximation proportional to the blood flow rather than to the capillaries' density. Therefore, the inflow of oxygen should increase much slower than the inflow of glucose, in the result of angiogenesis.

Radiosensitivity parameters are known to vary dramatically between various tumor cell lines [50] and, moreover, can significantly differ even for tumors of the same type [51]. Therefore, the parameters of tumor cells' radiosensitivity were varied within a physiologically justified range in order to investigate the potential for optimization of RT fractionation under various response of tumor to the treatment. At that, the alpha-beta ratio for tumor cells was kept constant and equal to 10 [3]. The alpha-beta ratio for normal tissue $(\alpha/\beta)_h$ was close to its experimental value for human muscle [3]. The value of the ratio of radiosensitivity of quiescent and proliferating tumor cells k for low malignant tumor was selected based on the fact that the radiosensitivity of quiescent and proliferating normal cells can differ five or more times [52]. Experimental data suggests that such difference should be leveled with the increase in tumor malignancy [53,54], therefore, the value of k increased up to unity with the increase in tumor malignancy. Other parameters of the optimization algorithm were adjusted manually in order to decrease the computational time as much as possible without noticeable distortion of the solution.

3. Results

3.1. Simulation of Tumor Growth and Radiotherapy

Figure 1 illustrates the single numerical simulation of the intermediate malignant tumor growth and radiotherapy (RT) with standard fractionation scheme $\mathbf{D}^{st}$ (see Equation (6)) and the values of tumor cells' radiosensitivity $\alpha = 10\beta = 0.1$. In the same way as it happened under other parameter values as well, after initial phase of exponential growth the living tumor cells concentrated at the tumor rim, closer to the source of nutrients, which in this model was considered to be proportional to the density of normal cells $h = 1 - n - m$. Due to the active consumption of glucose and oxygen by tumor cells, the proliferation rate of deeper located cells declined, and more deeper located cells began to die. Therefore, most of the volume of sufficiently large tumors was occupied by necrotic tissue, as Figure 1a demonstrates. Figure 1b corresponds to the day of the first irradiation. The radiosensitivity of tumor cells increased from the center of the tumor to its rim, where the most actively proliferating cells were situated and the concentration of oxygen was the highest throughout the tumor. During the first irradiation, approximately 79% of tumor cells survived in the nutrient-depleted regions, while only 12% of tumor cells survived in the outer layers of the tumor rim. The death of tumor cells due to this and following irradiations led to the gradual rise of the levels of nutrients that, in its turn, resulted in increase of radiosensitivity of tumor cells. Figure 1c demonstrates the 10th day of RT, by which eight irradiations were performed, the number of tumor cells decreased by 20 times and the levels of glucose and oxygen in the tumor center reached correspondingly 32% and 74% of their values in the normal tissue. Therefore, the effective radiosensivity of tumor cells sufficiently increased and became almost constant throughout the tumor. During the next irradiation, ≈10.8% of tumor cells died in the outer layers of the tumor rim, and ≈11.5%—in the deeply located regions.

Figure 1d shows the dynamics of the total number of tumor cells $N(t)$ and the number of proliferating tumor cells $N_p(t)$, estimated by the following formulas:

$$N(t) \equiv \hat{n}\hat{r}^3 \cdot 4\pi \int_0^X n(r,t)r^2 dr;$$
$$N_p(t) \equiv \hat{n}\hat{r}^3 \cdot 4\pi \int_0^X n(r,t)\frac{g(r,t)}{g(r,t)+g^*}r^2 dr. \quad (11)$$

In the considered simulation, the total number of tumor cells decreased during the RT course from ≈0.43 billion to ≈18 cells, at that the fraction of proliferating cells $N_p(t)/N(t)$ increased from ≈26% to ≈99.3%. Due to the nature of the model, tumor regrowth always happened after treatment, even if the total number of tumor cells that remained after RT, was less than one. Figure 1e shows the dynamics of the average oxygen pressure within the viable tumor rim, expressed in mmHg and estimated as

$$\langle pO_2 \rangle = 17.024 \cdot \Omega(t), \text{ where } \Omega(t) \equiv [\hat{r}\hat{r}^3 \cdot 4\pi \int_0^L w(r,t)n(r,t)r^2 dr]/N(t); \quad (12)$$

along with the dynamics of the oxygen pressure at the tumor center, that is, at the point $r = 0$. The former quantity cannot be straightforwardly measured in experiment, however, it can give a better estimation of the efficiency of the first irradiations. The latter quantity was up to five orders of magnitude smaller during the free tumor growth that $\langle pO_2 \rangle$, but their values became almost equal during RT.

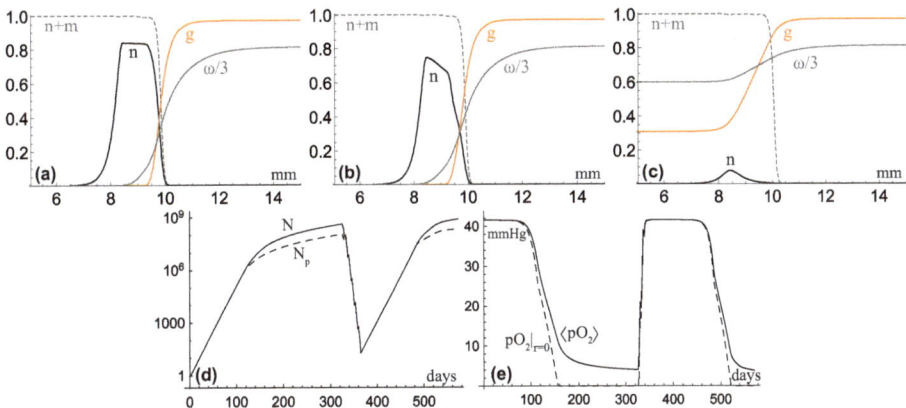

Figure 1. (a–c) The distributions of the variables, obtained in the numerical simulations of tumor growth with radiotherapy, administered by the standard scheme under the values of the model parameters, corresponding to intermediate malignant tumor and values of tumor cells' radiosensitivity $\alpha = 10\beta = 0.1$ on the 325th day, 326th day (the day when the first irradiation was performed) and 335th day of tumor growth. (d) The total number of tumor cells (solid line) and the estimated number of proliferating tumor cells (dashed line) during the same simulation. (e) The average oxygen pressure within the viable tumor rim (solid line) and the oxygen pressure in the tumor center (dashed line) during the same simulation.

3.2. Optimization of Radiotherapy Fractionation

Figure 2 shows the numerically obtained values of tumor control probability (TCP), estimated by Equation (10), for standard and optimized fractionation schemes for different types of tumor. Some of the optimized schemes are also shown, which were obtained under the designated values of tumor radiosensitivity $\alpha = 10\beta$. The solid lines interpolate the graphs of TCPs by the functions of the following form:

$$TCP(\alpha) = 0.5[1 + \tanh(\gamma\{\alpha - \alpha_{cr}\})], \quad (13)$$

where γ and α_{cr} are the fitting parameters. Let us introduce the following quantity as a high-level estimate of the effectiveness of RT fractionation optimization for different tumor types:

$$\Delta \alpha = \alpha_{cr}^{st} - \alpha_{cr}^{opt},$$

where α_{cr}^{st} and α_{cr}^{opt} are the fitting parameters in Equation (13) for standard and optimized RT fractionation schemes of the considered tumor type. Roughly speaking, this quantity denotes the increase in the curative range of the values of tumor radiosensitivity parameters due to the RT fractionation optimization.

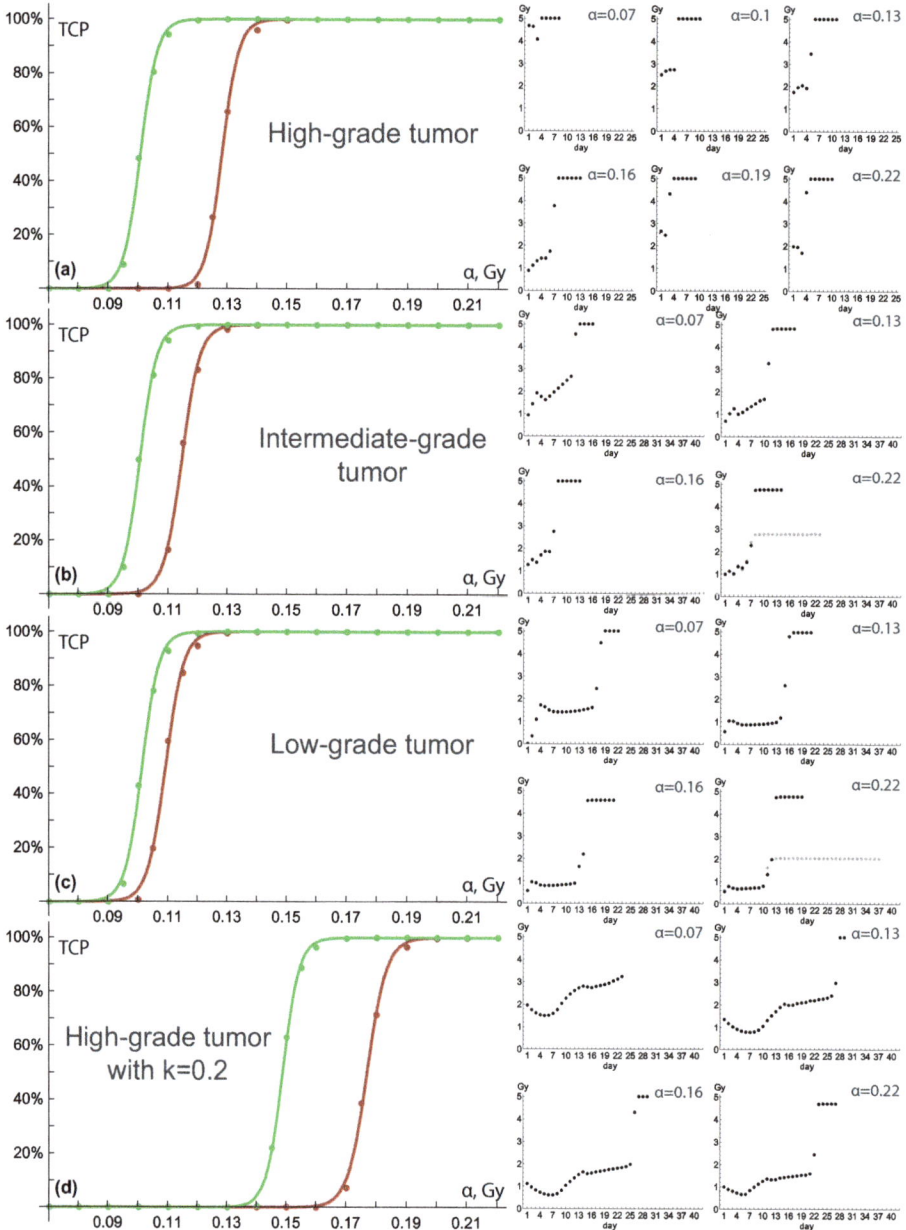

Figure 2. The tumor cure probability for fractionated radiotherapy of (**a**) high malignant tumor (**b**) intermediate malignant tumor (**c**) low malignant tumor and (**d**) high malignant tumor with decreased radiosensitivity of quiescent cells, under varied tumor radiosensitivity parameters $\alpha = 10\beta$ for standard fractionation scheme (red dots) and schemes, found by optimization procedures with Algorithm 1 (green dots). The solid lines are interpolations of data points. Inlets show the optimized schemes under designated values of tumor radiosensitivity. Some of the schemes, obtained without the use of step 4 of the algorithm, are denoted by gray dots.

Figure 2a–c correspond to high malignant (HM), intermediate malignant (IM) and low malignant (LM) tumors, the values of model parameters for which are listed in Table 1. The proliferation rates of these tumors' cells related as 4:2:1. This fact by itself stimulated increase in tumor cell number with the tumor malignancy. Other factors that contributed to the same effect were the decrease in tumor cells' death rate and the enhancement of nutrient inflow. However, at the moment, when these tumors reached the radii of 1 cm—at which the beginning of RT took place—their numbers of cells did not differ drastically and were equal to ≈0.51 billion, ≈0.43 billion and ≈0.43 billion correspondingly. This was due to the fact that tumor cells' rates of nutrients consumption also increased with the tumor malignancy, significantly reducing the pools of proliferating and alive cells. The presence of tumor cell motility in the cases of IM and HM tumors also led to a slight decrease in their numbers of tumor cells. However, it should be noted that in general case the variation of this parameter may have ambiguous effect on the total amount of tumor cells, which was discussed in our previous work [55].

The most prominent feature of all of the optimized RT fractionation schemes, obtained for these tumors for all values of α, is the fact that they could be clearly divided into two stages. The first stages were comprised of non-equal doses, which were noticeably less that the maximum fractional dose of D_{max} = 5 Gy. In every case they lasted until the following two quantities became smaller than one by no more than several percent: the first quantity was the ratio of the oxygen level inside the tumor to its value for the normal tissue $\Omega(t)/\omega(L,t)$ (see Equation (12)); the second quantity was the fraction of proliferating tumor cells $N_p(t)/N(t)$ (see Equation (11)). Thus, to the end of the first stages tumor cells radiosensitivity became close to its maximum level throughout all the pool of tumor cells. The aim of the second stages was to get advantage of the increased radiosensitivity of tumor cells. Therefore they represented a uniform sequence of doses, which were equal or close to the maximum fractional dose. The aim of the first stages was, therefore, to reach close to maximum sensitivity of tumor cells, reducing both the effective dose, delivered to normal tissues during the first stages (see Equation (7)), and their duration. The first aspect is crucial for the opportunity of increasing the number of more efficient irradiations during the second stage. The decrease of duration of the first stages, as well as of the whole courses, is crucial due to the process of tumor cells repopulation—that is, the shorter the treatment, the more effective it should be under the same amount of eradicated cells, since fewer acts of cell division should take place during its course.

For every considered value of α, the optimized RT schemes became longer with the decrease of tumor malignancy. This was due to the fact that quiescent tumor cells were more radioresistant in IM and especially LM tumors, because of the smaller value of the parameter k, which was equal to 1, 0.5 and 0.2 for HM, IM and LM tumors. Consequently, with the decrease of tumor malignancy it took longer time for radiation to eradicate enough tumor cells for the necessary increase in the level of glucose that would convert the remaining tumor cells into proliferative state, thus increasing their radiosensitivity. Therefore, the durations of the optimized RT courses were the shortest for HM tumors, which allowed to significantly decrease the influence of cell repopulation on the outcome of the treatments. Overall, the efficiency of RT fractionation optimization increased with tumor malignancy: the values of $\Delta \alpha$ for LM, IM and HM tumors were ≈0.008, ≈0.014 and ≈0.028 correspondingly. Quite surprisingly, the three interpolated functions of optimized TCPs turned out to be very close to each other. For every value of α the optimized treatments for each of the three tumor types yielded very close TCP values, differing by no more than 7%, and the corresponding values of α_{cr} for these tumors differed by less than 1%. On contrary, the efficiency of standard RT schemes significantly declined with the increase in tumor malignancy for every value of α. Of note, this happened despite the fact of greater radiosensitivity of HM tumor cells under glucose deficiency, and was due to their increased repopulation rate under normal level of glucose. It should be noted, however, that such qualitative outcome might change under different parameter values.

Since the presence of a radioresistant population within a malignant tumor may be of significant practical interest [56], we performed an analogical set of simulations for the fourth set of parameters, which corresponded to HM tumor with the only modification of decreased radiosensitivity of quiescent cells $k = 0.2$. The corresponding results are shown in Figure 2d. Expectedly, the standard fractionation schemes were the less effective for this case among the four considered parameter sets under any value of α. The length of the first stages of the schemes were significantly increased. Moreover, for the lowest values of α there were no second stages in the optimized schemes, since such weak therapies could not eradicate enough tumor cells for sufficient increase in the level of nutrients. The second stages became well-pronounced at $\alpha = 0.1$ and their lengths increased with the increase of α, as it happened under another parameter sets as well. However, the optimization procedures led to the value of $\Delta\alpha \approx 0.028$, close to its value for HM tumor with $k = 1$.

3.3. Efficiency of the Optimization Algorithm

As was already mentioned, Algorithm 1 can provide at best the local optimal schemes. It should be noted that for some of the parameter values slightly more effective schemes that the ones, shown in Figure 2, were obtained by manual manipulations or by replacing the scheme, produced by step 1, with other initial schemes for the following steps. Of especial importance is the fact that these manipulations did not lead to any noticeable change of the TCP graphs. However, such cases, in which further optimizations can be provided, are worth being noticed. The simplest case is removing of initial zero fractions produced by the algorithm, like in the depicted optimized scheme for LM tumor with $\alpha = 0.07$. Less trivial case took place when the search for the optimal uniform fractionation scheme, performed during step 1, yielded a bimodal distribution. In such case, setting another initial scheme for the following steps could optimize the result. For example, the optimized scheme for HM tumor with $\alpha = 0.19$ is obtained with the initial uniform scheme with 9 irradiations of ≈ 4.47 Gy, produced by step 1. Its replacement by the uniform scheme with 13 irradiations of ≈ 3.53 Gy, which by itself results in 30% greater minimum number of tumor cells that the previous uniform scheme, allowed to produce an optimized scheme, resulting in halved minimum number of tumor cells, compared to the depicted optimized scheme. It did not lead to a noticeable change of TCP, since it was already close to 100%, however, it should be noted that such manipulation might turn out to be important under some other used parameter values.

Another possibility for the slight optimization of some of the schemes is the increase of several doses in the second stage of the scheme, if they are less than D_{max}, by the expense of another doses. Such manipulation was not included in the algorithm for simplicity, since it was checked to provide only very slight improvements. However, the closeness of the doses of the second stages to the maximum fractional dose was by itself significant. For IM and LM tumors with sufficiently high values of α the first three steps of the algorithm by themselves yielded the schemes with doses of the second stage, equal to each other, but significantly less than D_{max}. Further slight variations of the fractionation, performed during steps 2 and 3, were ineffective. This result corresponds well to the findings, described previously in other studies, which considered homogeneous and constant radiosensitivity of tumor cells [15,16]. The aim of step 4 was to redistribute the fractions of the second stage, keeping them equal, which always resulted in close to maximum values of single doses. Two of the schemes, produced for IM and LM tumors with $\alpha = 0.22$ without the use of step 4, are shown in Figure 2 via gray dots. The improvements, introduced by step 4, allowed to decrease the minimum number of tumor cells more than threefold in both cases.

Of note, under neglect of the constraint on the maximum fractional dose (see Equation (8)), Algorithm 1 generally produced the optimized schemes, consisting of longer first stages with smaller doses and second stages, consisting of a few strong irradiations, which most frequently were a couple of doses close to 10 Gy. Furthermore, easing the constraint on normal tissue damage (see Equation (7)) by increasing the value of $(\alpha/\beta)_h$ resulted in more and more shorter optimized schemes with greater doses. These results are not surprising; moreover, Equation (1) immediately suggests that under absence of any time-dependent effects and any constraints increasing a single fraction should always be more efficient that its fractionation. Thus, in agreement with the clinical concepts, discussed in Section 1, the presented model indicates that dose fractionation is dictated by the constraints on normal tissue damage and by the alterations of tumor cells' radioresistance (which involves both reoxygenation and redistribution of proliferative states), while cell repopulation restricts the efficiency of the fractionation scheme.

4. Discussion

In this work, we presented a spatially-distributed continuous mathematical model of solid tumor growth and treatment by fractionated radiotherapy (RT). The model explicitly accounted for three factors that influence the efficiency of RT fractionation schemes—tumor cell repopulation, reoxygenation and redistribution of proliferative states. The main goal of this study was the search for optimized fractionation protocols that would increase the tumor cure probability under the constraints of maximum normal tissue damage and maximum fractional dose. For this goal, a special algorithm was developed. Its first step compared different uniformly fractionated RT schemes. By itself it showed that the length of an optimal treatment should grow with the decrease of the relative radiosensitivity of non-proliferating tumor cells, which occupied the main part of tumor at the beginning of RT. The next two steps of the algorithm represented an adaptation of the classical gradient descent method. They suggested dividing all the fractionation schemes in two stages. Fractionation during the first stages followed different trends, but their aim was always to spare the doses for the second stage, at that eradicating enough tumor cells for the levels of nutrients to increase close to their normal levels. Such approach brought the radiosensitivity of the remaining tumor cells close to the maximum level. If this goal was possible to achieve, the second stages began, which consisted of large equal doses. Of note, the qualitatively similar recommendation of a dose boost during the final part of the therapy was suggested in a previous theoretical study, which used another constitutive assumptions and considered treatment regimens, in which a patient is treated in several sessions, separated by weeks or months [57]. The aim of the fourth step of the algorithm was to optimize the number of fractions and the doses during the second stage, which previous steps could not do. Optimized RT fractionation schemes did not contain days-off, unlike standard clinical schemes. It should be noted that the current model neglected the change of radiosensitivity during the cell cycle for the proliferating cells, as well as the fact that cells do not die immediately due to irradiation [58]. The introduction of these aspects into the model may somehow alter the appearance of the optimized schemes, found by the used algorithm. However, they would hardly affect the main qualitative findings of this study.

We performed the optimization procedures using the objective function of minimum number of tumor cells during the treatment. Certainly, other objective functions can be incorporated within the introduced algorithm. One of them, which we implemented as well during the study, is the delay in tumor regrowth, which always happened during the simulations (see Section 3.1). The model simulations showed that only a rather moderate increase of it can be achieved for the considered parameter values under the restriction of maximum treatment duration of 6 weeks. However, the results suggested that sufficient tumor growth delays might be obtained for much longer treatments of slowly-proliferating tumors, which is in agreement with other theoretical studies [59,60]. Moreover, the presented model did not account for the drainage of necrotic tissue, which should be crucial for such problem. Furthermore, in this light, a very interesting augmentation of the model may be an introduction of concurrent antiangiogenic therapy, which not only influences the drainage of necrotic tissue [61], but also affects the intratumoral oxygen level in a complicated manner [62]. These factors should influence the outcome of combined radiotherapy and antiangiogenic therapy. Therefore, their consideration should provide insights into the ways of optimization of such treatment. This task lies within the scope of our future plans.

We tried to incorporate in the model the most basic features of malignant tumors, relevant for the considered task, and we varied the model parameters, assuming that some hallmarks of cancer should manifest themselves stronger with the increase in tumor malignancy. Certainly, this was a very general approach, and the results of this work are of purely qualitative nature. In our opinion, an important outcome of this study is the theoretical proof of concept that non-uniform RT fractionation schemes may be significantly more effective that uniform ones, due to the time and space-dependent effects. We hope that the presented algorithm would be useful for further, more specific, tasks. At that, one of the important aspects to be focused on is the consideration of a separate radioresistant population of cancer stem cells. Its determining role in optimization of RT treatment was already noticed in previous studies on mathematical modeling [18,63].

Supplementary Materials: The following are available at http://www.mdpi.com/2227-7390/8/8/1204/s1.

Author Contributions: Conceptualization, A.K.; methodology, A.K. and M.K.; software, M.K.; investigation, M.K.; writing—original draft preparation, M.K.; writing—review and editing, A.K.; visualization, M.K.; supervision, A.K.; funding acquisition, A.K. and M.K. All authors have read and agreed to the published version of the manuscript.

Funding: This work was supported by RFBR according to the research Project no. 19-01-00768. Numerical simulations were prepared with the support of the "RUDN University Program 5-100".

Conflicts of Interest: The authors declare no conflict of interest.

Abbreviations

The following abbreviations are used in this manuscript:

RT   radiotherapy
PDE partial differential equation
OER oxygen enhancement ratio
TCP tumor cure probability
HM   high malignant
IM   intermediate malignant
LM   low malignant

Appendix A. Choice of Discretization

The choice of discretization for the numerical simulations was dictated by two goals. The first goal was sparing computing resources, the importance of which was increased due to the fact that a lot of simulations of different treatment courses should have been preformed for a single optimization task. Namely, the optimization tasks, discussed in Section 3.2, required an average of ≈882 treatment simulations, with their minimum and maximum numbers of 66 and 3560. The second goal was providing sufficient accuracy of the solution. Since the model described biological objects, for which significant variability is natural, there was no need to pursue the degree of accuracy that would be necessary for solving, for example, physical problems, and the main aim was to capture the qualitative behavior of the model properly.

The considered low malignant tumor had zero cell motility, therefore, the only transport term for the tumor cells was the convective term in its case. As it was mentioned in Section 2.1.3, the corresponding equation was solved by the flux-corrected transport algorithm [37]. This algorithm is of indeterminate order, and the crucial condition for its workability is

$$|\mathbf{I}(r,t)\frac{dt}{dr}| < \frac{1}{2} \ \ \forall r \ \forall t,$$

where **I** is the field of the convective flow speed, dt and dr are the time and space steps. The flux-corrected transport algorithm consists of two stages. The first stage solves the convective equation, maintaining the total number of tumor cells and the non-negativity of their density profile. However, it introduces erroneous diffusion, reduction of which is the aim of the second stage. The simulations for the intermediate malignant and high malignant tumors included Crank-Nicholson method for the solution of tumor cell migration equation. Its main deficiency is the introduction of the spurious oscillations, which amplitude increases with the increase of $D_n dt/dr^2$, where D_n is tumor cell motility. Obviously, the decrease of the time step under constant space step should increase the accuracy of the Crank-Nicholson method. However, such action would play an ambiguous role on the accuracy of the flux-corrected transport algorithm, since more frequent calculations within a single unit of time should amplify its total erroneous diffusion.

Figure A1 shows the dependence of the low malignant, intermediate malignant and high malignant tumor growth speeds V, expressed in mm/week (which implies its multiplication by the factor of 16.8), on the time and space steps, equal to each other. The tumor growth speed was estimated via the method, described in our previous work [55], as the asymptotic value of the rate of change of the tumor radius, which was evaluated as the maximum space coordinate, at which $n + m \geq 0.1$. The low malignant tumor growth speed changed non-monotonically with the refinement of discretization, reflecting the increase of the erroneous diffusion, introduced by the flux-corrected transport algorithm. This effect was not pronounced for the intermediate malignant tumor. This tumor had non-zero cell motility, therefore, the migration equation was solved in its case, and its accuracy fell under such refinement of discretization. Nevertheless, the tumor cell motility was sufficiently low in this case for this effect to remain unnoticed on this graph. However, this effect was strongly pronounced for the high malignant tumor with sufficiently high cell motility. Moreover, the utilized numerical approach turned out to be unstable for high malignant tumor under $dr = dt = 0.01$. Overall, based on these graphs and on the amount of computing resources, spent under different discretizations, time and space steps $dr = dt = 0.1$ were chosen to be used for all the simulations.

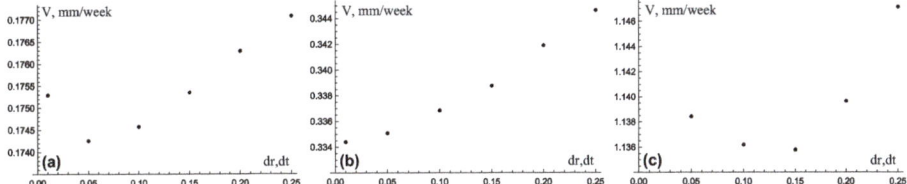

Figure A1. The values of the tumor growth speeds V, obtained numerically in the simulations of Equation (2) under designated space steps dr and time steps dt and the values of the model parameters, corresponding to: (**a**) low malignant tumor, (**b**) intermediate malignant tumor and (**c**) high malignant tumor.

References

1. Moding, E.J.; Kastan, M.B.; Kirsch, D.G. Strategies for optimizing the response of cancer and normal tissues to radiation. *Nat. Rev. Drug Discov.* **2013**, *12*, 526–542. [CrossRef]
2. Fowler, J.F. The linear-quadratic formula and progress in fractionated radiotherapy. *Br. J. Radiol.* **1989**, *62*, 679–694. [CrossRef]
3. Joiner, M.C.; Van der Kogel, A. *Basic Clinical Radiobiology*, 4th ed.; CRC Press: Boca Raton, FL, USA, 2009.
4. Hall, E.J. Intensity-modulated radiation therapy, protons, and the risk of second cancers. *Int. J. Radiat. Oncol. Biol. Phys.* **2006**, *65*, 1–7. [CrossRef]
5. Withers, H.R. The four R's of radiotherapy. *Adv. Radiat. Biol.* **1975**, *5*, 241–271.
6. Kodym, E.; Kodym, R.; Reis, A.E.; Habib, A.A.; Story, M.D.; Saha, D. The small-molecule CDK inhibitor, SNS-032, enhances cellular radiosensitivity in quiescent and hypoxic non-small cell lung cancer cells. *Lung Cancer* **2009**, *66*, 37–47. [CrossRef] [PubMed]
7. Powathil, G.; Kohandel, M.; Sivaloganathan, S.; Oza, A.; Milosevic, M. Mathematical modeling of brain tumors: Effects of radiotherapy and chemotherapy. *Phys. Med. Biol.* **2007**, *52*, 3291. [CrossRef] [PubMed]
8. Ahmed, K.A.; Correa, C.R.; Dilling, T.J.; Rao, N.G.; Shridhar, R.; Trotti, A.M.; Wilder, R.B.; Caudell, J.J. Altered fractionation schedules in radiation treatment: A review. *Semin. Oncol.* **2014**, *41*, 730–750. [CrossRef]
9. Ergun, A.; Camphausen, K.; Wein, L.M. Optimal scheduling of radiotherapy and angiogenic inhibitors. *Bull. Math. Biol.* **2003**, *65*, 407–424. [CrossRef]
10. Bertuzzi, A.; Bruni, C.; Papa, F.; Sinisgalli, C. Optimal solution for a cancer radiotherapy problem. *J. Math. Biol.* **2013**, *66*, 311–349. [CrossRef] [PubMed]
11. Badri, H.; Pitter, K.; Holland, E.C.; Michor, F.; Leder, K. Optimization of radiation dosing schedules for proneural glioblastoma. *J. Math. Biol.* **2016**, *72*, 1301–1336. [CrossRef]

12. Yang, Y.; Xing, L. Optimization of radiotherapy dose-time fractionation with consideration of tumor specific biology. *Med. Phys.* **2005**, *32*, 3666–3677. [CrossRef] [PubMed]
13. Leder, K.; Pitter, K.; LaPlant, Q.; Hambardzumyan, D.; Ross, B.D.; Chan, T.A.; Holland, E.C.; Michor, F. Mathematical modeling of PDGF-driven glioblastoma reveals optimized radiation dosing schedules. *Cell* **2014**, *156*, 603–616. [CrossRef] [PubMed]
14. Prokopiou, S.; Moros, E.G.; Poleszczuk, J.; Caudell, J.; Torres-Roca, J.F.; Latifi, K.; Lee, J.K.; Myerson, R.; Harrison, L.B.; Enderling, H. A proliferation saturation index to predict radiation response and personalize radiotherapy fractionation. *Radiat. Oncol.* **2015**, *10*, 159. [CrossRef] [PubMed]
15. Galochkina, T.; Bratus, A.; Pérez-García, V.M. Optimal radiation fractionation for low-grade gliomas: Insights from a mathematical model. *Math. Biosci.* **2015**, *267*, 1–9. [CrossRef] [PubMed]
16. Fernández-Cara, E.; Prouvée, L. Optimal control of mathematical models for the radiotherapy of gliomas: The scalar case. *Comput. Appl. Math.* **2018**, *37*, 745–762. [CrossRef]
17. Stamatakos, G.; Antipas, V.; Uzunoglu, N.; Dale, R. A four-dimensional computer simulation model of the in vivo response to radiotherapy of glioblastoma multiforme: Studies on the effect of clonogenic cell density. *Br. J. Radiol.* **2006**, *79*, 389–400. [CrossRef]
18. Enderling, H.; Park, D.; Hlatky, L.; Hahnfeldt, P. The importance of spatial distribution of stemness and proliferation state in determining tumor radioresponse. *Math. Model. Nat. Phenom.* **2009**, *4*, 117–133. [CrossRef]
19. Jalalimanesh, A.; Haghighi, H.S.; Ahmadi, A.; Soltani, M. Simulation-based optimization of radiotherapy: Agent-based modeling and reinforcement learning. *Math. Comput. Simul.* **2017**, *133*, 235–248. [CrossRef]
20. Kolobov, A.; Kuznetsov, M. Investigation of the effects of angiogenesis on tumor growth using a mathematical model. *Biophysics* **2015**, *60*, 449–456. [CrossRef]
21. Kuznetsov, M.B.; Kolobov, A.V. Mathematical investigation of antiangiogenic monotherapy effect on heterogeneous tumor progression. *Comput. Res. Model.* **2017**, *9*, 487–501. [CrossRef]
22. Kuznetsov, M.B.; Gubernov, V.V.; Kolobov, A.V. Analysis of anticancer efficiency of combined fractionated radiotherapy and antiangiogenic therapy via mathematical modelling. *Russ. J. Numer. Anal. Math. Model.* **2018**, *33*, 225–242. [CrossRef]
23. Kuznetsov, M.; Kolobov, A. Investigation of solid tumor progression with account of proliferation/migration dichotomy via Darwinian mathematical model. *J. Math. Biol.* **2020**, *80*, 601–626. [CrossRef] [PubMed]
24. Patra, K.C.; Hay, N. The pentose phosphate pathway and cancer. *Trends Biochem. Sci.* **2014**, *39*, 347–354. [CrossRef] [PubMed]
25. Vander Heiden, M.G.; Cantley, L.C.; Thompson, C.B. Understanding the Warburg effect: The metabolic requirements of cell proliferation. *Science* **2009**, *324*, 1029–1033. [CrossRef] [PubMed]
26. Phan, L.M.; Yeung, S.C.J.; Lee, M.H. Cancer metabolic reprogramming: Importance, main features, and potentials for precise targeted anti-cancer therapies. *Cancer Biol. Med.* **2014**, *11*, 1.
27. Sonveaux, P.; Végran, F.; Schroeder, T.; Wergin, M.C.; Verrax, J.; Rabbani, Z.N.; De Saedeleer, C.J.; Kennedy, K.M.; Diepart, C.; Jordan, B.F.; et al. Targeting lactate-fueled respiration selectively kills hypoxic tumor cells in mice. *J. Clin. Investig.* **2008**, *118*, 3930–3942. [CrossRef]
28. Mathew, R.; Karantza-Wadsworth, V.; White, E. Role of autophagy in cancer. *Nat. Rev. Cancer* **2007**, *7*, 961–967. [CrossRef]
29. Gatenby, R.A.; Gawlinski, E.T. A reaction-diffusion model of cancer invasion. *Cancer Res.* **1996**, *56*, 5745–5753.
30. Araujo, R.; McElwain, D. New insights into vascular collapse and growth dynamics in solid tumors. *J. Theor. Biol.* **2004**, *228*, 335–346. [CrossRef]
31. Raman, F.; Scribner, E.; Saut, O.; Wenger, C.; Colin, T.; Fathallah-Shaykh, H.M. Computational trials: Unraveling motility phenotypes, progression patterns, and treatment options for glioblastoma multiforme. *PLoS ONE* **2016**, *11*, e0146617. [CrossRef]
32. Levick, J.R. *An Introduction to Cardiovascular Physiology*; Butterworth-Heinemann: Oxford, UK, 2013.
33. Carreau, A.; Hafny-Rahbi, B.E.; Matejuk, A.; Grillon, C.; Kieda, C. Why is the partial oxygen pressure of human tissues a crucial parameter? Small molecules and hypoxia. *J. Cell. Mol. Med.* **2011**, *15*, 1239–1253. [CrossRef] [PubMed]
34. Brown, W.; Hill, A. The oxygen-dissociation curve of blood, and its thermodynamic basis. *Proc. R. Soc. Lond. Ser. B Contain. Pap. Biol. Character* **1923**, *94*, 297–334.

35. Kuznetsov, M.B.; Kolobov, A.V. Transient alleviation of tumor hypoxia during first days of antiangiogenic therapy as a result of therapy-induced alterations in nutrient supply and tumor metabolism—Analysis by mathematical modeling. *J. Theor. Biol.* **2018**, *451*, 86–100. [CrossRef] [PubMed]
36. Freyer, J.; Sutherland, R. A reduction in the in situ rates of oxygen and glucose consumption of cells in EMT6/Ro spheroids during growth. *J. Cell. Physiol.* **1985**, *124*, 516–524. [CrossRef]
37. Boris, J.P.; Book, D.L. Flux-corrected transport. I. SHASTA, a fluid transport algorithm that works. *J. Comput. Phys.* **1973**, *11*, 38–69. [CrossRef]
38. Press, W.H.; Teukolsky, S.A.; Vetterling, W.T.; Flannery, B.P. *Numerical Recipes 3rd Edition: The Art of Scientific Computing*; Cambridge University Press: Cambridge, UK, 2007.
39. Wouters, B.G.; Brown, J.M. Cells at intermediate oxygen levels can be more important than the "hypoxic fraction" in determining tumor response to fractionated radiotherapy. *Radiat. Res.* **1997**, *147*, 541–550. [CrossRef]
40. Hanahan, D.; Weinberg, R.A. The hallmarks of cancer. *Cell* **2000**, *100*, 57–70. [CrossRef]
41. Izuishi, K.; Kato, K.; Ogura, T.; Kinoshita, T.; Esumi, H. Remarkable tolerance of tumor cells to nutrient deprivation: Possible new biochemical target for cancer therapy. *Cancer Res.* **2000**, *60*, 6201–6207.
42. Swanson, K.R.; Alvord, E.C., Jr.; Murray, J. A quantitative model for differential motility of gliomas in grey and white matter. *Cell Prolif.* **2000**, *33*, 317–329. [CrossRef]
43. Baker, P.G.; Mottram, R. Metabolism of exercising and resting human skeletal muscle, in the post-prandial and fasting states. *Clin. Sci. Mol. Med.* **1973**, *44*, 479–491. [CrossRef]
44. Casciari, J.; Sotirchos, S.; Sutherland, R. Mathematical modelling of microenvironment and growth in EMT6/Ro multicellular tumour spheroids. *Cell Prolif.* **1992**, *25*, 1–22. [CrossRef]
45. Tuchin, V.; Bashkatov, A.; Genina, E.; Sinichkin, Y.P.; Lakodina, N. In vivo investigation of the immersion-liquid-induced human skin clearing dynamics. *Tech. Phys. Lett.* **2001**, *27*, 489–490. [CrossRef]
46. Richardson, R.S.; Duteil, S.; Wary, C.; Wray, D.W.; Hoff, J.; Carlier, P.G. Human skeletal muscle intracellular oxygenation: The impact of ambient oxygen availability. *J. Physiol.* **2006**, *571*, 415–424. [CrossRef] [PubMed]
47. Pittman, R.N. Regulation of tissue oxygenation. In *Colloquium Series on Integrated Systems Physiology: From Molecule to Function*; Morgan & Claypool Life Sciences: San Rafael, CA, USA, 2011; Volume 3, pp. 1–100.
48. Clerbaux, T.; Gustin, P.; Detry, B.; Cao, M.; Frans, A. Comparative study of the oxyhaemoglobin dissociation curve of four mammals: Man, dog, horse and cattle. *Comp. Biochem. Physiol. Part A Physiol.* **1993**, *106*, 687–694. [CrossRef]
49. Androjna, C.; Gatica, J.E.; Belovich, J.M.; Derwin, K.A. Oxygen diffusion through natural extracellular matrices: Implications for estimating "critical thickness" values in tendon tissue engineering. *Tissue Eng. Part A* **2008**, *14*, 559–569. [CrossRef] [PubMed]
50. Skarsgard, L.D.; Skwarchuk, M.W.; Wouters, B.G.; Durand, R.E. Substructure in the radiation survival response at low dose in cells of human tumor cell lines. *Radiat. Res.* **1996**, *146*, 388–398. [CrossRef]
51. Rockne, R.; Rockhill, J.; Mrugala, M.; Spence, A.; Kalet, I.; Hendrickson, K.; Lai, A.; Cloughesy, T.; Alvord, E., Jr.; Swanson, K. Predicting the efficacy of radiotherapy in individual glioblastoma patients in vivo: A mathematical modeling approach. *Phys. Med. Biol.* **2010**, *55*, 3271. [CrossRef]
52. Zhao, L.; Wu, D.; Mi, D.; Sun, Y. Radiosensitivity and relative biological effectiveness based on a generalized target model. *J. Radiat. Res.* **2017**, *58*, 8–16. [CrossRef]
53. Grönvik, C.; Capala, J.; Carlsson, J. The non-variation in radiosensitivity of different proliferative states of human glioma cells. *Anticancer Res.* **1996**, *16*, 25–31.
54. Onozato, Y.; Kaida, A.; Harada, H.; Miura, M. Radiosensitivity of quiescent and proliferating cells grown as multicellular tumor spheroids. *Cancer Sci.* **2017**, *108*, 704–712. [CrossRef]
55. Kuznetsov, M. Mathematical Modeling Shows That the Response of a Solid Tumor to Antiangiogenic Therapy Depends on the Type of Growth. *Mathematics* **2020**, *8*, 760. [CrossRef]
56. Pajonk, F.; Vlashi, E.; McBride, W.H. Radiation resistance of cancer stem cells: The 4 R's of radiobiology revisited. *Stem Cells* **2010**, *28*, 639–648. [CrossRef] [PubMed]
57. Unkelbach, J.; Craft, D.; Hong, T.; Papp, D.; Ramakrishnan, J.; Salari, E.; Wolfgang, J.; Bortfeld, T. Exploiting tumor shrinkage through temporal optimization of radiotherapy. *Phys. Med. Biol.* **2014**, *59*, 3059. [CrossRef] [PubMed]

58. Lorenzo, G.; Pérez-García, V.M.; Mariño, A.; Pérez-Romasanta, L.A.; Reali, A.; Gomez, H. Mechanistic modelling of prostate-specific antigen dynamics shows potential for personalized prediction of radiation therapy outcome. *J. R. Soc. Interface* **2019**, *16*, 20190195. [CrossRef] [PubMed]
59. Pérez-García, V.M.; Pérez-Romasanta, L.A. Extreme protraction for low-grade gliomas: Theoretical proof of concept of a novel therapeutical strategy. *Math. Med. Biol. J. IMA* **2016**, *33*, 253–271. [CrossRef] [PubMed]
60. Henares-Molina, A.; Benzekry, S.; Lara, P.C.; García-Rojo, M.; Pérez-García, V.M.; Martínez-González, A. Non-standard radiotherapy fractionations delay the time to malignant transformation of low-grade gliomas. *PLoS ONE* **2017**, *12*, e0178552. [CrossRef]
61. Jain, R.K.; Di Tomaso, E.; Duda, D.G.; Loeffler, J.S.; Sorensen, A.G.; Batchelor, T.T. Angiogenesis in brain tumours. *Nat. Rev. Neurosci.* **2007**, *8*, 610–622. [CrossRef]
62. Dings, R.P.; Loren, M.; Heun, H.; McNiel, E.; Griffioen, A.W.; Mayo, K.H.; Griffin, R.J. Scheduling of radiation with angiogenesis inhibitors anginex and Avastin improves therapeutic outcome via vessel normalization. *Clin. Cancer Res.* **2007**, *13*, 3395–3402. [CrossRef]
63. Alfonso, J.C.L.; Jagiella, N.; Núñez, L.; Herrero, M.A.; Drasdo, D. Estimating dose painting effects in radiotherapy: A mathematical model. *PLoS ONE* **2014**, *9*, e89380.

© 2020 by the authors. Licensee MDPI, Basel, Switzerland. This article is an open access article distributed under the terms and conditions of the Creative Commons Attribution (CC BY) license (http://creativecommons.org/licenses/by/4.0/).

Article

Efficient Methods for Parameter Estimation of Ordinary and Partial Differential Equation Models of Viral Hepatitis Kinetics

Alexander Churkin [1],*, Stephanie Lewkiewicz [2], Vladimir Reinharz [3], Harel Dahari [4] and Danny Barash [5],*

1. Department of Software Engineering, Sami Shamoon College of Engineering, Beer-Sheva 8410501, Israel
2. Department of Mathematics, University of California at Los Angeles, Los Angeles, CA 90095, USA; slewkiewicz@math.ucla.edu
3. Department of Computer Science, Université du Québec à Montréal, Montreal, QC H3C 3P8, Canada; reinharz.vladimir@uqam.ca
4. Program for Experimental and Theoretical Modeling, Division of Hepatology, Department of Medicine, Stritch School of Medicine, Loyola University Medical Center, Maywoood, IL 60153, USA; hdahari@luc.edu
5. Department of Computer Science, Ben-Gurion University, Beer-Sheva 8410501, Israel
* Correspondence: alexach3@sce.ac.il (A.C.); dbarash@cs.bgu.ac.il (D.B.); Tel.: +972-8-647-5281 (A.C.); +972-8-647-2714 (D.B.)

Received: 30 June 2020; Accepted: 24 August 2020; Published: 2 September 2020

Abstract: Parameter estimation in mathematical models that are based on differential equations is known to be of fundamental importance. For sophisticated models such as age-structured models that simulate biological agents, parameter estimation that addresses all cases of data points available presents a formidable challenge and efficiency considerations need to be employed in order for the method to become practical. In the case of age-structured models of viral hepatitis dynamics under antiviral treatment that deal with partial differential equations, a fully numerical parameter estimation method was developed that does not require an analytical approximation of the solution to the multiscale model equations, avoiding the necessity to derive the long-term approximation for each model. However, the method is considerably slow because of precision problems in estimating derivatives with respect to the parameters near their boundary values, making it almost impractical for general use. In order to overcome this limitation, two steps have been taken that significantly reduce the running time by orders of magnitude and thereby lead to a practical method. First, constrained optimization is used, letting the user add constraints relating to the boundary values of each parameter before the method is executed. Second, optimization is performed by derivative-free methods, eliminating the need to evaluate expensive numerical derivative approximations. The newly efficient methods that were developed as a result of the above approach are described for hepatitis C virus kinetic models during antiviral therapy. Illustrations are provided using a user-friendly simulator that incorporates the efficient methods for both the ordinary and partial differential equation models.

Keywords: parameter estimation; constrained optimization; derivative free optimization; multiscale models; differential equations; viral hepatitis

1. Introduction

Chronic viral hepatitis (hepatitis C, hepatitis B, and hepatitis D) is a major public health concern. Approximately 500 million individuals worldwide are living with chronic viral hepatitis; above a million of those who are infected die each year, primarily from cirrhosis or liver cancer resulting from their hepatitis infection [1–3]. Deaths related to chronic hepatitis are as many as those due to human immunodeficiency virus (HIV) infection, tuberculosis, or malaria [4], and are projected to exceed the combined mortality associated with HIV infection, tuberculosis, and malaria by 2040 [5]. Only a small subset of patients are cured with currently available drugs for hepatitis B and hepatitis D. As such, a deeper understanding of hepatitis B and D infection dynamics is needed to enable the development of more curative therapeutics. Despite the significant advances in hepatitis C therapy, it is widely acknowledged that cost remains a major barrier for achieving global elimination. Thus, there still exists a need for affordable therapy with similar high efficacy and with much shorter treatment durations and vaccine development.

Mathematical models have been developed to provide insights into viral hepatitis and host dynamics during infection and the pathogenesis of infection [6–11]. The standard biphasic model for viral infection is a set of three ordinary differential equations (ODEs) with three variables. This ODE model has been used to study hepatitis C virus (HCV), hepatitis B virus (HBV), and hepatitis D virus (HDV) kinetics during antiviral treatment. It contributed to the assessment of antivirals efficacy and to our understanding of their mechanism of action [9,12–18]. The ODE model can be further simplified (termed biphasic mode) by assuming that target cells remain constant during antiviral treatment. The biphasic model has been extensively used for modeling HCV [19–27], HBV [28–31], or HDV [32–35] kinetics during antiviral treatment. Notably, we recently showed in a proof-of-concept pilot study that using the biphasic model in real time (i.e., on treatment) can shorten HCV treatment duration (and cost) with direct-acting antivirals without compromising efficacy or patient safety [36], which confirmed our retrospective biphasic modeling reports in more than 250 patients [37–41].

Modeling efforts using ODEs for understanding the intracellular viral hepatitis genome dynamics have been done in [7,42–46]. Recently, partial differential equation (termed PDE, age-structured or multiscale) models for HCV infection and treatment were developed [47–50]. These PDE models are an extension to the classical biphasic models in which the infected cell is a "black box", producing virions but without any consideration of the intracellular viral RNA replication and degradation within the infected cell [42,43,51]. The multiscale models consider the intracellular viral RNA in an additional equation for the variable (R), with the introduction of age-dependency in addition to time-dependency, making it a PDE model. They are considerably more difficult to solve and to perform parameter estimation on compared to the biphasic model. Unlike the construction of numerical schemes in other applications, for example in the nonlinear diffusion of digital images [52–54] where accuracy can be limited, herein it is advisable to construct a stable and efficient scheme that belongs to the Runge–Kutta family with a higher accuracy than in nonlinear diffusion. Our numerical solution strategy was outlined in [55–57] and herein we continue [57] by providing an efficient parameter estimation method that follows this strategy.

Parameter estimation (or calibration) of multiscale HCV models with HCV kinetic data measured in treated patients is challenging. To overcome this, several strategies have been employed. The first strategy, employed in [48], utilizes an analytical solution named long-term approximation for solving the model equations along with calling the Levenberg–Marquardt [58,59] as a canned method for performing the fitting. The second strategy, employed in [60], transforms the multiscale model to a system of ODEs and, as such, simple parameter estimation methods can be used in the same manner as the biphasic model. The third strategy, employed in [50] that also deals with spatial models of intracellular virus replication, is based on the method of lines and utilizes canned methods for both the numerical solution of the resulting equations (Matlab's *ode45*) and for performing the fitting (Matlab's *fmincon*). While these strategies are adequate for specific cases, they rely on canned methods and are problematic when it comes to the user's capability to access and control them. For these reasons, we have developed our own open source code released free of charge for the benefit of the community that allows the user to make modifications to the model and provides prospects for future development, while ensuring that it is practical in running time and enabling the user to insert constraints for the parameters that need to be estimated. In contrast to these approaches, our strategy does not rely on any canned method but fully implements our own optimization routine, thus making it suitable to other multiscale model equations by modifications inside the routine and an early preparation of the multiscale model equations by taking derivatives with respect to the parameters before the optimization procedure.

The general ideas that have led to [57], including the parameter estimation procedure described in this reference, have been laid out in order to remain self-contained. The motivation of the present work is to develop a tool that can provide similar calibration values in significantly less time. More specifically, the main contribution herein is as follows. Because of precision problems in [57] encountered with Levenberg–Marquardt that caused the parameter estimation procedure to become highly non-efficient, we developed an efficient constrained optimization procedure that is based on damped Gauss–Newton instead such that we avoid problematic use of derivatives, while alternatively offering the possibility to apply Powell's Constrained optimization by linear approximation (COBYLA) [61] for the optimization procedure. In the following sections, we describe the model and the optimization procedure that is used in our HCVMultiscaleFit simluator. Illustrations of simulations using HCVMultiscaleFit are provided and the efficiency and practicality relative to the initial version put forth in [57] are discussed.

2. Methods

2.1. Development of Mathematical Models

2.1.1. The Standard Biphasic Model

The three variables this model keeps track of are the target cells T, in Equation (1), the infected cells I in Equation (2), and the free virus V in Equation (3). The target cells T are produced at constant rate s, die at per capita rate d, and are infected by virus V at constant rate β. The infected cells I increase with the new infections at rate $\beta V(t)T(t)$ and die at constant rate δ. The virus V is produced at rate p by each infected cell and is cleared at constant rate c. The ϵ term denotes the effectiveness of the anti-viral treatment that decreases the production from p to $(1-\epsilon)p$. Formally, the ensemble of ODEs for this model is:

$$\frac{dT(t)}{dt} = s - \beta V(t)T(t) - dT(t) \tag{1}$$

$$\frac{dI(t)}{dt} = \beta V(t)T(t) - \delta I(t) \tag{2}$$

$$\frac{dV(t)}{dt} = (1-\epsilon)pI(t) - cV(t). \tag{3}$$

From the mathematical perspective, the standard biphasic model is relatively much simpler than the multiscale model. Although it is nonlinear, it can be solved analytically when assuming that T is constant (target cells remain constant during antiviral treatment).

2.1.2. The Multiscale HCV Model

A multiscale PDE model for HCV infection and treatment dynamics was introduced in [47–49]. Intracellular HCV RNA plays a biologically significant role during the HCV replication and multiscale models are considering it by additional equations for the RNA that are age-dependent, with the most complete model to date that was recently put forth in [50].

The multiscale model [47–49] can be formulated as follows:

$$\frac{dT(t)}{dt} = s - dT(t) - \beta V(t)T(t) \tag{4}$$

$$\frac{\partial I(a,t)}{\partial t} + \frac{\partial I(a,t)}{\partial a} = -\delta I(a,t) \tag{5}$$

$$\frac{dV(t)}{dt} = (1-\varepsilon_s)\int_0^\infty \rho R(a,t)I(a,t)da - cV(t) \tag{6}$$

$$\frac{\partial R(a,t)}{\partial t} + \frac{\partial R(a,t)}{\partial a} = (1-\varepsilon_\alpha)\alpha e^{-\gamma t} - ((1-\varepsilon_s)\rho + \kappa\mu)R(a,t), \tag{7}$$

with the initial and boundary conditions $T(0) = \bar{T}$, $V(0) = \bar{V}$, $I(0,t) = \beta V(t)T(t)$, $I(a,0) = \bar{I}(a)$, $R(0,t) = 1$, and $R(a,0) = \bar{R}(a)$. The initial condition $R(0,t) = 1$ reflects the assumption that a cell is infected by a single virion and therefore there is only one vRNA in an infected cell at age zero.

The four variables this model keeps track of are the target cells T in Equation (4), the infected cells I in Equation (5), the free virus V in Equation (6), and the intracellular viral RNA R in an infected cell in Equation (7).

The target cells T are produced at constant rate s, and decrease by the number of cells infected by virus in blood V at constant rate β and their death rate d. The infected cells I die at constant rate δ. The quantity of intracellular viral RNA R depends on its production α and its degradation μ and expulsion from the cell ρ. The quantity of free virus V depends on the number of assembled and released virions and their clearance rate c. The parameter γ represents the decay of replication template under therapy. The decrease in viral RNA synthesis is represented by ε_α, the reduction in secretion by ε_s, and the increase in viral degradation by $\kappa \geq 1$.

The parameters that were used in the multiscale model described in [48] are depicted in Table 1. The model forms an example of our parameter estimation calibration method for PDE models developed herein that can easily be extended to include additional parameters.

Table 1. The 12 parameters of the model.

Parameter	Description
s (cells mL^{-1})	Influx rate of new hepatocytes
d (d^{-1})	Target cell loss/death rate constant
β (mL d^{-1} virion^{-1})	Infection rate constant
δ (d^{-1})	HCV-infected cell loss/death rate constant
ρ (d^{-1})	Virion assembly/secretion rate constant
c (d^{-1})	Virion clearance rate constant
α (vRNAd^{-1})	vRNA synthesis rate
μ (d^{-1})	vRNA degradation
κ	Enhancement of intracellular viral RNA degradation
γ (d^{-1})	Loss rate of vRNA replication complexes
ε_s	Treatment vs. secretion/assembly effectiveness
ε_α	Treatment vs. production effectiveness

An important consideration in this model is that the treatment starts after the infection has reached its steady state. The steady states of the different variables are $\bar{R}(a,t)$, $\bar{I}(a,t)$, $\bar{V}$, and $\bar{T}$. The term N represents the total number of virions produced by infected cells.

These values have been previously derived in [48] and can be expressed as follows:

$$\bar{T} = c/\beta N \tag{8}$$

$$\bar{V} = (\beta Ns - dc)/(\beta c) \tag{9}$$

$$\bar{I}(a) = \beta \bar{V}\bar{T}e^{-\delta a} \tag{10}$$

$$\bar{R}(a) = \frac{\alpha}{\rho+\mu} + \left(1 - \frac{\alpha}{\rho+\mu}\right)e^{-(\rho+\mu)(a)} \tag{11}$$

$$N = \frac{\rho(\alpha+\delta)}{\delta(\rho+\mu+\delta)} \tag{12}$$

It has been shown that the equations for $I(a,t)$ and $R(a,t)$ can be solved by the method of characteristics to yield:

$$I(a,t) = \begin{cases} \beta V(t-a)T(t-a)e^{-\delta a} & a < t \\ I(a-t)e^{-\delta t} = \beta \tilde{V}\tilde{T}e^{-\delta(t-a)} = (\beta Ns - dc)/(\beta N)e^{-\delta a} & a > t \end{cases} \quad (13)$$

and

$$R(a,t) = \begin{cases} \frac{(1-\varepsilon_a)\alpha e^{-\gamma t}}{(1-\varepsilon_s)\rho+\kappa\mu-\gamma} + \left(1 - \frac{(1-\varepsilon_a)\alpha e^{-\gamma(t-a)}}{(1-\varepsilon_s)\rho+\kappa\mu-\gamma}\right)e^{-((1-\varepsilon_s)\rho+\kappa\mu)a} & a < t \\ \\ \frac{(1-\varepsilon_a)\alpha e^{-\gamma t}}{(1-\varepsilon_s)\rho+\kappa\mu-\gamma} \\ + \left(\frac{\alpha}{\rho+\mu} + \left(1 - \frac{\alpha}{\rho+\mu}\right)e^{-(\rho+\mu)(a-t)}\right) & a > t \\ - \frac{(1-\varepsilon_a)\alpha}{(1-\varepsilon_s)\rho+\kappa\mu-\gamma} \Big)e^{-((1-\varepsilon_s)\rho+\kappa\mu)t} \end{cases} \quad (14)$$

whereas the equations for $V(a,t)$ and $T(a,t)$ cannot be solved analytically without any approximations. The equations for $V(a,t)$ and $T(a,t)$ when using the short-term and long-term approximations can be found in [48].

2.2. Data Description

Calibration of the model was performed with data from treated patients by [48]. The data points to fit the model and on which the error is computed are only V. We assume that we start at a steady state and begin by computing the steady state given the initial parameters by using Equation (8). While the raw data are not available, we used the freely accessible tool of [62] to retrieve it from the figures directly. A visual example for one patient is available at [57].

In our method, we mostly use the default parameters from [48] that are shown in Table 2. The main difference concerns parameter s. The pre-treatment steady state viral load $\tilde{V}$ in each patient is different. Since $\tilde{V}$ is a necessary value in computing the long-term approximation, it was approximated as the pre-treatment viral load observed per patient. In the full model that we are implementing, we do not directly use $\tilde{V}$. Instead, we have from Equation (9) that $\tilde{V}$ is a function of many parameters, in particular s which is not present in the long term approximation that was outlined in [48]. Inspired by the method of [48], we chose to also fix $\tilde{V}$. The counterpart in our method is that s changes per patient being, by Equation (9), equal to $(\tilde{V}\beta c + dc) / (\beta N)$, where N is from Equation (12).

More details about preparing the system with data from patients and the model parameters are available in [57]. Herein, the methods are different from [57] and are significantly more efficient, but the model parameters and the system preparation are exactly the same.

Table 2. Default parameters that are used herein. Parameter s comes from Equation (9), taking $\tilde{V}$ as the max Virions value.

α	40 d^{-1}	β	5×10^{-8} mL d^{-1}
c	22.3 d^{-1}	δ	0.14 d^{-1}
μ	1 d^{-1}	d	0.01 d^{-1}
ρ	8.18 d^{-1}	s	$(\tilde{V}\beta c + dc) / (\beta N)$ cells/mL

2.3. Solving the Model Equations

In [48], the multiscale model equations were solved by analytical approximations but, as discussed in [56], those analytical approximations have limitations that should be alleviated. The long-term approximation is an underestimate of the PDE model since some infection events are being ignored. Moreover, for each multiscale model, the long-term approximation needs to be derived analytically, which is not a trivial task. Thus, numerical solutions provide an attractive alternative and could be easier to adjust when introducing changes to the model. A more general and comprehensive approach to parameter fitting without relying on analytical approximations would be useful. In addition, although it was shown recently that it is possible to transform the PDE multiscale model to a system of ODEs [60], this transformation problematically introduces some of the boundary conditions, e.g., ζ, as new parameters inside the model equations. A numerical approach to parameter fitting of multiscale models was recently put forth and described in [50], by the use of the method of lines and canned methods that are available in Matlab. Our new numerical approach that originated in [56] and described in [57] in detail does not rely on canned methods, with considerable benefits.

For the numerical solution of the multiscale model equations, properties such as approximation, stability, and convergence were discussed in [56] and numerical robustness was discussed in [55,56]. Future work should expand towards the advanced treatment of properties as covered in [63,64]. Concerning the numerical solution itself, we showed in [56] that the full implementation of the Rosenbrock method is preferable over the use of a canned solver in terms of efficiency and stability. Therefore, the Rosenbrock method has been implemented for the purpose of our parameter fitting method as well. In order to apply the Rosenbrock method, it is simplest to represent the system to be solved as a vector f of two functions:

$$y' = f(t,y) = \left[\frac{dT}{dt}, \frac{dV}{dt}\right]^T = \Big[s - dT - \beta VT, \\ (1-\varepsilon_s)\int_0^\infty \rho R(a,t)I(a,t)da - cV\Big]^T, \tag{15}$$

where y is a vector with the values of $[T,V]^T$ and the transpose symbol can be omitted from now on for brevity. This representation has originated in [56] for convenience with formulating the numerical schemes described in that reference. This function depends on three variables, t, V and T. While V and T are the values at the time point we are evaluating, inside the equation of I, the function $V(t-a)$ and $T(t-a)$ do depend on t directly. In our implementation, when computing the integral, we need to divide into two cases. If $a > t$, we analytically determine the values of $R(a,t)$ and $I(a,t)$ for small time steps a. When $a < t$, the system was previously solved at times $\tau_0, \ldots, \tau_n$. Therefore, we evaluate the integrals at times $a_0 = t - \tau_0, \ldots, a_n = t - \tau_n$, ensuring that the required values of $V(t-a)$ and $T(t-a)$ are already known, following the scheme presented in [56].

The Rosenbrock method additionally requires the Jacobian matrix, denoted by f'. As was shown in [57], the Jacobian can be controlled and, with some proper computational simplifications to avoid singularities that were shown to yield correct results in [57], we can implement the Rosenbrock method convincingly for both solution and parameter estimation of the multiscale models.

2.4. Parameter Estimation

2.4.1. Preliminaries

As outlined in [48], the HCV multiscale model has 12 parameters (Table 1) and the nonlinear differential equations that comprise it are stiff [56]. In addition, the integral term in the equation complicates matters, as described in [56,57]. Parameter fitting is known to be a difficult problem in general and for multiscale models, in particular, one needs to approach it carefully with the use of robust techniques for the optimization, but, at the same time, these techniques can be made highly efficient for practical computations. The novelty in this work is described next.

For efficiency reasons, we revert from the Levenberg–Marquardt method for optimization that was used in albeit different ways in both [48,57] and implement significant improvements. Already in [57], we have noticed that more difficult fitting cases take several hours to perform, and this situation needs to be remedied for a practical use of our simulator. The reason for the lengthy running times was non-trivial and only after a considerable period of time, having tried the simplest numerical method for the solution of the equations (the Euler method instead of the Rosenbrock method) and not noticing a significant time reduction in the parameter estimation calculation, we began to understand that the problem lies in the optimization method being used. We then examined interior point methods for performing constrained optimization instead of the Levenberg–Marquardt method we used in [57] and found out that the Hessian calculations in these interior point methods are problematic, causing precision problems near the parameter boundaries that are the source of running time accumulations. There was definitely a need to avoid the use of derivatives and therefore two alternative approaches were taken. The first was to try a constrained damped Gauss–Newton strategy, which can also be looked at as a simple version of Levenberg–Marquardt without gradient descent, or alternatively Levenberg–Marquardt is a pseudo second-order method with added derivatives to approximate the Hessian and thereby adds complications that should better be avoided. While in general Levenberg–Marquardt is considered more robust than Gauss–Newton, for our constrained application, the simplicity of the damped Gauss–Newton in terms of derivative calculations relative to Levenberg–Marquardt, in which also the Lagrange parameter needs to be calculated at each step, makes the damped Gauss–Newton significantly preferable. The second approach taken was that, while developing our own damped Gauss–Newton method for the constrained application, we also examined a completely derivative-free approach based on COBYLA (Constrained Optimization by Linear Approximation). These two approaches turned out to be complementary to each other as by default the quicker and sometimes somewhat more accurate damped Gauss–Newton can be tried first, but, when it fails, COBYLA can provide a good alternative or it can even be used from the start and all along a research study as the difference in the calculated error that has been minimized is quite small. This contribution allows for reaching an overall procedure for parameter estimation that is practical and by orders of magnitude less demanding in computing time relative to [57], which provides a technical breakthrough from the computational standpoint.

Thus, two newly developed methods have been introduced to perform constrained optimization for this application in an efficient manner: LSF (Least Squares Fitter using Gauss–Newton) with a flowchart shown in Figure 1 and Powell's COBYLA (Constrained Optimization by Linear Approximation) with a pseudocode shown in Algorithm 1. The latter is a derivative-free optimization method that solves the constrained optimization by linear programming. The former is a constrained optimization that performs linearization in the manner described herein.

In both approaches, the objective function to be minimized is described as follows. The objective consists of adjusting the parameters of a model function to best fit a data set. A simple data set consists of n points (data pairs) (x_i, y_i), $i = 1, \ldots, n$, where x_i is an independent variable and y_i is a dependent variable whose value is found by observation. The model function has the form $f(x, p)$, where m adjustable parameters are held in the vector p. The goal is to find the parameter values for the model that best fits the data. The fit of a model to a data point is measured by its residual, defined as the difference between the actual value of the dependent variable and the value predicted by the model:

$$r_i = y_i - f(x_i, p) \tag{16}$$

The least-squares method obtains the optimal parameter values by minimizing the sum of squared residuals:

$$S = \sum_{i=1}^{n}(r_i)^2 = \sum_{i=1}^{n}(y_i - f(x_i, p))^2 \tag{17}$$

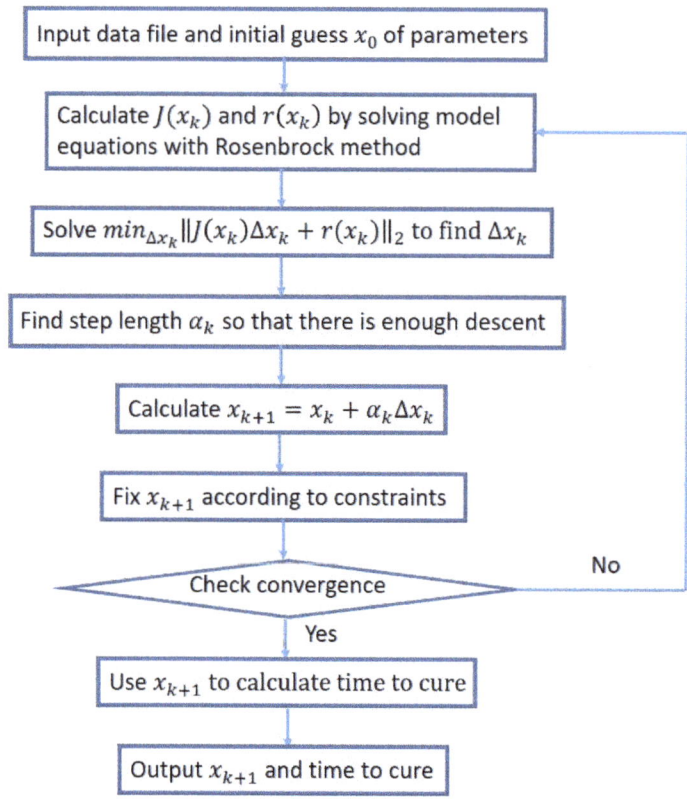

Figure 1. A flowchart of our constrained damped Gauss–Newton method.

2.4.2. Optimization by a Constrained Version of Nonlinear Least Squares (Gauss–Newton Method)

If we assume that $f(x)$ is twice continuously differentiable, then we can utilize Newton's method to solve the system of nonlinear equations:

$$\nabla f(x) = J(x)^T r(x) = 0, \tag{18}$$

which provides local stationary points for $f(x)$, where $r(x)$ is the vector of residuals associated with data points as functions of parameter vector x and J is the Jacobian. Written in terms of derivatives of $r(x)$ and starting from an initial guess x_0, this version of the Newton iteration scheme takes the form:

$$x_{k+1} = x_k - \left[J(x_k)^T J(x_k) + S(x_k)\right]^{-1} J(x_k)^T r(x_k), \quad k = 0, 1, 2, \ldots \tag{19}$$

where $S(x_k)$ denotes the matrix:

$$S(x_k) = \sum_{i=1}^{m} r_i(x_k) \nabla^2 r_i(x_k). \tag{20}$$

In order to obtain the correction $\Delta x_k = x_{k+1} - x_k$, a linear system is solved by a direct or iterative method:

$$\left[J(x_k)^T J(x_k) + S(x_k)\right] \Delta x_k = -J(x_k)^T r(x_k). \tag{21}$$

For our application, we use the Gauss–Newton method, which neglects the second term $S(x_k)$ of the Hessian, and the computation of the step Δx_k involves the solution of the linear system:

$$\left[J(x_k)^T J(x_k)\right] \Delta x_k = -J(x_k)^T r(x_k), \tag{22}$$

and $x_{k+1} = x_k + \Delta x_k$.

In our application, we use the following steps that comprise a damped Gauss–Newton strategy:

- Start with an initial guess x_0 and iterate for $k = 0, 1, 2, \ldots$
- Solve $min_{\Delta x_k} \|J(x_k)\Delta x_k + r(x_k)\|_2$ to compute the correction Δx_k.
- Choose a step length α_k so that there is enough descent.
- Calculate the new iterate $x_{k+1} = x_k + \alpha_k \Delta x_k$.
- Check for convergence.

We choose α_k to be 1.0 at the beginning of the algorithm and decrease it by dividing by two each time the error increases relative to the previous iteration. More sophisticated damping strategies such as the Armijo–Goldstein step-length principle are not suitable in our application because of constraints violation that is described next. We also extend the Damped Gauss–Newton method to be constrained in the following way: in the case that one of the parameters, during the convergence process, exceeds its bounds (constraints provided in the GUI by the user), the algorithm assigns to this parameter its corresponding bound value instead. For this reason, we cannot apply the Armijo–Goldstein condition and need to revert to a simple damping strategy that is suitable with our constrained modification of a damped Gauss–Newton method. A flowchart of our method is shown in Figure 1.

2.4.3. Optimization by Derivative-Free Methods (COBYLA Method)

Should the Gauss–Newton method fail to carry out the optimization of Equation (18), a helpful alternative is the COBYLA algorithm, a derivative-free simplex method originally developed by Powell [65]. The parameters in the algorithm have mathematical meanings that are outside the scope of the model employed, as will be shown herein, and a pseudocode of the algorithm is available in Algorithm 1. In general, a simplex method seeks to minimize an objective function using *simplices*, where *simplex* refers to the convex hull of a set of $n + 1$ points in n-dimensional space. Such an algorithm begins by evaluating the objective function at the vertices of an initial simplex, and then strategically adjusting the simplex so that the objective function attains generally smaller values at the vertices of the new simplex than it did at those of the previous simplex. At each iteration, a vertex of the simplex may be altered, or the simplex itself rescaled, so as to guide the simplex into a region at which the objective function is minimized. When sufficient accuracy is attained, the vertex of the final simplex at which the objective function is smallest is returned as the function's minimizer.

A major benefit of both the Gauss–Newton and COBYLA algorithms is in reducing and even abolishing the use of derivatives of the objective function. In our model, the Hessian matrix associated with our objective function imposes a heavy computational burden on the optimization problem, and methods that do not require it are preferable. Numerical results indicate that COBYLA is generally very effective when the Gauss–Newton method fails; the latter, however, is quicker and more accurate than COBYLA. By default, we use the Gauss–Newton method, and, when it fails, the user is prompted to initiate COBYLA. The details of the COBYLA algorithm are described in Appendix A, beginning with a description of the Nelder and Mead simplex method from which it is derived.

Algorithm 1: COBYLA method.

```
1  begin
2      ρ ← ρ_beg
3      μ ← 0
4      Branch ← (∗)
5      Form the initial simplex
6      loop
7          Ensure that x_0 is the optimal vertex and that Flag = ON iff the simplex is acceptable
8          if Branch = (∗) or Flag = ON then
9              Generate x*
10             if ‖x* − x_0‖_2 ≥ 0.5ρ then
11                 if μ is not large enough then
12                     Revise μ
13                     if x_0 is not optimal vertex then
14                         Continue
15                     end
16                 end
17                 Calculate f(x*) and { c_i(x*) : i = 1, 2, ..., m }
18                 if Φ(x*) < Φ(x_0) or the change may help acceptability then
19                     Revise the simplex
20                 end
21                 if (Φ(x*) − Φ(x_0))/(Φ̂(x*) − Φ̂(x_0)) ≥ 0.1 then
22                     Branch ← (∗)
23                     Continue
24                 end
25             end
26             if Flag = ON then
27                 if ρ ≤ ρ_end then
28                     Break
29                 else
30                     Branch ← (∗)
31                     Update ρ and μ
32                 end
33             else
34                 Branch ← (Δ)
35             end
36         else
37             Calculate x^Δ, f(x^Δ) and { c_i(x^Δ) : i = 1, 2, ..., m }
38             Make x^Δ a vertex of the simplex
39             Branch ← (∗)
40         end
41     endloop
42  end
```

2.5. Method Scope and Other Approaches

The strategy that was introduced in [57] and also implemented herein prepares the multiscale model equations for parameter fitting by working on them directly as an initial step. This strategy is beneficial in postponing approximations to later steps and ensuring full control of the user during the whole fitting procedure. It should be noted that, for each parameter introduced in future multiscale models, the derivative with respect to the new parameter needs to be taken and more equations need to be derived, as illustrated in this section. However, this technical procedure is significantly less complicated than deriving analytical approximations to a modified model with a change in the parameters. In our package, the code is written in Java and at present the method is hard coded for the model; therefore, some technical expertise is needed if a new model is given and the method needs to be hard coded in Java for the new model. In future work we plan to separate the model from the method and make it generic, which needs to be done only once and then it can easily handle various modifications to the model and become modeler friendly. Until that time, we do rely on some amount of expert knowledge, but, overall, it should still be easier than deriving analytical approximations to a modified model.

The importance of parameter estimation to the model was already noted in previous studies. It was addressed in [48] and attempts to come up with improved strategies were tried thereafter in [60] and in [50]. Here, we briefly relate to each of these approaches in order to remain self-contained. More information can be found in [57].

2.5.1. Parameters Change When Transforming a PDE Multiscale Model to a System of ODEs

An approach taken in [60] showed how a PDE multiscale model of hepatitis C virus can be transformed to a system of ODEs. In principle, parameter estimation should then become easier, avoiding the complications in dealing with the PDE multiscale model. However, there are side effects introduced in such a transformation, as can be noticed in Equation (9) of [60] where the boundary condition $R(t,0) = \zeta$ gets inside the differential equations. Consequently, as admitted in the discussion of that reference, all parameters in Equations (7)–(10) must be estimated including ζ. The inclusion of boundary conditions as new parameters inside the model equations is a drawback compared to parameter estimation performed on the original multiscale model equations before the transformation. Another drawback from the perspective of parameters change is the fact that the simplest PDE multiscale model appearing in [47] was used in the transformation to ODEs, but important additions such as the inclusion of parameter γ as in [48] are not taken into account. It is not obvious how to include the parameter γ and other developments to the multiscale model inside the system of ODEs. Finally, any information regarding the age of the cell since infection is lost. Thus, if one would wish, for example, to vary the parameter α from infection to a certain time; this is not possible. In summary, while the transformation works for the simplest multiscale model, it is limited in considering developments to the multiscale model and the parameters in the system of ODEs are not the same as the parameters in the multiscale model.

2.5.2. Problematic Issues in Strategies Relying on Canned Methods

The previous approaches for parameter fitting of the multiscale model with age are all relying on canned methods. The two main strategies are the ones worked out in [48,50]. In [48], the long-term approximation is used for the solution of the multiscale model equations and Levenberg–Marquardt is used as a canned method. One drawback of such an approach is that it is limited to the multiscale model under treatment. In addition, the analytical approximation would change when various multiscale models are introduced and the elaborative derivations would need to be carried for each one, with restrictions that are incorporated by the approximation being used. Finally, as elaborated in [57], the use of a canned method is distancing the user away from having control over the main optimization procedure and the ability to tune it from the programming standpoint.

3. Results

Having described the newer and significantly more efficient methods for parameter estimation relative to [57], we present the new results obtained for both the biphasic model [26] and multiscale models [47–50]. We first provide a basic illustration with the mutliscale model in which run-time and performance comparisons between methods are generated. Then, in Appendix B, for each type of model, some examples are described. The results are presented using a newer (efficient) version of the user-friendly simulator that we have initially developed in [55,57] for both biphasic and multiscale models. We start from the biphasic model in Appendix B and end with the multiscale model in Appendix C. The simulator with a GUI is freely available at http://www.cs.bgu.ac.il/~dbarash/Churkin/SCE/Efficient/Parameter_Estimation (the efficient version, with the option to either select the biphasic or the multiscale model).

For a basic comparison between all relevant parameter estimation methods, we apply our new methods on the difficult case of the retrieved data points that was also used for this purpose in [57] to compare our efficient methods with the previous ones. Results were obtained after a few minutes instead of the several hours that was reported in [57], making our tool practical also for difficult cases. As in [57], we fitted the four treatment parameters κ, ε_s, ε_α and γ and all other parameters were selected with the values of Table 2.

We show in Table 3 the different values of those four parameters and sum of squared-errors fitted with the various methods (new efficient ones vs. previously published ones) to the data emanating from a patient. In the rightmost column, we fitted the long-term approximation with the retrieved data points using the scipy.optimize.curve_fit method, which is a Python implementation of a simple Levenberg–Marquardt scheme as a canned method. The next column to the right are the values obtained previously by the use of Levenberg–Marquardt along with the numerical method to solve the model equations as outlined in [57]. In the left columns are the values obtained by our new efficient methods. The small differences assure us that the significant efficiency achieved, thereby making our simulator a practical and useful tool, did not result in less accuracy.

Table 3. Values of the parameters when fitted to the patient digitized data. The rightmost column has the values when the retrieved data points are fitted to the long-term approximation as in [57]. The left columns contain the fitted parameter values by our efficient methods. Except for the rightmost column, all methods are combined with the Rosenbrock numerical scheme. The fixed parameters have the values shown in Table 2. Run-time comparison is reported in seconds in the last row.

	Gauss–Newton (LSF)	COBYLA	Levenberg–Marquardt	Long-Term
ε_s	0.609	0.598	0.602	0.600
ε_α	0.995	0.994	0.995	0.994
κ	6.210	6.375	6.219	6.160
γ (d^{-1})	0.137	0.177	0.139	0.140
accuracy (sum error^2)	0.538	0.582	0.538	0.587
run-time (s)	194	3698	70118	<1

To further illustrate the tool we provide, we show in Figure 2 the starting configuration after the data was inserted as input. The shown fitting curve is the one for default parameters (not considering data points) before running any fitting method. In Figures 3 and 4, the final results are shown when selecting LSF and COBYLA, respectively. In Figures 5 and 6, we present the curves of all methods shown in the same simulator window and in a separate graph, to which Table 3 corresponds.

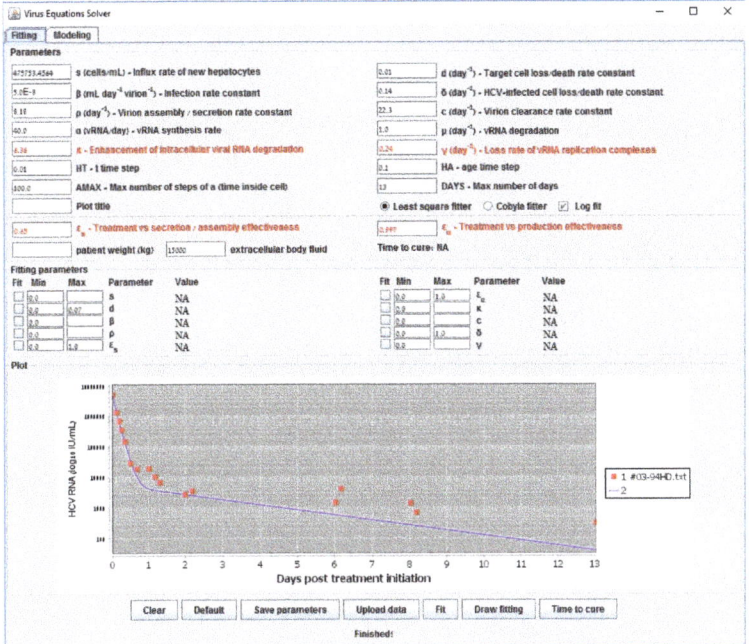

Figure 2. Start fit that emanates from data of a patient reported in [48]. The fitting curve corresponds to default parameters before fitting with our methods. The multiscale model is used.

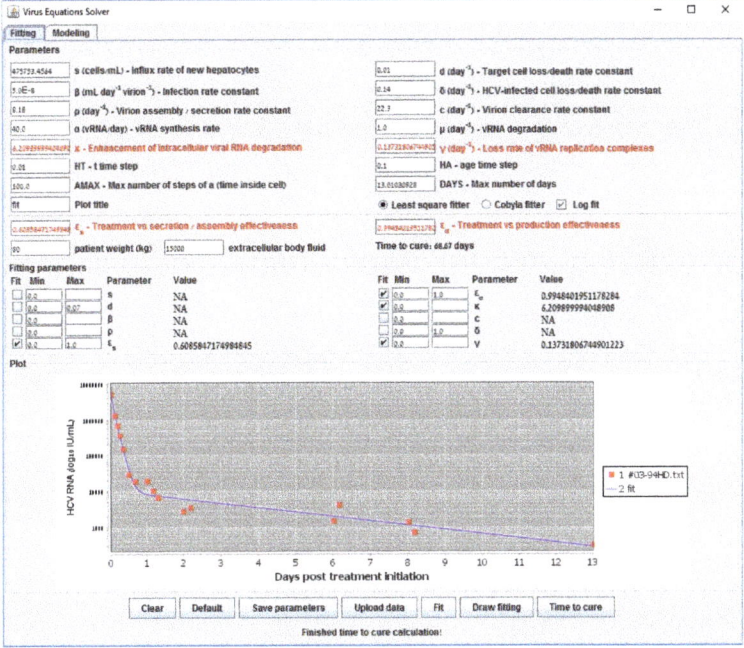

Figure 3. End fit using Gauss–Newton (LSF) that emanates from data of a patient reported in [48].

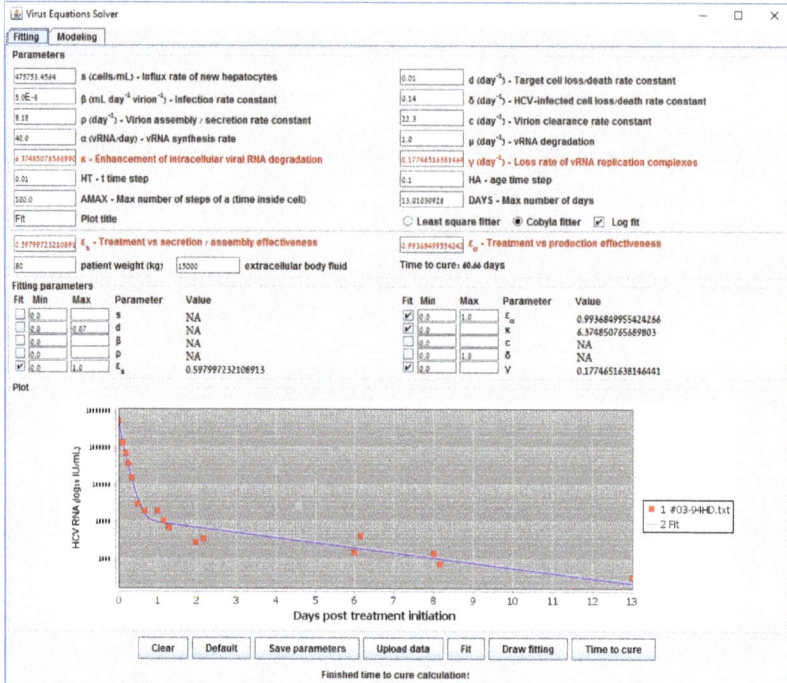

Figure 4. End fit using COBLYA that emanates from data of a patient reported in [48].

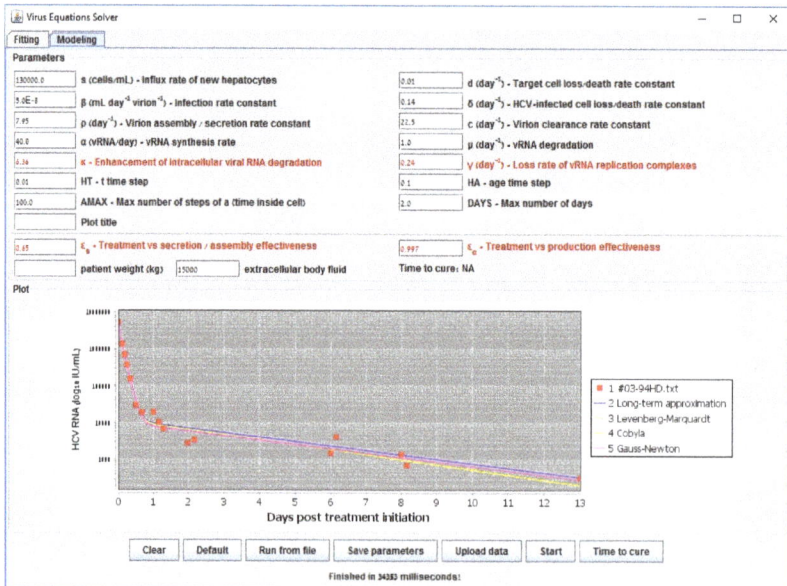

Figure 5. Comparison between the line fits of different methods inside the simulator window for the retrieved data points of patient HD that was reported in [48].

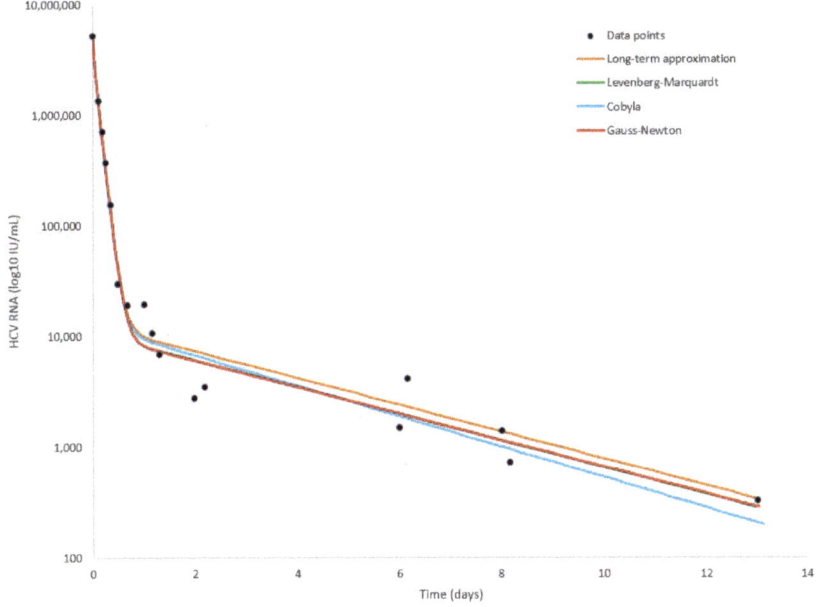

Figure 6. Comparison between the line fits of different methods for the retrieved data points of patient HD that was reported in [48].

4. Discussion

A practical and user-guided automatic procedure for parameter estimation is an important goal to achieve for mathematical models that are based on differential equations. It enables users to test a variety of fitting scenarios, either for the model calibration or model calibration with validation, by inserting different available data points of patients used for the fitting and fixed parameter values. The motivation is to use the parameters obtained by the fitting procedure to perform successful predictions for other data, where other data are data of new patients that form initial conditions to the model and successful predictions mean that the solution of the model equations yields a correct extraction of important quantities such as time to cure. In the context of viral dynamic models, even a simple model such as the biphasic model [26] that is beneficial to be tested by users requires a nonlinear method for the least squares minimization because a linear method is not sufficient [57]. The development of more complicated models such as viral dynamic models that consider intracellular viral RNA replication, namely age-structured PDE multiscale models to study viral hepatitis dynamics during antiviral therapy [47–50], presents a need for even more sophisticated strategies that perform parameter estimation while solving the model equations simultaneously. Efficient methods as developed herein are crucial such that the parameter estimation can be performed in a reasonable time.

From the parameter estimation standpoint, as previously outlined [57] and briefly mentioned in the Introduction, multiscale models are even more challenging than the biphasic model. Not only is conducting a search in at least a 10-parameter search space more difficult than in a 4-parameter search space, but also the task of solving the model equations themselves and how to connect the equations solution to the optimization procedure requires more sophistication. Previously, this was approached in [48] by using the long-term approximation along with a canned method for Levenberg–Marquardt, and in [50] by the method of lines and then employing Matlab's 4th order Runge–Kutta solver along with a canned method available in Matlab called *fmincon* for the optimization. While these strategies work sufficiently well for specific cases because of their use of canned methods, they are problematic from the standpoint of the user's capability to access and control them. Thus far, to the best of our knowledge, no specific source code for viral hepatitis kinetics besides our initial attempt at [57] (more general software such as DUNE, DuMuX, and UG4 for the solution of PDE models are available at [66–68] and would be worthwhile exploring in the future) has been released free of charge for the benefit of the community and while these strategies were described coherently in the context of presenting multiscale models, they were not intended to provide to the user a comprehensive solution of their own. There is clearly a need to provide the user with a free of charge simulator that is effortless to operate and a code that can be accessed for dissemination and future development. Furthermore, it should be practical in running time and allow inserting constraints for the parameters that need be estimated, which is not available in our initial attempt of [57] because of reverting to the standard non-constrained Levenberg–Marquardt method for the optimization and encountering numerical precision problems that were difficult to detect when developing the complete strategy for parameter estimation in our initial attempt.

The strategy we presented herein is a direct continuation to [57] and requires no canned methods utilization. It works directly on the multiscale model equations, preparing them in advance for the optimization procedure by taking their derivatives with respect to the parameters, in contrast to solving them first by an analytical approximation or performing the method of lines as a first step. For the solution of the model equations, the Rosenbrock method described in [55] is employed, as was shown to be advantageous in comparison to other solution schemes in [56]. For the constrained optimization procedure, as a departure from [57], either the Gauss–Newton or COBYLA are employed in full (not as a canned method) such that the user has access to the source code at each point in the procedure. Both Gauss–Newton and COBYLA are significantly more efficient in their constrained optimization procedure relative to the Levenberg–Marquardt employed in [57]. More complicated patient cases that took several hours of run time simulation in [57] (19.48 h reported in Table 3 for Levenberg–Marquardt) are now calculated in a few minutes (3.23 min reported in Table 3 for Gauss–Newton) on a standard PC, and simpler cases that took several minutes are now performed in seconds. Thus, the obtained results are much faster to compute than the existing solutions without sacrificing accuracy. The whole method is provided in a form of a model simulator with a user-friendly GUI, letting the user insert parameter constraints.

We note by passing that the aforementioned general software DUNE, DuMuX, and UG4 [66–68] are written in C++ using the MPI library allowing for massively parallel evaluations in the context of HPC, which might allow significantly more extended data sets to consider in the future. Thus, High Performance Computing (HPC) might also be an option for future development.

The code is open source and is divided into several packages: two fitter packages and a third default package with solver (Solver.java) class and GUI (GUI.java) class. The default package also contains different helper classes, like a class with all parameters and adapter classes to define the objective function for the fitters. The code is flexible and it is easy to add any new model solver class or any new parameters fitting class, library, or package. We use adapter design pattern to connect between the model solver and the parameters fitting algorithm. Thus, to add a new solver or fitter, one should add the solver/fitter code to the project and implement the adapter class that matches the interface of the model solver to the objective function interface of the parameters fitting method. In addition, one should change 2–3 rows in the GUI.java class to make use of a new solver/fitter from the GUI interface.

5. Conclusions and Future Work

The efficient methods described herein make the simulator a practical tool that is distributed free of charge for the benefit of the community and the dissemination of viral hepatitis models. Furthermore, the methods for parameter estimation employed can conceptually be used in other mathematical models in biomedicine.

Future work would include the development of the code in several directions. First, the code can be made more modular such that the modeler can easily implement the method for a different model or a modified version of one of these models. In this way, portability of the method to other models can be achieved such that a significant modification of the code is not needed as a consequence of a change in the model, ensuring that the modification is relatively straightforward. Second, at present, individual fits to individual time course profiles is available, which is useful when one wants to describe viral dynamics within one patient. The code can be developed for use also for fitting the in vitro time course profiles of pooled patient datasets. Thus, as future work, having the option to import and fit the models to repeated/multiple measurements would be useful.

From the numerical perspective, it might be possible to try weak methods for the solution of the model equations such as finite elements, finite volumes, or discontinuous Galerkin as described in [63,64]. The two time scales might present different challenges as compared to PDEs that are dependent on time and space in their partial derivatives. Independently, much of the computations are present in the optimization stage as compared to the solution stage and therefore efforts centered on the model equations solution could focus on simplified strategies, if at all possible, for the benefit of gaining more efficiency.

Finally, machine learning methods can be used to improve parameter estimation. There are already enough patient cases, as more than 250 patients have been modeled, which can be used to prepare the data for the parameter estimation of our simulator. Machine learning can then be used for outliers' removal, replacement of the incorrect and missed data with the correct one (currently done manually), and correction of the data for the parameters and time to cure estimation. The machine learning algorithm can then be integrated with the parameter estimation method to yield an overall improved procedure.

Author Contributions: Conceptualization, D.B.; methodology, A.C., S.L., V.R., H.D., and D.B.; software, A.C., V.R.; investigation, A.C., S.L., V.R., H.D., and D.B.; writing—original draft preparation, A.C., S.L., V.R., H.D., and D.B.; writing—review and editing, A.C., S.L., V.R., H.D., and D.B.; supervision, D.B.; funding acquisition, A.C., D.B., V.R., and H.D. All authors have read and agreed to the published version of the manuscript.

Funding: This research was funded by the U.S. National Institutes of Health Grant Nos. R01-AI078881, R01-AI144112, R01-AI146917, and R01-GM121600.

Conflicts of Interest: The authors declare no conflict of interest. The funders had no role in the design of the study; in the collection, analyses, or interpretation of data; in the writing of the manuscript, or in the decision to publish the results.

Appendix A. Details of the COBYLA Method

The original simplex method, devised by Spendley, Hext, and Himsworth [61,69], seeks to advance a simplex towards a minimizer of $f(x) : \mathbb{R}^n \to \mathbb{R}$ by reflecting vertices at which f is large across opposite faces of the simplex in the hopes of reducing f. If the initial simplex has vertices $x_0, x_1, \ldots, x_n$, ordered so that $f(x_0) \leq f(x_1) \leq \ldots \leq f(x_n)$, then x_n is the vertex at which f is largest, and is considered for revision. The vector $\hat{x}$, defined below, is proposed as a replacement:

$$\hat{x} := \frac{2}{n}\left(\sum_{i=0}^{n-1} x_i\right) - x_n, \tag{A1}$$

$$= \frac{1}{n}\left(\sum_{i=0}^{n-1} x_i\right) + \left(\frac{1}{n}\left(\sum_{i=0}^{n-1} x_i\right) - x_n\right), \tag{A2}$$

$$:= \bar{x} + (\bar{x} - x_n). \tag{A3}$$

Note that the vector $\bar{x}$ is the centroid of the convex hull of the points $x_0, x_1, \ldots, x_{n-1}$, so that $\hat{x}$ is the reflection of the vertex through the face of the simplex opposite x_n. As such, the volume of the simplex is preserved if the vertex exchange $x_n \to \hat{x}$ is made, with the swap occurring if $f(\hat{x}) < f(x_{n-1})$.

If $f(\hat{x})$ is comparable to $f(x_n)$, the assumption is made that f is minimized in the area between x_n and $\hat{x}$, so that the change $x_n \to \hat{x}$ simply bypasses this region. To access this interior region, the simplex is rescaled without replacing any of the vertices with $\hat{x}$. The optimal vertex, x_0, is left alone, and each vertex x_k, for $k = 1, \ldots, n$, is replaced with $(1/2)(x_0 + x_k)$. Both such changes to the simplex are illustrated in the case $n = 2$ in Figure A1.

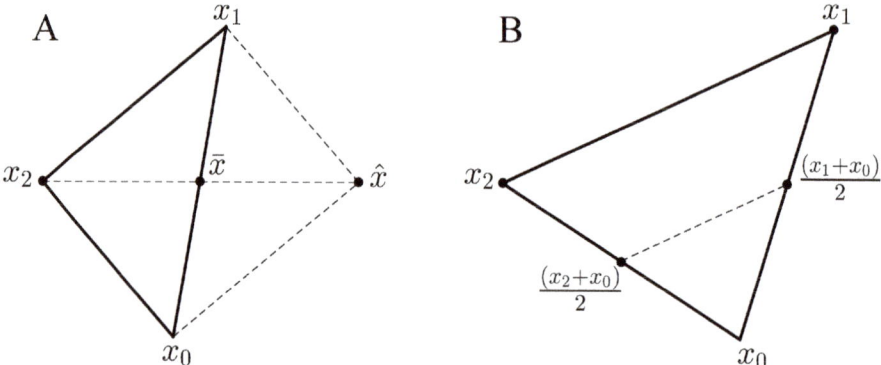

Figure A1. Illustration of the original simplex method. The points x_0, x_1, and x_2 form the initial simplex. (**A**) The point $\bar{x}$ is the midpoint of the line joining x_0 and x_1, and $\hat{x}$ is the reflection of x_2 through this line. If $f(\hat{x}) < f(x_1)$, x_2 is replaced by $\hat{x}$, shifting the location of the simplex. (**B**) If $f(\hat{x}) \geq f(x_1)$, x_2 is replaced with $(1/2)(x_2 + x_0)$ and x_1 is replaced with $(1/2)(x_1 + x_0)$, reducing the volume of the simplex.

The Nelder–Mead method [70] expands on this basic simplex method, removing much of the inefficiency that arises when rescalings result in a small simplex that takes longer to converge to the function's minimizing region. It does so by moving the vertex x_n to a new point along the line joining x_n and $\hat{x}$, strategically chosen to reduce f as much as possible. A generic expression for the new vertex, x_{new}, is now,

$$x_{\text{new}} := \bar{x} + \theta(\bar{x} - x_n), \tag{A4}$$

where $\theta > 0$ may be different at each iteration. In the case of a linear f, we have that

$$f(x_{\text{new}}) = f(\bar{x}) + \theta(f(\bar{x}) - f(x_n)) \leq f(\bar{x}), \tag{A5}$$

where the last equality above follows from the fact that $f(x_n) \geq f(x_k)$ for all $k = 0, 1, \ldots, n-1$, which implies $f(\bar{x}) \leq f(x_n)$. Regarding the linear case as a proxy for the more general case, we expect such a vector x_{new} to decrease the value of f, if θ is chosen well. One particular implementation bases the choice of θ—and thus the new vertex—on the value of $f(\hat{x})$ relative to the value of f at other vertices. The new vertex, $\check{x}$, is defined as such:

$$\begin{cases} 2\hat{x} - \bar{x} & \text{if } f(\hat{x}) < f(x_0), & \text{(A6a)} \\ \frac{1}{2}(\bar{x} + \hat{x}) & \text{if } f(x_0) \leq f(\hat{x}) < f(x_{n-1}), & \text{(A6b)} \\ \frac{1}{2}(x_n + \bar{x}) & \text{if } f(x_{n-1}) \leq f(\hat{x}). & \text{(A6c)} \end{cases}$$

The choices in Equations (A6a)–(A6c) are obtained by taking $\theta = 2, (1/2), (-1/2)$, respectively, in Equation (A4), and are depicted in Figure A2. If $f(\hat{x}) < f(x_0)$, as in Equation (A6a), then f decreases so significantly along the line from x_n to $\hat{x}$ that choosing a new vertex, $2\hat{x} - \bar{x}$, even further down the line is presumed to result in a greater reduction. If $f(x_0) \leq f(\hat{x}) < f(x_{n-1})$, as in Equation (A6b), the reduction in f is less significant, and the new vertex is placed between $\hat{x}$ and the face of the simplex opposite x_n. If $f(x_{n-1}) \leq f(\hat{x})$, as in Equation (A6c), the reduction in f at $\hat{x}$ is minimal, and $\check{x}$ is placed between x_n and the face of the opposite simplex, as placement of the vertex near $\hat{x}$ is not warranted.

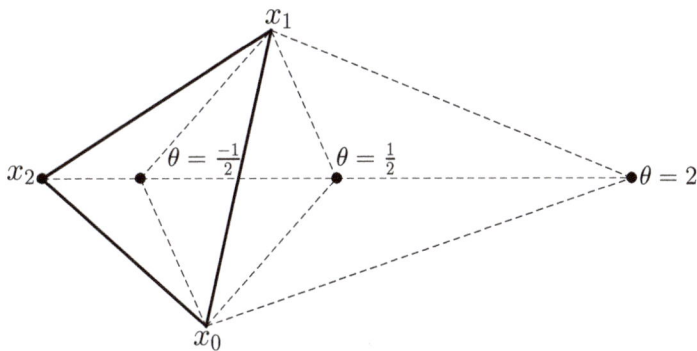

Figure A2. Illustration of the Nelder–Mead method. The points x_0, x_1 and x_2 form the initial simplex. The vertex x_2 is replaced with a vertex of the form $x_{\text{new}} = \bar{x} + \theta(\bar{x} - x_2)$. If $f(\hat{x}) < f(x_0)$, $\theta = 2$, if $f(x_0) \leq f(\hat{x}) < f(x_{n-1})$, $\theta = 1/2$, and if $f(x_{n-1}) \leq f(\hat{x})$, $\theta = -1/2$.

The COBYLA algorithm is used to minimize an objective function $f(x) : \mathbb{R}^n \to \mathbb{R}$ subject to the set of $m \in \mathbb{N}$ constraints,

$$\{c_i(x) \geq 0 : i = 1, 2, \ldots, m\}, \tag{A7}$$

where $c_i(x) : \mathbb{R}^n \to \mathbb{R}$ for each $i = 1, \cdots, m$. As derivatives are omitted from the algorithm entirely, no smoothness assumptions are required for the functions f, c_i; they must simply be well-defined on $\mathbb{R}^n$. After generating the initial simplex from an initial guess as to the location of the minimizer, the optimal vertex is identified and labeled x_0. In this case, a vertex of the simplex is considered optimal if $\Phi(x_0) \leq \Phi(x_k)$, for $k = 1, \cdots, n$, where the x_k are the other n vertices of the simplex, and $\Phi(x)$ is defined by

$$\Phi(x) = f(x) + \mu[\max\{-c_i(x) : i = 1, \cdots, m\}]_+, \tag{A8}$$

with $[x]_+ = \max\{x, 0\}$, and $\mu \geq 0$ a constant parameter. The optimality of a point x is thus affected by both the value of $f(x)$ and how closely it satisfies the constraints in Equation (A7). If $c_i(x) \geq 0$ for all $i = 1, \cdots, m$, then $\Phi(x) = f(x)$, but if at least one constraint is violated, then $\Phi(x) > f(x)$, lessening the "worth" of the point x as an approximation to the minimizer. From there, each iteration of the algorithm generates a new candidate vertex, designed to either replace an existing vertex with one that decreases $\Phi(x)$ or improves the shape of the simplex. The shape of the simplex is particularly crucial in this algorithm because its vertices are used to define linear programming problems, from which new candidate vertices designed to improve the optimality condition are derived. Specifically, if $\{x_k : k = 0, \cdots, n\}$ are the vertices of the current simplex, we let $\hat{f}(x) : \mathbb{R}^n \to \mathbb{R}$ be the unique affine function that passes through points $(x_k, f(x_k)) \in \mathbb{R}^{n+1}$, and, analogously, we let $\hat{c}_i(x) : \mathbb{R}^n \to \mathbb{R}$ be the unique affine function that passes through the points $(x_k, c_i(x_k))$. Should the shape of the simplex be "acceptable"—in a way to be defined later—a new candidate vertex is chosen to improve optimality by minimizing $\hat{f}(x)$ subject to the constraints $\{\hat{c}_i(x) \geq 0\}$. Should the simplex be of an unacceptable shape, the linear programming problem may be ill-defined or fail to provide a reasonable approximation to the functions $f(x), c_i(x)$. If this newly generated vector improves the value of Φ, it replaces a vertex of the simplex. This process continues until a pre-determined final trust region radius, which represents the desired accuracy of the approximation to the minimizer, is achieved.

The algorithm takes as inputs an initial guess, x_0, as to the location of the minimizer, and the constants $\rho_{beg}, \rho_{end} > 0$, which represent the initial and final trust region radii. Additionally, μ is set to zero. At the start, the initial simplex is generated from x_0 and ρ_{beg}. The vector x_0 is one vertex, with the other vertices, $x_k, k = 1, \ldots, n$, defined by $x_k = x_0 + \rho_{beg} e_k$, for $k = 1, \ldots, n$. Here, e_k is the k-th coordinate vector. After each x_k is generated, $f(x_k)$ is computed, and the labels of the vectors x_0 and x_k are swapped if $f(x_k) < f(x_0)$, to ensure that f is minimized at x_0. After the initial simplex is defined, the algorithm proceeds to advance the simplex at each iteration by either generating a new candidate vertex, denoted x^*, to decrease $\Phi(x)$, or an alternate vertex, denoted x^Δ, to improve the shape of the simplex. Each iteration begins by ensuring that x_0 is the optimal vertex–and relabeling vertices if it is not–before assessing the suitability of the simplex. For this, denote by σ^k the Euclidean distance from the vertex x_k to the face of the simplex opposite x_k, and, by η^k, the length of the segment joining x_k to x_0. The simplex is deemed to have an acceptable shape if and only if $\sigma^k \geq \alpha\rho$ and $\eta^k \leq \beta\rho$ for all $k = 1, \ldots, n$, where α, β satisfy $0 < \alpha < 1 < \beta$. These conditions prevent the development of flat, degenerate simplices, which would result in poorly-formulated linear programming problems. If the simplex is of an unacceptable shape, a vertex x^Δ is generated; otherwise, a candidate vertex x^* is computed.

The iteration immediately following formation of the initial simplex always generates a vertex x^*, as the initial simplex satisfies $\sigma^k = \eta^k = \rho_{beg}$ for all $k = 1, \ldots, n$, and is thus always acceptable. The loop that generates such an x^* in general is described below; in the very first iteration, $\rho = \rho_{beg}$. The vector x^* is generated by minimizing $\hat{f}(x)$ subject to the constraints in Equation (A7) and the trust region condition,

$$||x - x_0||_2 \leq \rho, \tag{A9}$$

as illustrated in Figure A3. Should the constraints in Equations (A7) and (A9) be inconsistent with one another, the candidate x^* is chosen by minimizing the "greatest constraint violation" function, $\hat{M}(x)$, defined by,

$$\hat{M}(x) := \max\{-\hat{c}_i(x) : i = 1, \ldots, m\} \tag{A10}$$

subject to the trust region constraint in Equation (A9). This process is illustrated in Figure A4. If there exist multiple x^* that satisfy $\hat{M}(x^*) = \min\{\hat{M}(x) : ||x - x^{(0)}||_2 \leq \rho\}$, then the x^* that minimizes $\hat{f}(x)$ is chosen from among those that minimized $\hat{M}(x)$. If there are multiple such minimizers of $\hat{f}(x)$, then the one that minimizes $||x - x_0||_2$ is chosen.

When the appropriate x^* is identified, the condition $||x^* - x_0||_2 < \frac{1}{2}\rho$ is tested. If $||x^* - x_0||_2 \geq \frac{1}{2}\rho$, the relative "size" of the parameter μ is evaluated. To this end, denote by $\bar{\mu}$ the smallest value of μ for which $\hat{\Phi}(x^*) \leq \hat{\Phi}(x_0)$, where

$$\hat{\Phi}(x) = \hat{f}(x) + \mu[\max\{-\hat{c}_i(x) : i = 1, \ldots, m\}]_+ = \hat{f}(x) + \mu[\hat{M}(x)]_+. \tag{A11}$$

If Equations (A7) and (A9) are consistent with one another, then x^* had been chosen to minimize $\hat{f}(x)$, and it follows that $\hat{\Phi}(x^*) = \hat{f}(x^*) \leq \hat{f}(x_0) = \hat{\Phi}(x_0)$. In this case, $\bar{\mu} = 0$. Should Equations (A7) and (A9) be inconsistent, then x^* had been chosen to minimize $\hat{M}(x)$ as in Equation (A10), implying that $\hat{M}(x_0) - \hat{M}(x^*) \geq 0$. If $\hat{M}(x_0) - \hat{M}(x^*) = 0$, then $\hat{M}$ has at least two minimizers, and x^* had been chosen to minimize $\hat{f}(x)$, implying $\hat{\Phi}(x^*) \leq \hat{\Phi}(x_0)$ from Equation (A11). If $\hat{M}(x_0) - \hat{M}(x^*) > 0$, there exists a $\bar{\mu}$ sufficiently large to guarantee $\hat{f}(x_0) - \hat{f}(x^*) + \mu\left(\hat{M}(x_0) - \hat{M}(x^*)\right) > 0$, even if $\hat{f}(x_0) - \hat{f}(x^*) < 0$, implying $\hat{\Phi}(x^*) \leq \hat{\Phi}(x_0)$ for $\mu \geq \bar{\mu}$. If the current value of μ satisfies $\mu \geq \frac{2}{3}\bar{\mu}$, then μ is considered "sufficiently large" and its value is left alone. Otherwise, μ is increased to 2μ.

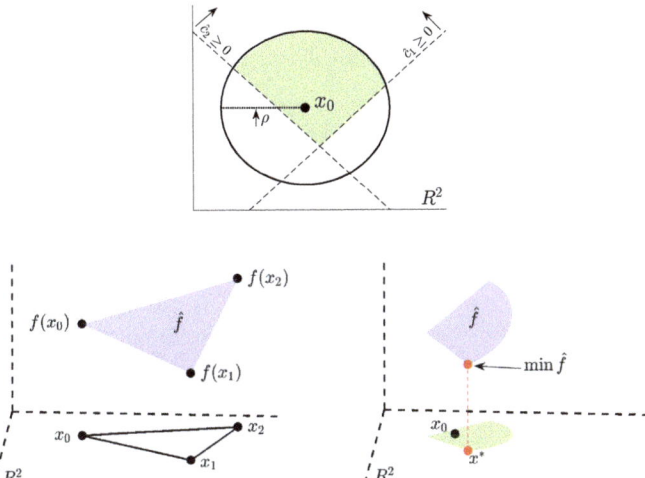

Figure A3. Minimization of $\hat{f}$ The candidate vertex x^* is computed by minimizing $\hat{f}(x)$ subject to the constraints $\hat{c}_1, \hat{c}_2 \geq 0$ within the trust region $||x - x_0||_2 \leq \rho$. (top) The region of optimization (green) is the intersection of the trust region $||x - x_0||_2 \leq \rho$ with the half planes defined by the affine constraints $\hat{c}_1 \geq 0$ and $\hat{c}_2 \geq 0$. (bottom left) The function, $\hat{f}$, to be minimized is represented graphically by the plane (blue) passing through points $(x_0, f(x_0)), (x_1, f(x_1)), (x_2, f(x_2))$. (bottom right) The vertex x^* is defined to be the point within the region of optimization (green) at which $\hat{f}$ (blue) is minimized.

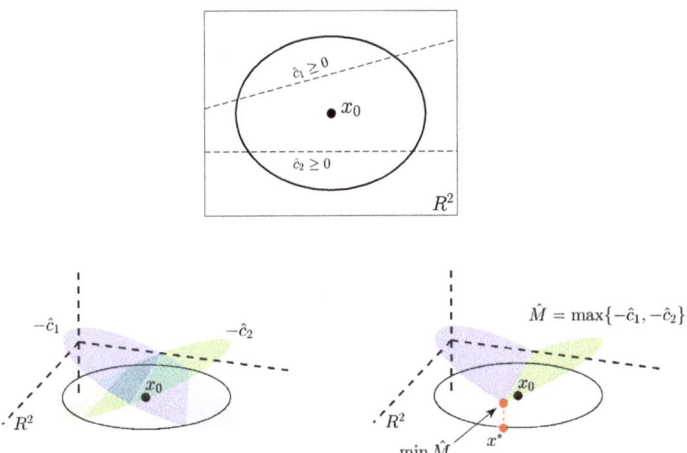

Figure A4. Minimization of $\hat{M}$ Should the constraints $\hat{c}_i(x) \geq 0$ be inconsistent with one another within the trust region $||x - x_0||_2 \leq \rho$, the candidate vertex x^* is chosen to minimize $\hat{M} := \max\{-\hat{c}_i(x) : i = 1, \ldots, m\}$. (top) The constraints $\hat{c}_1(x) \geq 0$ and $\hat{c}_2(x) \geq 0$ are inconsistent within the region $||x - x_0||_2 \leq \rho$. (bottom left) Graphs of the affine functions $-\hat{c}_1(x)$ (blue) and $-\hat{c}_2(x)$ (green). (bottom right) The vertex x^* is defined to be the point within the trust region (black circle) at which $\hat{M}$ is minimized.

If μ is increased, it may no longer be the case that x_0 is optimal, in the sense that there may exist some k between 1 and n for which $\Phi(x_k) < \Phi(x_0)$. From the form of Φ in Equation (A8), this reversal of the original order relation $\Phi(x_k) \geq \Phi(x_0)$ as μ increases can only occur if $[M(x_0)]_+ > [M(x_k)]_+$, with $M(x)$ defined as:

$$M(x) := \max\{-c_i(x) : i = 1, \ldots, m\}. \tag{A12}$$

If x_0 is no longer optimal, the process returns to the start of the loop, from which it first rearranges the labeling of the vertices so as to label the optimal one x_0. No change was made to the vertices of the simplex beyond this relabeling, and if the simplex had been acceptable previously, it will remain acceptable. In this case, another candidate vertex x^* will be computed with respect to the new x_0. The process of generating x^*, increasing μ, and then relabeling the vertices can only happen a finite number of times; the labels of the vertices x_0 and x_k are only switched if $M(x_0) > M(x_k)$, meaning that the value of M decreases each time a vertex exchange is made. Thus, increasing μ can only change the order relation between the values $\Phi(x_k)$ until the optimal vertex x_0 also satisfies $M(x_0) = \min\{M(x_k) : k = 0, \ldots, n\}$, at the latest.

If μ is sufficiently large and x_0 is optimal, or once these conditions have been achieved, the possible replacement of one vertex with the new candidate vertex x^* is considered. The values of $f(x^*)$ and $c_i(x^*)$ are computed, and the simplex is revised to incorporate x^* as a vertex if $\Phi(x^*) < \Phi(x_0)$, or if such a revision will improve acceptability of the simplex. The choice of which existing vertex should be replaced with x^* is determined by the predicted effect of the change on the acceptability conditions and the volume of the simplex. For this, consider the quantity $\bar{\sigma}^k$, defined to be the Euclidean distance from x^* to the face of the simplex opposite x_k. If x_k is replaced with x^*, but all other vertices are left unchanged, the volume, V^*, of the new simplex is related to the volume, V^k, of the original simplex by the formula,

$$V^* = \frac{\bar{\sigma}^k}{\sigma^k} V^k. \tag{A13}$$

This follows from the standard formula for the volume of a simplex in $\mathbb{R}^n$, which yields $V^k = (1/n)V^H \sigma^k$, $V^* = (1/n)V^H \bar{\sigma}^k$, where V^H is the volume of the convex hull of the n points $x_0, x_1, \ldots, x_{j-1}, x_{j+1}, \ldots, x_n$. It is considered advantageous for a change in the vertices to increase the volume of the simplex, and, as such, vertices x_k for which $\bar{\sigma}^k > \sigma^k$ are sought for replacement by x^*. Specifically, consider the set J defined by

$$J := \{j : \bar{\sigma}^j \geq \sigma^j\} \cup \{j : \bar{\sigma}^j \geq \alpha\rho\}, \tag{A14}$$

thus consisting of indices j for which x^j could be replaced with x^* either without decreasing the volume of the simplex or without disturbing the acceptability condition $\sigma^j \geq \alpha\rho$. To consider the effect of a change on the acceptability condition $\eta^k \leq \beta\rho$, note that, if a vertex is replaced by x^*, the optimal vertex of the new simplex will be either the previous optimal vertex, x_0, or the vertex x^* itself. Denoting the new optimal vertex by $\bar{x}_0$, if J is nonempty, let $\ell \in \mathbb{N}$ be defined as

$$\ell := \min\{k \in \mathbb{N} \cap [1, n] : \|x_k - \bar{x}_0\|_2 = \max\{\|x_j - \bar{x}_0\|_2 : j \in J\}\}, \tag{A15}$$

so that $x_{(\ell)}$ is the vertex with smallest index at a maximal distance from the optimal vertex. If $\|x^\ell - \bar{x}_0\|_2 > \delta\rho$, for $1 < \delta \leq \beta$, then x_k is replaced by x^*. If these conditions are not met, the simplex is revised regardless, as long as either $\Phi(x^*) < \Phi(x_0)$ or there exists an index k for which $\bar{\sigma}^k > \sigma^k$—that is, if Φ is smaller at the new candidate vertex, or a change increases the simplex's volume. In this case, x_0 is replaced by x_ℓ, where ℓ is now defined as

$$\ell := \min\{k \in \mathbb{N} \cap [1, n] : \bar{\sigma}^k / \sigma^k = \max\{\bar{\sigma}^j / \sigma^j : j \in \mathbb{N} \cap [1, n]\}\}. \tag{A16}$$

Note that, even though the simplex is updated whenever $\Phi(x^*) < \Phi(x_0)$, the vertex to discard is chosen based on the effect the change will have on the volume and acceptability of the simplex, with no regard for the values of Φ at different vertices. This process almost always results in an update to the simplex, with an update failing to occur only if Φ is not decreased at x^*, no vertex swap would increase the volume of the simplex, and either J is empty or $\|x_j - \tilde{x}_0\|_2 \leq \delta\rho$ for all $j = 1,\ldots,n$. The set J is empty if each hypothetical vector swap actually decreases the volume of the simplex and fails to maintain the first acceptability condition, $\bar{\sigma}^j \geq \alpha\rho$.

After potentially making a vector replacement, the progress made in reducing Φ as the simplex advances is examined. If sufficient progress is not made, the trust region radius, ρ, will be reduced. To determine if such a reduction should be made, the change in Φ at x^* is compared to the change in $\hat{\Phi}$. The condition

$$\frac{\Phi(x^*) - \Phi(x_0)}{\hat{\Phi}(x^*) - \hat{\Phi}(x_0)} \leq 0.1 \tag{A17}$$

is tested. If Cond. (A17) fails, then the improvement to Φ is at least 10% of the improvement to $\hat{\Phi}$. This improvement is considered significant enough, and the iteration returns to the start of the loop to generate a successive x^*, assuming the simplex is still acceptable. If Cond. (A17) fails, the improvement to x^* will be less than 10% of the improvement to $\hat{\Phi}$, and a decrease in the trust region radius is called for. Before the trust region radius is assessed, the acceptability of the simplex is checked. If the simplex is unacceptable, the iteration returns to the start of the loop, and generates an alternate vertex x^Δ. If the simplex is acceptable, the condition $\rho \leq \rho_{\text{end}}$ is tested. If $\rho \leq \rho_{\text{end}}$, the final trust region radius has been reached, and the algorithm terminates. If $\rho > \rho_{\text{end}}$, further advances to the simplex are required to reduce the trust region radius to its final value. The iteration returns to the start of the loop, and will generate a new candidate, x^*, as the acceptability of the simplex has already been verified. Before this, ρ and μ are updated. If $\rho > 3\rho_{\text{end}}$, then ρ is decreased by half. If $\rho \leq 3\rho_{\text{end}}$, then ρ is set to ρ_{end} itself.

It is considered sensible to update μ whenever ρ is updated, as the value of μ can become quite large. A constraint function $c_i(x)$ is regarded as "significant" to Φ if $i \in I$, where

$$I := \{i \in \mathbb{N} \cap [1,m] : c_i^{\min} < (1/2)c_i^{\max}\}, \tag{A18}$$

with $c_i^{\min}$ and $c_i^{\max}$ representing the minimum and maximum values of c_i at the vertices of the current simplex. If I is empty, $\mu = 0$, and if I is nonempty, μ is set to the value,

$$\frac{\{\max_{k=0,1,\ldots,n} f(x_k) - \min_{k=0,1,\ldots,n} f(x_k)\}}{\min\{[c_i^{\max}]_+ - c_i^{\min} : i = 1,\ldots,m\}}, \tag{A19}$$

assuming that the quantity in Equation (A19) is less than the current value of μ.

If after generating the candidate vector x^* the condition $\|x^* - x_0\|_2 < \frac{1}{2}\rho$ is met, much of the previously described process is omitted. The algorithm proceeds as it did after advancing the simplex and verifying Cond. (A17), by first checking the acceptability of the simplex and revising, if necessary, and then checking the condition $\rho \leq \rho_{\text{end}}$. As before, if $\rho \leq \rho_{\text{end}}$, the algorithm terminates, and, if $\rho > \rho_{\text{end}}$, ρ and μ are updated as previously described and the process returns to the start of the loop to generate a new x^*.

It only remains to describe the process of updating the simplex to improve its acceptability. This is done by generating an alternate vertex, x^Δ, to replace one of the vertices of the simplex. Recall that, if the simplex is unacceptable, then there either exists a $j \in \mathbb{N} \cap [1,n]$ such that $\sigma^k < \alpha\rho$ or such that $\eta^k > \beta\rho$. If the latter is true, define $\ell \in \mathbb{N} \cap [1,n]$ to be the index that satisfies

$$\eta^\ell = \max\{\eta^k : k = 1,\ldots,n\}. \tag{A20}$$

Otherwise, define ℓ to be the index that satisfies

$$\sigma^\ell = \min\{\sigma^k : k = 1,\ldots,n\}. \tag{A21}$$

The vertex x_ℓ is then one that violates one of the acceptability conditions most egregiously, either by being the closest to the optimal vertex or the farthest from its opposite face. The vertex x_ℓ will be replaced by, x^Δ, where x^Δ is defined by

$$x^\Delta = x_0 \pm \gamma \rho v_\ell, \tag{A22}$$

where v_ℓ is the unit vector perpendicular to the face of the simplex opposite the outgoing vertex x_ℓ, and $\gamma \in (\alpha, 1)$. The + or − is chosen to minimize $\hat{\Phi}$. The new vertex x_Δ, illustrated in Figure A5, maintains the general position of x_ℓ relative to the opposite face of the simplex, while satisfying $\sigma^\Delta = \eta^\Delta = \|x^\Delta - x_0\|_2 = \gamma \rho \in (\alpha \rho, \beta \rho)$. Even though the acceptability condition is not violated at the new vertex x^Δ, acceptability may still be violated at other vertices. After replacing x_ℓ with x_Δ, the process always generates a vertex of the type x^*, as opposed to conducting several replacements to improve acceptability. As described previously, the algorithm continues to advance the simplex by replacing current vertices with improvements of the form x^* and x^Δ until the condition $\rho \leq \rho_{\text{end}}$ is achieved.

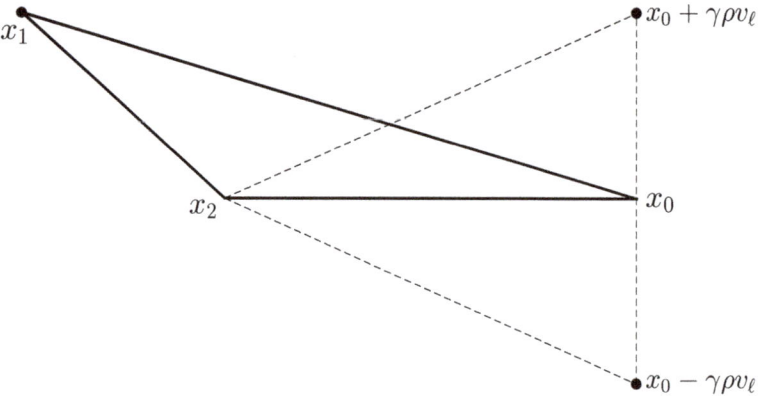

Figure A5. Illustration of the new vector x^Δ, generated to improve the shape of the simplex. The vertex x_1 is replaced with either $x^\Delta = x_0 + \gamma \rho v_\ell$ or $x^\Delta = x_0 - \gamma \rho v_\ell$, whichever point results in a smaller value of $\hat{\Phi}$.

Appendix B. Parameter Estimation in the Biphasic Model

We begin with the standard model for HCV dynamics, the biphasic model of Neumann et al. [26]. Although the model is nonlinear, it can be solved analytically when assuming that the target cells T variable is constant. We incorporated the analytical solution described in [26] to our simulator and performed parameter estimation using a constrained optimization Gauss–Newton solver to solve the minimum least squares problem, which is more efficient and stable than the non-constrained nonlinear solver with a damping factor described in [57]. There can be instances in which the Gauss–Newton solver fails, although it is the simplest and most efficient, in which case the COBYLA solver should be used instead by selecting its option in the simulator or it can be used in all instances from the start. Figures A6 and A7 present fitting results from two patients (Pts). Figure A6 corresponds to Pt3 who was treated with mavyret [36] where LSF was selected for the optimization (default). In a case that corresponds to Pt285003 who was treated with epclusa [36] where LSF was selected for the optimization (default), a warning appeared because of a failure, after which Figure A7 corresponds to the same case, but this time COBYLA was selected for the optimization and succeeded to yield a fit.

Our current method has recently been used in [71,72]. A webpage with user instructions is available at http://www.cs.bgu.ac.il/~dbarash/Churkin/SCE/Efficient/Parameter_Estimation/Biphasic.

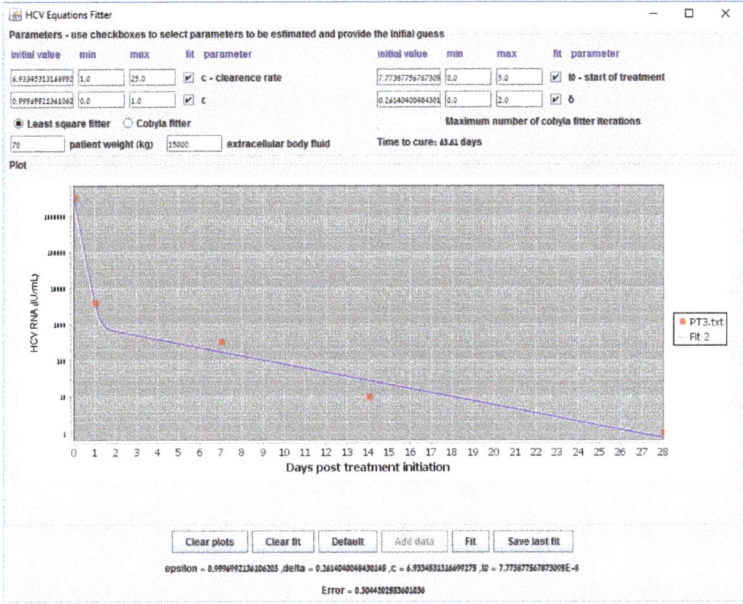

Figure A6. Biphasic model fitting example with data taken from [36] of a patient who was treated with mavyret. The LSF method (default) is recommended for use.

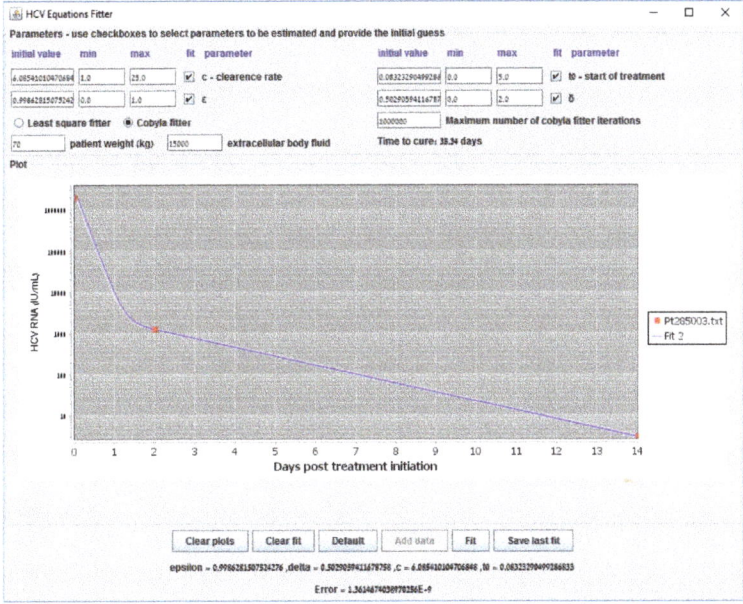

Figure A7. Biphasic model fitting example with data taken from [36] of a patient who was treated with epclusa. In this particular case, COBYLA was selected instead of LSF and succeeded to yield a fit.

Appendix C. Parameter Estimation in the Multiscale Model

Extending to the multiscale model for HCV dynamics, we perform parameter estimation on the model taken from [48]. As previously described, in our simulator, we solved the model equations using the Rosenbrock method [55] and performed parameter estimation after preparing the derivative equations using a full implementation of our developed Gauss–Newton (LSF) and COBYLA methods that are suitable to our application domain without reverting to canned methods.

To illustrate the tool we provide, we first show in Figure A8 the result of fitting to generated data points from the default values of the parameters c ρ, when starting away from their real values (with initial guesses of $\rho = 7.95, c = 22.5$) and selecting the LSF method. The predicted values after running LSF ($\rho = 5.0, c = 31.0$) are very close to their real values (error of $2.58 \cdot 10^{-6}$) and run-time was 40 s. In Figure A9, we show the result of the selecting the COBYLA method for the case as in the previous figure. The predicted values after running COBYLA are even slightly closer to their real values (error of $1.34 \cdot 10^{-8}$) for this particular case and run-time was 109 s. Thus, starting from initial guesses that are far from the real values can be handled, indicating the robustness of our methods. Run-times of our methods were significantly faster than a run-time of 1680 s (with even a larger error of around 0.5 showing much less robustness) when using the Levenberg–Marquardt method that was implemented in [57]. A webpage with user instructions is available at http://www.cs.bgu.ac.il/~dbarash/Churkin/SCE/Efficient/Parameter_Estimation/Multiscale.

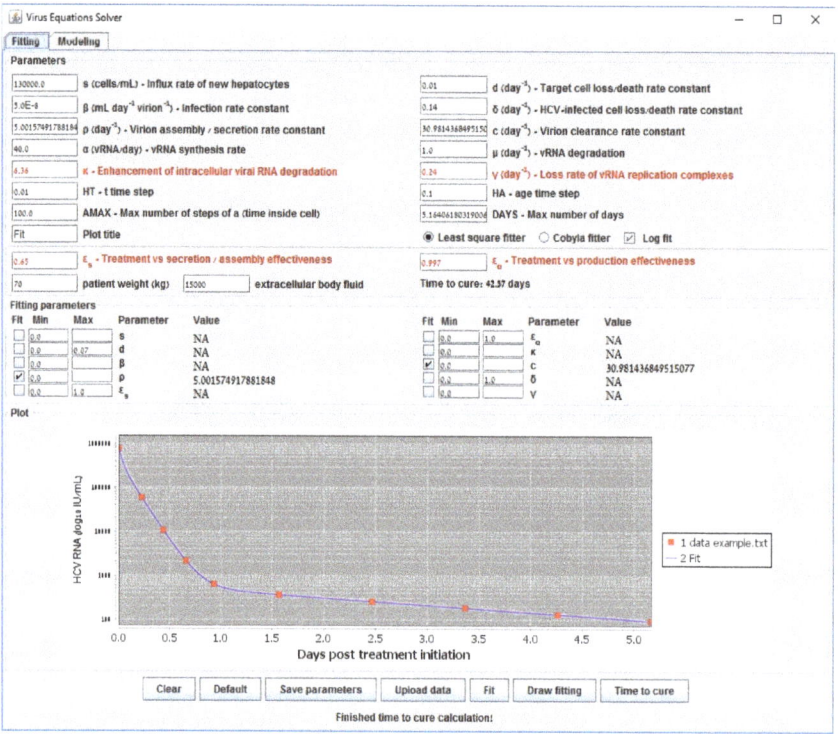

Figure A8. Fitting the parameters c and ρ of the multiscale model to generated data points using the LSF method.

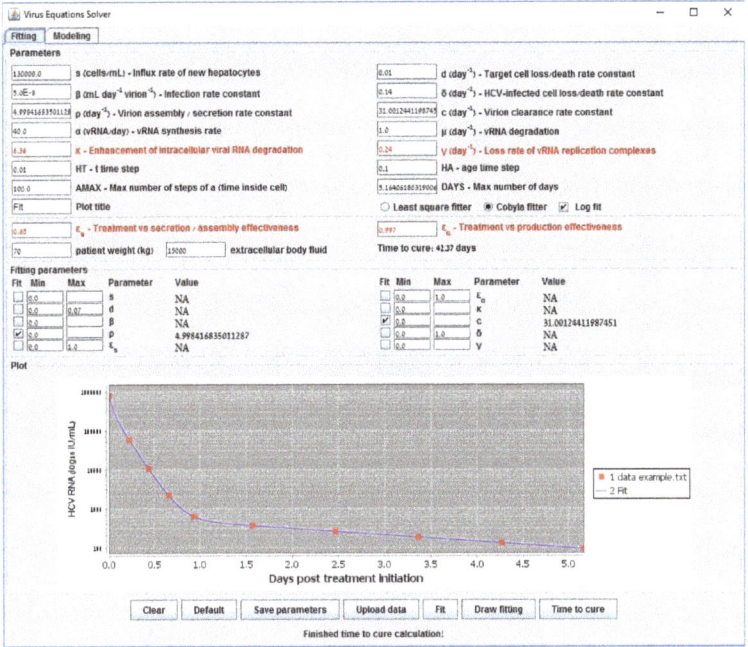

Figure A9. Fitting the parameters c and ρ of the multiscale model to generated data points using the COBYLA method.

References

1. World Health Organization. *Global Hepatitis Report 2017: Web Annex A: Estimations of Worldwide Prevalence of Chronic Hepatitis B Virus Infection: A Systematic Review of Data Published between 1965 and 2017*; Technical Report; World Health Organization: Geneva, Switzerland, 2018.
2. Gilman, C.; Heller, T.; Koh, C. Chronic hepatitis delta: A state-of-the-art review and new therapies. *World J. Gastroenterol.* **2019**, *25*, 4580. [CrossRef] [PubMed]
3. Blach, S.; Zeuzem, S.; Manns, M.; Altraif, I.; Duberg, A.S.; Muljono, D.H.; Waked, I.; Alavian, S.M.; Lee, M.H.; Negro, F.; et al. Global prevalence and genotype distribution of hepatitis C virus infection in 2015: A modelling study. *Lancet Gastroenterol. Hepatol.* **2017**, *2*, 161–176. [CrossRef]
4. Stanaway, J.D.; Flaxman, A.D.; Naghavi, M.; Fitzmaurice, C.; Vos, T.; Abubakar, I.; Abu-Raddad, L.J.; Assadi, R.; Bhala, N.; Cowie, B.; et al. The global burden of viral hepatitis from 1990 to 2013: Findings from the Global Burden of Disease Study 2013. *Lancet* **2016**, *388*, 1081–1088. [CrossRef]
5. Foreman, K.J.; Marquez, N.; Dolgert, A.; Fukutaki, K.; Fullman, N.; McGaughey, M.; Pletcher, M.A.; Smith, A.E.; Tang, K.; Yuan, C.W.; et al. Forecasting life expectancy, years of life lost, and all-cause and cause-specific mortality for 250 causes of death: Reference and alternative scenarios for 2016–2040 for 195 countries and territories. *Lancet* **2018**, *392*, 2052–2090. [CrossRef]
6. Ciupe, S.M. Modeling the dynamics of hepatitis B infection, immunity, and drug therapy. *Immunol. Rev.* **2018**, *285*, 38–54. [CrossRef] [PubMed]
7. Means, S.; Ali, M.A.; Ho, H.; Heffernan, J. Mathematical Modeling for Hepatitis B Virus: Would Spatial Effects Play a Role and How to Model It? *Front. Physiol.* **2020**, *11*, 146. [CrossRef]
8. Dahari, H.; Major, M.; Zhang, X.; Mihalik, K.; Rice, C.M.; Perelson, A.S.; Feinstone, S.M.; Neumann, A.U. Mathematical modeling of primary hepatitis C infection: Noncytolytic clearance and early blockage of virion production. *Gastroenterology* **2005**, *128*, 1056–1066. [CrossRef]
9. Dahari, H.; Layden-Almer, J.E.; Kallwitz, E.; Ribeiro, R.M.; Cotler, S.J.; Layden, T.J.; Perelson, A.S. A mathematical model of hepatitis C virus dynamics in patients with high baseline viral loads or advanced liver disease. *Gastroenterology* **2009**, *136*, 1402–1409. [CrossRef]

10. Goyal, A.; Murray, J.M. Dynamics of in vivo hepatitis D virus infection. *J. Theor. Biol.* **2016**, *398*, 9–19. [CrossRef]
11. Goyal, A.; Ribeiro, R.M.; Perelson, A.S. The role of infected cell proliferation in the clearance of acute HBV infection in humans. *Viruses* **2017**, *9*, 350. [CrossRef]
12. Neumann, A.U.; Phillips, S.; Levine, I.; Ijaz, S.; Dahari, H.; Eren, R.; Dagan, S.; Naoumov, N.V. Novel mechanism of antibodies to hepatitis B virus in blocking viral particle release from cells. *Hepatology* **2010**, *52*, 875–885. [CrossRef] [PubMed]
13. Dahari, H.; de Araujo, E.S.A.; Haagmans, B.L.; Layden, T.J.; Cotler, S.J.; Barone, A.A.; Neumann, A.U. Pharmacodynamics of PEG-IFN-α-2a in HIV/HCV co-infected patients: Implications for treatment outcomes. *J. Hepatol.* **2010**, *53*, 460–467. [CrossRef] [PubMed]
14. Dahari, H.; Ribeiro, R.M.; Perelson, A.S. Triphasic decline of hepatitis C virus RNA during antiviral therapy. *Hepatology* **2007**, *46*, 16–21. [CrossRef] [PubMed]
15. Dahari, H.; Lo, A.; Ribeiro, R.M.; Perelson, A.S. Modeling hepatitis C virus dynamics: Liver regeneration and critical drug efficacy. *J. Theor. Biol.* **2007**, *247*, 371–381. [CrossRef] [PubMed]
16. Dahari, H.; Shudo, E.; Ribeiro, R.M.; Perelson, A.S. Modeling complex decay profiles of hepatitis B virus during antiviral therapy. *Hepatology* **2009**, *49*, 32–38. [CrossRef] [PubMed]
17. Koh, C.; Dubey, P.; Han, M.A.T.; Walter, P.J.; Garraffo, H.M.; Surana, P.; Southall, N.T.; Borochov, N.; Uprichard, S.L.; Cotler, S.J.; et al. A randomized, proof-of-concept clinical trial on repurposing chlorcyclizine for the treatment of chronic hepatitis C. *Antivir. Res.* **2019**, *163*, 149–155. [CrossRef]
18. Dubey, P.; Koh, C.; Surana, P.; Uprichard, S.L.; Han, M.A.T.; Fryzek, N.; Kapuria, D.; Etzion, O.; Takyar, V.K.; Rotman, Y.; et al. Modeling hepatitis delta virus dynamics during ritonavir boosted lonafarnib treatment-the LOWR HDV-3 study. *Hepatology* **2017**, *66*, 21A.
19. Pawlotsky, J.M.; Dahari, H.; Neumann, A.U.; Hezode, C.; Germanidis, G.; Lonjon, I.; Castera, L.; Dhumeaux, D. Antiviral action of ribavirin in chronic hepatitis C. *Gastroenterology* **2004**, *126*, 703–714. [CrossRef]
20. Neumann, A.U.; Lam, N.P.; Dahari, H.; Davidian, M.; Wiley, T.E.; Mika, B.P.; Perelson, A.S.; Layden, T.J. Differences in viral dynamics between genotypes 1 and 2 of hepatitis C virus. *J. Infect. Dis.* **2000**, *182*, 28–35. [CrossRef]
21. Dahari, H.; Shudo, E.; Ribeiro, R.M.; Perelson, A.S. Mathematical modeling of HCV infection and treatment. *Methods Mol. Biol.* **2009**, *510*, 439–453.
22. DebRoy, S.; Hiraga, N.; Imamura, M.; Hayes, C.N.; Akamatsu, S.; Canini, L.; Perelson, A.S.; Pohl, R.T.; Persiani, S.; Uprichard, S.L.; et al. hepatitis C virus dynamics and cellular gene expression in uPA-SCID chimeric mice with humanized livers during intravenous silibinin monotherapy. *J. Viral Hepat.* **2016**, *23*, 708–717. [CrossRef] [PubMed]
23. Canini, L.; DebRoy, S.; Mariño, Z.; Conway, J.M.; Crespo, G.; Navasa, M.; D'Amato, M.; Ferenci, P.; Cotler, S.J.; Forns, X.; et al. Severity of liver disease affects HCV kinetics in patients treated with intravenous silibinin monotherapy. *Antivir. Ther.* **2015**, *20*, 149. [CrossRef] [PubMed]
24. Goyal, A.; Lurie, Y.; Meissner, E.G.; Major, M.; Sansone, N.; Uprichard, S.L.; Cotler, S.J.; Dahari, H. Modeling HCV cure after an ultra-short duration of therapy with direct acting agents. *Antivir. Res.* **2017**, *144*, 281–285. [CrossRef] [PubMed]
25. Dahari, H.; Shteingart, S.; Gafanovich, I.; Cotler, S.J.; D'Amato, M.; Pohl, R.T.; Weiss, G.; Ashkenazi, Y.J.; Tichler, T.; Goldin, E.; et al. Sustained virological response with intravenous silibinin: individualized IFN-free therapy via real-time modelling of HCV kinetics. *Liver Int.* **2015**, *35*, 289–294. [CrossRef]
26. Neumann, A.U.; Lam, N.P.; Dahari, H.; Gretch, D.R.; Wiley, T.E.; Layden, T.J.; Perelson, A.S. Hepatitis C viral dynamics in vivo and the antiviral efficacy of interferon-α therapy. *Science* **1998**, *282*, 103–107. [CrossRef] [PubMed]
27. Guedj, J.; Dahari, H.; Pohl, R.T.; Ferenci, P.; Perelson, A.S. Understanding silibinin's modes of action against HCV using viral kinetic modeling. *J. Hepatol.* **2012**, *56*, 1019–1024. [CrossRef]
28. Lewin, S.R.; Ribeiro, R.M.; Walters, T.; Lau, G.K.; Bowden, S.; Locarnini, S.; Perelson, A.S. Analysis of hepatitis B viral load decline under potent therapy: complex decay profiles observed. *Hepatology* **2001**, *34*, 1012–1020. [CrossRef]

29. Ribeiro, R.M.; Germanidis, G.; Powers, K.A.; Pellegrin, B.; Nikolaidis, P.; Perelson, A.S.; Pawlotsky, J.M. hepatitis B virus kinetics under antiviral therapy sheds light on differences in hepatitis B e antigen positive and negative infections. *J. Infect. Dis.* **2010**, *202*, 1309–1318. [CrossRef]
30. Nowak, M.A.; Bonhoeffer, S.; Hill, A.M.; Boehme, R.; Thomas, H.C.; McDade, H. Viral dynamics in hepatitis B virus infection. *Proc. Natl. Acad. Sci. USA* **1996**, *93*, 4398–4402. [CrossRef]
31. Tsiang, M.; Rooney, J.F.; Toole, J.J.; Gibbs, C.S. Biphasic clearance kinetics of hepatitis B virus from patients during adefovir dipivoxil therapy. *Hepatology* **1999**, *29*, 1863–1869. [CrossRef]
32. Canini, L.; Koh, C.; Cotler, S.J.; Uprichard, S.L.; Winters, M.A.; Han, M.A.T.; Kleiner, D.E.; Idilman, R.; Yurdaydin, C.; Glenn, J.S.; et al. Pharmacokinetics and pharmacodynamics modeling of lonafarnib in patients with chronic hepatitis delta virus infection. *Hepatol. Commun.* **2017**, *1*, 288–292. [CrossRef] [PubMed]
33. Koh, C.; Canini, L.; Dahari, H.; Zhao, X.; Uprichard, S.L.; Haynes-Williams, V.; Winters, M.A.; Subramanya, G.; Cooper, S.L.; Pinto, P.; et al. Oral prenylation inhibition with lonafarnib in chronic hepatitis D infection: A proof-of-concept randomised, double-blind, placebo-controlled phase 2A trial. *Lancet Infect. Dis.* **2015**, *15*, 1167–1174. [CrossRef]
34. Guedj, J.; Rotman, Y.; Cotler, S.J.; Koh, C.; Schmid, P.; Albrecht, J.; Haynes-Williams, V.; Liang, T.J.; Hoofnagle, J.H.; Heller, T.; et al. Understanding early serum hepatitis D virus and hepatitis B surface antigen kinetics during pegylated interferon-alpha therapy via mathematical modeling. *Hepatology* **2014**, *60*, 1902–1910. [CrossRef] [PubMed]
35. Shekhtman, L.; Cotler, S.J.; Hershkovich, L.; Uprichard, S.L.; Bazinet, M.; Pantea, V.; Cebotarescu, V.; Cojuhari, L.; Jimbei, P.; Krawczyk, A.; et al. Modelling hepatitis D virus RNA and HBsAg dynamics during nucleic acid polymer monotherapy suggest rapid turnover of HBsAg. *Sci. Rep.* **2020**, *10*, 1–7. [CrossRef]
36. Etzion, O.; Dahari, H.; Yardeni, D.; Issachar, A.; Nevo-Shor, A.; Naftaly-Cohen, M.; Uprichard, S.L.; Arbib, O.S.; Munteanu, D.; Braun, M.; et al. Response-Guided Therapy with Direct-Acting Antivirals Shortens Treatment Duration in 50% of HCV Treated Patients. *Hepatology* **2018**, *68*, 1469A–1470A.
37. Dahari, H.; Canini, L.; Graw, F.; Uprichard, S.L.; Araújo, E.S.; Penaranda, G.; Coquet, E.; Chiche, L.; Riso, A.; Renou, C.; et al. HCV kinetic and modeling analyses indicate similar time to cure among sofosbuvir combination regimens with daclatasvir, simeprevir or ledipasvir. *J. Hepatol.* **2016**, *64*, 1232–1239. [CrossRef]
38. Canini, L.; Imamura, M.; Kawakami, Y.; Uprichard, S.L.; Cotler, S.J.; Dahari, H.; Chayama, K. HCV kinetic and modeling analyses project shorter durations to cure under combined therapy with daclatasvir and asunaprevir in chronic HCV-infected patients. *PLoS ONE* **2017**, *12*, e0187409. [CrossRef]
39. Gambato, M.; Canini, L.; Lens, S.; Graw, F.; Perpiñan, E.; Londoño, M.C.; Uprichard, S.L.; Mariño, Z.; Reverter, E.; Bartres, C.; et al. Early HCV viral kinetics under DAAs may optimize duration of therapy in patients with compensated cirrhosis. *Liver Int.* **2019**, *39*, 826–834. [CrossRef]
40. Sandmann, L.; Manns, M.P.; Maasoumy, B. Utility of viral kinetics in HCV therapy—It is not over until it is over? *Liver Int.* **2019**, *39*, 815–817. [CrossRef]
41. Deng, B.; Lou, S.; Dubey, P.; Etzion, O.; Chayam, K.; Uprichard, S.; Sulkowski, M.; Cotler, S.; Dahari, H. Modeling time to cure after short-duration treatment for chronic HCV with daclatasvir, asunaprevir, beclabuvir and sofosbuvir: the FOURward study. *J. Viral Hepat.* **2018**, *25*, 58.
42. Dahari, H.; Sainz, B.; Perelson, A.S.; Uprichard, S.L. Modeling subgenomic hepatitis C virus RNA kinetics during treatment with alpha interferon. *J. Virol.* **2009**, *83*, 6383–6390. [CrossRef] [PubMed]
43. Dahari, H.; Ribeiro, R.M.; Rice, C.M.; Perelson, A.S. Mathematical modeling of subgenomic hepatitis C virus replication in Huh-7 cells. *J. Virol.* **2007**, *81*, 750–760. [CrossRef] [PubMed]
44. Murray, J.M.; Goyal, A. In silico single cell dynamics of hepatitis B virus infection and clearance. *J. Theor. Biol.* **2015**, *366*, 91–102. [CrossRef] [PubMed]
45. Murray, J.M.; Wieland, S.F.; Purcell, R.H.; Chisari, F.V. Dynamics of hepatitis B virus clearance in chimpanzees. *Proc. Natl. Acad. Sci. USA* **2005**, *102*, 17780–17785. [CrossRef]
46. Packer, A.; Forde, J.; Hews, S.; Kuang, Y. Mathematical models of the interrelated dynamics of hepatitis D and B. *Math. Biosci.* **2014**, *247*, 38–46. [CrossRef]
47. Guedj, J.; Dahari, H.; Rong, L.; Sansone, N.D.; Nettles, R.E.; Cotler, S.J.; Layden, T.J.; Uprichard, S.L.; Perelson, A.S. Modeling shows that the NS5A inhibitor daclatasvir has two modes of action and yields a shorter estimate of the hepatitis C virus half-life. *Proc. Natl. Acad. Sci. USA* **2013**, *110*, 3991–3996. [CrossRef]

48. Rong, L.; Guedj, J.; Dahari, H.; Coffield, D.J.J.; Levi, M.; Smith, P.; Perelson, A.S. Analysis of hepatitis C virus decline during treatment with the protease inhibitor danoprevir using a multiscale model. *PLoS Comput. Biol.* **2013**, *9*, e1002959. [CrossRef]
49. Rong, L.; Perelson, A.S. Mathematical analysis of multiscale models for hepatitis C virus dynamics under therapy with direct-acting antiviral agents. *Math. Biosci.* **2013**, *245*, 22–30. [CrossRef]
50. Quintela, B.M.; Conway, J.M.; Hyman, J.M.; Guedj, J.; dos Santos, R.W.; Lobosco, M.; Perelson, A.S. A New Age-Structured Multiscale Model of the Hepatitis C Virus Life-Cycle During Infection and Therapy with Direct-Acting Antiviral Agents. *Front. Microbiol.* **2018**, *9*, 601. [CrossRef]
51. Guedj, J.; Neumann, A.U. Understanding hepatitis C viral dynamics with direct-acting antiviral agents due to the interplay between intracellular replication and cellular infection dynamics. *J. Theor. Biol.* **2010**, *267*, 330–340. [CrossRef]
52. Weickert, J.; ter Haar Romeny, B.; Viergever, M. Efficient and reliable schemes for nonlinear diffusion filtering. *IEEE Trans. Imag. Proc.* **1998**, *7*, 398–410. [CrossRef]
53. Barash, D.; Israeli, M.; Kimmel, R. An Accurate Operator Splitting Scheme for Nonlinear Diffusion Filtering. In Proceedings of the 3rd International Conference on ScaleSpace and Morphology, Vancouver, BC, Canada, 7–8 July 2001; pp. 281–289.
54. Barash, D. Nonlinear Diffusion Filtering on Extended Neighborhood. *Appl. Num. Math.* **2005**, *52*, 1–11. [CrossRef]
55. Reinharz, V.; Churkin, A.; Dahari, H.; Barash, D. A Robust and Efficient Numerical Method for RNA-mediated Viral Dynamics. *Front. Appl. Math. Stat.* **2017**, *3*, 20. [CrossRef] [PubMed]
56. Reinharz, V.; Dahari, H.; Barash, D. Numerical schemes for solving and optimizing multiscale models with age of hepatitis C virus dynamics. *Math. Biosci.* **2018**, *300*, 1–13. [CrossRef]
57. Reinharz, V.; Churkin, A.; Lewkiewicz, S.; Dahari, H.; Barash, D. A Parameter Estimation Method for Multiscale Models of hepatitis C Virus Dynamics. *Bull. Math. Biol.* **2019**, *81*, 3675–3721. [CrossRef] [PubMed]
58. Levenberg, K. A method for the solution of certain nonlinear problems in least squares. *Q. Appl. Math.* **1944**, *2*, 164–168. [CrossRef]
59. Marquardt, D.W. An algorithm for least-squares estimation of nonlinear parameters. *J. Soc. Ind. Appl. Math.* **1963**, *11*, 431–441. [CrossRef]
60. Kitagawa, K.; Nakaoka, S.; Asai, Y.; Watashi, K.; Iwami, S. A PDE Multiscale Model of hepatitis C virus infection can be transformed to a system of ODEs. *J. Theor. Biol.* **2018**, *448*, 80–85. [CrossRef]
61. Powell, M.J.D. A View of Algorithms for Optimization Without Derivatives. *Math. Today* **2007**, *43*, 170–174.
62. Rohatgi, A. WebPlotDigitizer: Web Based Tool to Extract Data from Plots, Images, and Maps. V 4.1. 2018. Available online: https://automeris.io/WebPlotDigitizer (accessed on 22 August 2020).
63. Quarteroni, A.; Valli, A. *Springer Series in Computational Mathematics, 1994*; Springer: Berlin/Heidelberg, Germany, 1994.
64. Knabner, P.; Angermann, L. *Numerical Methods for Elliptic and Parabolic Partial Differential Equations: An Applications-Oriented Introduction*; Springer: Berlin/Heidelberg, Germany, 2004.
65. Powell, M.J.D. A Direct Search Optimization Method That Models Objective and Constraint Functions by Linear Interpolation. In *Advances in Optimization and Numerical Analysis. Mathematics and its Applications*; Gomez, S., Hennart, J., Eds.; Springer: Dordrecht, The Netherlands, 1994; Volume 275.
66. Bastian, P.; Heimann, F.; Marnach, S. Generic implementation of finite element methods in the distributed and unified numerics environment (DUNE). *Kybernetika* **2010**, *46*, 294–315.
67. Flemisch, B.; Darcis, M.; Erbertseder, K.; Faigle, B.; Lauser, A.; Mosthaf, K.; Müthing, S.; Nuske, P.; Tatomir, A.; Wolff, M.; et al. DuMux: DUNE for multi-{phase, component, scale, physics,...} flow and transport in porous media. *Adv. Water Resour.* **2011**, *34*, 1102–1112. [CrossRef]
68. Vogel, A.; Reiter, S.; Rupp, M.; Nägel, A.; Wittum, G. UG 4: A novel flexible software system for simulating PDE based models on high performance computers. *Comput. Vis. Sci.* **2013**, *16*, 165–179. [CrossRef]
69. Spendley, W.; Hext, G.R.; Himsworth, F.R. Sequential Application of Simplex Designs in Optimisation and Evolutionary Operation. *Technometrics* **1962**, *4*, 441–461. [CrossRef]
70. Nelder, J.; Mead, R. A Simplex Method for Function Minimization. *Comput. J.* **1965**, *7*, 308–313. [CrossRef]

71. Dasgupta, S.; Imamura, M.; Gorstein, E.; Nakahara, T.; Tsuge, M.; Churkin, A.; Yardeni, D.; Etzion, O.; Uprichard, S.L.; Barash, D.; et al. Modeling-Based Response-Guided Glecaprevir-Pibrentasvir Therapy for Chronic Hepatitis C to Identify Patients for Ultrashort Treatment Duration. *J. Infect. Dis.* **2020**, jiaa219. [CrossRef] [PubMed]
72. Gorstein, E.; Martinello, M.; Churkin, A.; Dasgupta, S.; Walsh, K.; Applegate, T.; Yardeni, D.; Etzion, O.; Uprichard, S.L.; Barash, D.; et al. Modeling based response guided therapy in subjects with recent hepatitis C infection. *Antivir. Res.* **2020**. [CrossRef]

© 2020 by the authors. Licensee MDPI, Basel, Switzerland. This article is an open access article distributed under the terms and conditions of the Creative Commons Attribution (CC BY) license (http://creativecommons.org/licenses/by/4.0/).

Article

Nonlocal Reaction–Diffusion Model of Viral Evolution: Emergence of Virus Strains

Nikolai Bessonov [1], Gennady Bocharov [2,3], Andreas Meyerhans [2,4,5], Vladimir Popov [6] and Vitaly Volpert [2,6,7,8,*]

1. Institute of Problems of Mechanical Engineering, Russian Academy of Sciences, 199178 Saint Petersburg, Russia; nickbessonov1@gmail.com
2. Marchuk Institute of Numerical Mathematics, Russian Academy of Sciences, 199333 Moscow, Russia; bocharov@m.inm.ras.ru (G.B.); andreas.meyerhans@upf.edu (A.M.)
3. Key Center of Excellence on Experimental immunophysiology and immunochemistry, Ural Federal University, 620002 Ekaterinburg, Russia
4. Institució Catalana de Recerca i Estudis Avançats (ICREA), Pg. Lluis Companys 23, 08003 Barcelona, Spain
5. Infection Biology Laboratory, Universitat Pompeu Fabra, 08003 Barcelona, Spain
6. Peoples' Friendship University of Russia (RUDN University), 6 Miklukho-Maklaya St, 117198 Moscow, Russia; popov-va@rudn.ru
7. Institut Camille Jordan, UMR 5208 CNRS, University Lyon 1, 69622 Villeurbanne, France
8. INRIA Team Dracula, INRIA Lyon La Doua, 69603 Villeurbanne, France
* Correspondence: volpert@math.univ-lyon1.fr

Received: 2 December 2019; Accepted: 9 January 2020; Published: 12 January 2020

Abstract: This work is devoted to the investigation of virus quasi-species evolution and diversification due to mutations, competition for host cells, and cross-reactive immune responses. The model consists of a nonlocal reaction–diffusion equation for the virus density depending on the genotype considered to be a continuous variable and on time. This equation contains two integral terms corresponding to the nonlocal effects of virus interaction with host cells and with immune cells. In the model, a virus strain is represented by a localized solution concentrated around some given genotype. Emergence of new strains corresponds to a periodic wave propagating in the space of genotypes. The conditions of appearance of such waves and their dynamics are described.

Keywords: virus density distribution; genotype; virus infection; immune response; resistance to treatment; nonlocal interaction; quasi-species diversification

MSC: 35K57; 92C20

1. Introduction

Human infections with rapidly evolving viruses such as the human immunodeficiency virus (HIV) or the hepatitis C virus (HCV) remain a challenge for health-care systems. Infections are usually initiated by one or few virions that then replicate and generate a swarm of progeny viruses with distinct but related genomes [1,2]. Collectively these swarms of viruses are called a virus quasi-species [3–7]. This quasi-species nature enables viruses to rapidly evolve within an infected host organism and adapt to constraints mediated by immune responses or antiviral drugs [8,9]. It also allows viruses to broaden their host cell tropism and to spread to diverse tissues [10]. Well studied examples for virus adaptation are the development of drug resistance or the generation of variants within virus-specific cytotoxic T lymphocyte (CTL) epitopes that diminish immune recognition and destruction of infected cells [11–13]. Since the immune system can also adapt to respective virus changes [14], an increase in the number of CTL target regions over time of infection as well as successive shifts in the hierarchy of immunodominance have been observed [15,16].

Gaining a mechanistic understanding of the dynamic interplay between the processes of virus replication, mutation, and elimination by immune responses and drug-based treatment requires the development of mathematical models which could be used to predict the generation of viral variants that escape the immune recognition and confer resistive to antiviral drugs. The existing models of virus evolution are based on the concept of quasi-species. i.e., an ensemble of related genomes [4]. The models can be formulated either as deterministic high-dimensional systems of ODEs, describing the densities of individual strain [15,17] or stochastic models with genetic algorithms [18]. The cooperative interactions in viral populations are considered to be key for linking the quasi-species dynamics in a changing virus-host environment with the genetic markers of viral evolution and the disease pathogenesis [10,19]. This implies that nonlocal interactions between the quasi-species in the genotype space need to be considered to predict the evolution of viruses to form distinct phenotypes.

Nonlocal reaction–diffusion equations represent an appropriate framework to describe evolution of biological species [20–22]. These equations take into account nonlocal consumption of resources characterizing intraspecific competition and possibly leading to the emergence of multi-modal population density distributions. Considered to be depending on a morphological characteristic and on time, localized in-space distributions can be interpreted as biological species, and the emergence of multi-modal distributions corresponds to the appearance of new species. In this work we will study virus quasi-species and will analyze the emergence of new strains in the space of genotypes. We consider the nonlocal reaction–diffusion equation

$$\frac{\partial u}{\partial t} = D\frac{\partial^2 u}{\partial x^2} + ru(1 - qJ(u)) - uf(S(u_\tau)) - \sigma(x)u \qquad (1)$$

introduced in the previous work [23] devoted to the existence and dynamics of virus strains, but not to the emergence of new strains since this question is different both from the biological and modelling points of view. Here $u(x,t)$ is a dimensionless virus density distribution depending on its genotype x considered to be a continuous variable and on time t. The diffusion term in the right-hand side of this equation characterizes virus mutations, and the other terms describe virus reproduction, its elimination by immune response and by genotype-dependent mortality, either natural or caused by an antiviral treatment.

We describe the virus reproduction and immune response terms in more detail. We begin with the diffusion term. Assuming that there is a sequence of reversible mutations with consecutive genotypes x_i, we can write the equation for the density u_i of virus with genotype x_i:

$$\frac{du_i}{dt} = \mu(u_{i-1} - u_i) + \mu(u_{i+1} - u_i),$$

where μ is the frequency of mutations. This equation represents a discretization of the diffusion equation with the diffusion coefficient proportional to μ.

The virus reproduction rate is conventionally considered either proportionally to its density u or, if we take into account the limitation on the quantity of the host cells where virus can multiplicate, as a logistic term $ru(1 - qu)$. Here r is a proportionality coefficient, and q is a positive constant corresponding to the inverse carrying capacity in population dynamics. The latter case implicitly signifies that there is one-to-one correspondence between the virus genotype and the type of infected cells. In a more general and biologically realistic setting we should accept that viruses with different genotypes can infect the same cells. In this case, we replace the conventional logistic term by the term $ru(1 - qJ(u))$, where $J(u) = \int_{-\infty}^{\infty} \phi(x - y)u(y,t)dy$. The kernel $\phi(x - y)$ in this integral characterizes the efficacy of host cell infection depending on the difference in genotypes. In general, it is a decreasing function of the modulus of its argument. Its exact form in the applications is not known, and different examples will be considered below. The integral is taken in the infinite limits for convenience of presentation. It implies that the genotype space is sufficiently large, and it can be mathematically approximated as a real line.

The reproduction rate with the integral $J(u)$ corresponds to nonlocal consumption of resources in population dynamics [22,24]. If we replace the kernel $\phi(x)$ by the δ-function, then we obtain the previous "local" case. Finally, if cell contamination is independent of virus genotype, then we have the integral $I(u) = \int_{-\infty}^{\infty} u(y,t)dy$ corresponding to the global consumption of resources. Behavior of solutions of Equation (1) can be essentially different in these three cases.

Virus elimination by immune cells is proportional to the virus density u and to the concentration of immune cells C. Since immune response is stimulated by the antigen (virus), then the concentration of immune cells can be considered to be a function of virus density $C = f(u)$. The function $f(u)$ characterizes the intensity of immune response. It is positive and growing for some limited interval of values of u where the antigen stimulates the immune response, and it is decreasing for u sufficiently large due to the exhaustion and death of immune cells provoked by large virus concentrations [25–27]. The qualitative form of this function is well described by the dependence $f(u) = (k_1 u + k_2)e^{-k_3 u}$ with some positive constants k_1, k_2, and k_3. The approximation of the concentration of immune cells as a function of virus density can be derived from a more complete model under the assumption of large reaction rate constants [28].

Clonal expansion of immune cells requires several cell proliferations and differentiations, and it usually takes 3–4 days. Therefore, the rate of virus elimination by immune cells can take into account this time delay if it is not negligible in the time scale of virus evolution. In this case, the corresponding term becomes $uf(u_\tau)$, where $u_\tau = u(x, t-\tau)$. Furthermore, similar to virus reproduction, virus elimination is also nonlocal in the genotype space. In [23] it was described by the term $\int_{-\infty}^{\infty} \theta(x-y) f(S(u_\tau)(y,t)) dy$, where $S(u_\tau)(y,t) = \int_{-\infty}^{\infty} \psi(y-z) u(z, t-\tau) dz$. The inner integral S characterizes a cross-reactive stimulation of immune response by different antigens, while the outer integral describes a cross-reactive virus elimination by different immune cells. Both assumptions are biologically justified. However, this model becomes excessively complex, and we will restrict ourselves here only to the inner integral assuming the outer term is local, i.e., that the kernel $\theta(x)$ is replaced by the δ-function.

The last term in the right-hand side of Equation (1) describes virus mortality with the rate depending on its genotype. The viability interval, i.e., the rate of genotypes where virus multiplication rate exceeds its mortality can depend on its intrinsic features and on an antiviral treatment.

Some particular cases of Equation (1) are studied in the literature. Considered without immune response and genotype-dependent mortality, this nonlocal reaction–diffusion equation and some its variations were widely studied in relation to various applications [29–32] and from the point of view of their mathematical properties [33–36]. One of the main features of this equation is that its homogeneous in-space stationary solution can become unstable leading to the emergence of periodic in-space solutions. We will return to this question below. The local equation ($J(u) \to u, S(u) \to u$) with time delay in the immune response term was suggested in [23,26,27,37] as a model of virus spread in tissues. The presence of time delay can lead to complex patterns of wave propagation.

In our previous work [23] we studied the existence and dynamics of virus strains considered to be localized solutions (pulses) in the space of genotypes, with the understanding that a virus strain can be characterized by its most frequent genotype with a narrow density distribution around it. The existence and stability of such solutions is not a priori given. In the local bistable equation such solutions exist but they are not stable. In the local monostable equation such solutions do not exist. It was previously shown that stable pulses exist for the nonlocal bistable equation [38]. In [23], it was revealed that persistent virus strains can exist due to the interaction of nonlocal (global) virus reproduction with immune response or with genotype-dependent mortality rate. This modelling approach allows us to investigate the competition of different strains and the emergence of resistant strains due to treatment.

In this work we will study the question about the emergence of new strains. From the modelling point of view, these two cases, existence of stable strains and emergence of new strains are complementary, they are not observed at the same time. The former corresponds to stable pulses while the latter to periodic travelling waves. A nonlocal reaction–diffusion equation describing the emergence of biological species was suggested in [22]. It represents a particular case of Equation (1) without immune response and genotype-dependent mortality. We will show that immune response plays an important role in the dynamics of virus quasi-species.

2. Bifurcations of Periodic Structures

An important property of nonlocal reaction–diffusion equations is that the homogeneous in-space stationary solution can lose its stability with respect to spatial perturbations leading to the emergence of periodic solutions. This property was revealed and studied in some models of population dynamics [22,39]. Here we will study it for the models of infection development described by the equation

$$\frac{\partial u}{\partial t} = D\frac{\partial^2 u}{\partial x^2} + ru(1 - qJ(u)) - uf(S(u_\tau)) \tag{1}$$

similar to Equation (1) but without the genotype-dependent mortality term. We will analyze various particular cases of this equation.

2.1. Single Nonlocal Term

We consider the reaction–diffusion equation

$$\frac{\partial u}{\partial t} = D\frac{\partial^2 u}{\partial x^2} + ru(1 - qJ(u)) - f(u)u \tag{2}$$

for all real x, where

$$J(u) = \int_{-\infty}^{\infty} \phi(x-y)u(y,t)dy,$$

the function $\phi(x)$ is bounded and non-negative. We will suppose that $\int_{-\infty}^{\infty} \phi(x)dx = 1$. In what follows, we set $r = q = 1$. Let $u_0 > 0$ be a solution of the equation

$$f(u) = 1 - u. \tag{3}$$

Then u_0 is a stationary solution of Equation (2). We will study stability of this stationary solution. To linearize Equation (2) about $u = u_0$, we look for its solution in the form $u = u_0 + ve^{\lambda t}$, where v is a small perturbation and we obtain the eigenvalue problem

$$Dv'' - u_0(J(v) + f'(u_0)v) = \lambda v. \tag{4}$$

Applying the Fourier transform, we get

$$\lambda = -D\xi^2 - u_0(\tilde{\phi}(\xi) + f'(u_0)), \tag{5}$$

where $\tilde{\phi}(x)$ is the Fourier transform of the function $\phi(x)$. If we replace $\phi(x)$ by the δ-function, instead of (5) we have

$$\lambda = -D\xi^2 - u_0(1 + f'(u_0)). \tag{6}$$

Assuming that

$$1 + f'(u_0) > 0, \tag{7}$$

we conclude from (6) that $\lambda < 0$ for all real ξ.

Let us now analyze equality (5). Since $\tilde{\phi}(0) = 1$, then $\lambda(0) < 0$. Suppose that $\phi(x)$ is an even function, $\phi(x) = \phi(-x)$ for all x. Then

$$\tilde{\phi}(\xi) = \int_{-\infty}^{\infty} \phi(x)\cos(\xi x)dx,$$

$\tilde{\phi}(0) = 1$, and $\tilde{\phi}(\xi) < 1$ for all $\xi \neq 0$. Assuming that $\lambda = 0$ in (5) for some ξ, we obtain the stability boundary:
$$\tilde{\phi}(\xi) = -f'(u_0) - D\xi^2/u_0. \tag{8}$$

This equality should be satisfied for some real ξ. Its value is related to the wave number of the corresponding eigenfunction. If we consider a bounded interval with periodic boundary conditions, then $\xi = 2\pi k/L$, where L is the length of the interval and $k = 1, 2, 3 \ldots$

2.2. Examples

Consider the functions

$$\phi_1(x) = \sqrt{\frac{a}{\pi}} e^{-ax^2}, \quad \phi_2(x) = \frac{a}{2} e^{-a|x|}, \quad \phi_3(x) = \begin{cases} 1/(2N) &, \quad |x| \leq N \\ 0 &, \quad |x| > N \end{cases}.$$

Then

$$\tilde{\phi}_1(\xi) = e^{-\xi^2/(4a)}, \quad \tilde{\phi}_2(\xi) = \frac{a^2}{a^2 + \xi^2}, \quad \tilde{\phi}_3(\xi) = \frac{1}{\xi N} \sin(\xi N).$$

Figure 1 (left) shows a graphical solution of Equation (8) for the function $\phi_3(x)$. The curves corresponding to the functions in the left-hand side and in the right-hand side of this equation touch each other. The corresponding values of parameters belong to the stability boundary. For lesser values of the diffusion coefficient, the homogeneous in-space stationary solution u_- loses its stability resulting in the emergence of a periodic in-space solution.

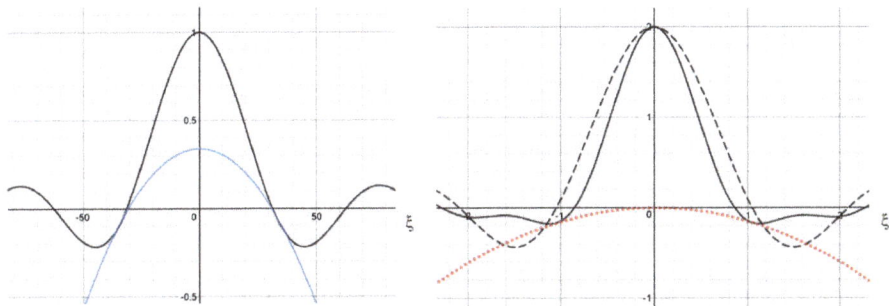

Figure 1. (**Left**) Graphical solution of Equation (8): functions $(\sin(\xi N))/(\xi N)$ and $-f'(u_0) - D\xi^2/u_0$, where $N = 0.1$, $f'(u_0) = -0.34$, $D = 0.00033$, $u_0 = 0.93$. (**Right**) Graphical solution of Equation (11): the function $\phi(\xi) + b\tilde{\psi}(\xi)$ for the values of parameters $b = 1$, $N_1 = 3$, $N_2 = 5$ (solid line), $N_1 = 3$, $N_2 = 3$ (dashed line), and the function $-D/u_-\xi^2$ with $D/u_- = 0.1508$ (point line).

Let us note that Fourier transforms of the functions $\phi_1(x)$ and $\phi_2(x)$ are positive. Therefore, if $f'(u_-) \geq 0$, then Equation (8) does not have solution, and solution u_- is stable. If $f'(u_-) < 0$, it has a solution for sufficiently small values of the diffusion coefficient. Thus, emergence of periodic solutions is determined by the interaction of virus mutations, nonlocal competition for host cells and immune response. In terms of virus population distribution in the space of genotypes, these periodic solutions correspond to different virus strains.

2.3. Double Nonlocal Equation

Consider Equation (1) with two nonlocal terms $J(u)$ and $S(u)$ and without time delay ($\tau = 0$). For simplicity of presentation, we set $f(u) = bu$. In the stationary case, we obtain the equation

$$Du'' + u(1 - J(u) - bS(u)) = 0.$$

Let us recall that $J(u) = \int_{-\infty}^{\infty} \phi(x-y) u(y) dy$, $S(u) = \int_{-\infty}^{\infty} \psi(x-y) u(y) dy$. Assuming that $\int_{-\infty}^{\infty} \phi(y) dy = \int_{-\infty}^{\infty} \psi(y) dy = 1$, we get a homogeneous in-space stationary solution $u_- = 1/(1+b)$ of this equation. Linearizing the equation about this stationary solution, we obtain the eigenvalue problem:

$$Dv'' - u_-(J(v) + bS(v)) = \lambda v. \tag{9}$$

Applying the Fourier transform to (9), we get

$$\lambda = -D\xi^2 - u_-\left(\tilde{\phi}(\xi) + b\tilde{\psi}(\xi)\right). \tag{10}$$

Consider the functions

$$\phi(x) = \begin{cases} 1/(2N_1) &, |x| \leq N_1 \\ 0 &, |x| > N_1 \end{cases}, \quad \psi(x) = \begin{cases} 1/(2N_2) &, |x| \leq N_2 \\ 0 &, |x| > N_2 \end{cases}$$

Then

$$\tilde{\phi}(\xi) = \frac{1}{\xi N_1} \sin(\xi N_1), \quad \tilde{\psi}(\xi) = \frac{1}{\xi N_2} \sin(\xi N_2).$$

Stability boundary is determined by the equation

$$\tilde{\phi}(\xi) + b\tilde{\psi}(\xi) = -D\xi^2/u_- \tag{11}$$

obtained from Equation (10) with $\lambda = 0$. An example of graphical solution of this equation is shown in Figure 1. The stability boundary corresponds to the case where the functions in the left-hand side and in the right-hand side of this equation touch each other (solid and point lines). If they intersect (dashed and point lines), then the stationary solution is unstable. For the fixed values of N_1, N_2 and b, stability conditions are determined by the diffusion coefficient.

Proposition 1. *For any positive values N_1, N_2 and b there exists a critical value D_c of the diffusion coefficient such that the stationary solution $u_- = 1/(1+b)$ is stable for $D > D_c$ and unstable for $D < D_c$.*

The proof of this proposition is straightforward. It is sufficient to note that the function $\tilde{\phi}(\xi) + b\tilde{\psi}(\xi)$ has negative values for any values of parameters.

Dependence of the stability conditions on N_1 and N_2 is more complex. Both can have stabilizing or destabilizing effect on the solution. In the example in Figure 1, decreasing the value N_2 leads to the instability of the solution (solid and dashed lines).

2.4. Delay Equation

Stability of solutions of the delay equation

$$\frac{\partial u}{\partial t} = D\frac{\partial^2 u}{\partial x^2} + ru(1-qu) - f(u_\tau)u \tag{12}$$

without nonlocal terms was studied in [37]. Spatial perturbations of the homogeneous in-space solution can lead to a complex spatiotemporal behavior with standing wave, travelling waves and aperiodic dynamics.

Nonlocal delay equation.

Consider now Equation (1) with a single nonlocal term and with time delay:

$$\frac{\partial u}{\partial t} = D\frac{\partial^2 u}{\partial x^2} + ru(1-qJ(u)) - uf(u_\tau). \tag{13}$$

In what follows, we set $r = q = 1$. Linearizing this equation about a stationary solution $u = u_0$, we obtain the eigenvalue problem

$$Dv'' - u_0\left(J(v) + f'(u_0)e^{-\lambda\tau}v\right) = \lambda v.$$

Applying the Fourier transform, we get

$$\lambda = -D\xi^2 - u_0\left(\tilde{\phi}(\xi) + f'(u_0)e^{-\lambda\tau}\right).$$

For $\lambda = 0$ we obtain Equation (8) for the stability boundary. Therefore, as it can be expected, time delay does not influence the bifurcation of stationary solution. Consider now the bifurcation of time periodic solution. We set $\lambda = i\nu$, where ν is a real number. Then separating the real and imaginary part in the last equation we obtain:

$$\nu = u_0 f'(u_0)\sin(\nu\tau), \quad D\xi^2 + u_0\tilde{\phi}(\xi) + u_0 f'(u_0)\cos(\nu\tau) = 0.$$

Set $z = \nu\tau$. Then

$$\cos z = -\frac{D\xi^2/u_0 + \tilde{\phi}(\xi)}{f'(u_0)}, \quad \tau = \frac{z}{u_0 f'(u_0)\sin z}. \tag{14}$$

We find the value of z from the first equation and the value of τ from the second equation. The first equation has a solution if and only if

$$\left|\frac{D\xi^2/u_0 + \tilde{\phi}(\xi)}{f'(u_0)}\right| \leq 1. \tag{15}$$

In the case of the local equation where $\tilde{\phi}(\xi) = 1$, the minimum of the numerator is reached at $\xi = 0$. Therefore, the loss of stability occurs with the space independent perturbations assuming that $|f'(u_0)| < 1$. For the nonlocal equation the loss of stability can occur for $\xi \neq 0$ and for $|f'(u_0)| \geq 1$.

Consider the case where the function $f(u)$ is linear, $f(u) = k_1 u$. If $k_1 > 1$ and N_1 is sufficiently small, then the homogeneous in-space solution u_0 can lose its stability with respect to temporal perturbation for the time delay τ large enough. If τ is less than a critical value then temporal and spatial perturbations decay (Figure 2, left). If we increase N for the other parameters fixed, then the constant solution u_0 can lose its stability with respect to spatiotemporal perturbations. Figure 2 (right) shows the emergence of a spatiotemporal pattern at the center of the interval. It propagates and gradually fills the whole spatial domain. Let us now take $k_1 < 1$. Then time oscillations in the local problem ($N_1 = 0$) decay for any time delay. In the nonlocal problem and N_1 sufficiently large various spatiotemporal patterns can be observed (Figure A1 in Appendix A).

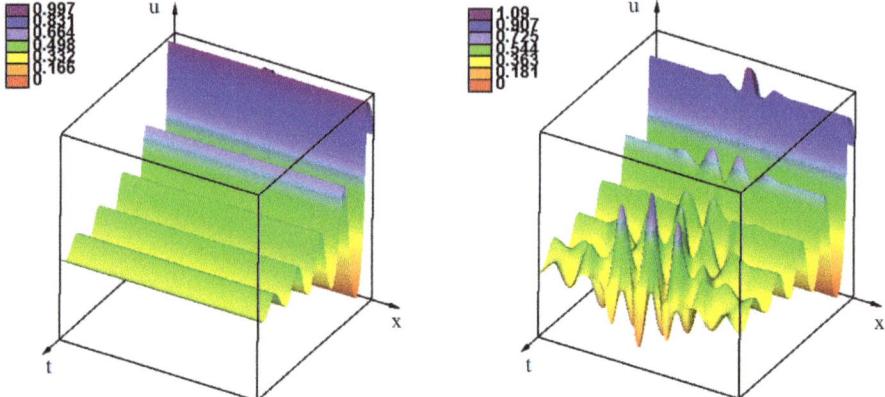

Figure 2. Numerical simulations of Equation (13) for the linear function $f(u) = k_1 u$. Spatial and temporal perturbation decay if the solution u_- is stable (**left**). The spatial perturbation at the center of the interval leads to the emergence of a spatiotemporal pattern propagating from the center and gradually filling the whole spatial domain (**right**). The value of parameters: $r = 1, q = 1, k_1 = 1.5, D = 10^{-5}, \tau = 3$, $N_1 = 0.01$ (**left**) and $N_1 = 0.1$ (**right**), $t = 50$. Here and in all figures below, $L = 1$, unless another value is indicated.

3. Emergence of Strains as Periodic Wave Propagation

3.1. Propagation Of Waves

We study in this section propagation of described by Equation (1) assuming for simplicity that the second integral term $S(u)$ becomes local, i.e., the kernel $\psi(x)$ is the replaced by the δ-function:

$$\frac{\partial u}{\partial t} = D\frac{\partial^2 u}{\partial x^2} + ru(1 - qJ(u)) - uf(u_\tau). \tag{1}$$

Local and delay equations.

To study the behavior of solutions, we begin with the local case. Then we get conventional reaction–diffusion equation

$$\frac{\partial u}{\partial t} = D\frac{\partial^2 u}{\partial x^2} + F(u), \tag{2}$$

where the function $F(u) = ru(1 - qu) - uf(u)$ can have different numbers of zeros depending on the values of parameters. Besides the zero $u_+ = 0$, there can exist up to three positive zeros, the maximal zero u_- and possibly one or two intermediate zeros u_1 and u_2.

Monostable case. If there is only one positive zero u_-, then $F'(u_+) > 0$, $F'(u_-) < 0$. The $[u_+, u_-]$-waves, i.e., the waves with the limits $u(\pm\infty) = u_\pm$ at infinity, exist for all values of the speed greater than or equal to some minimal speed c_0. These waves are stable in appropriate weighted spaces [24].

Bistable case. In the bistable case, $F'(u_\pm) < 0$, and there is an additional zero $u_1 \in (u_+, u_-)$. The $[u_+, u_-]$-wave exists for a single value of speed c_1, and this wave is globally asymptotically stable.

Monostable–bistable case. In this case, there are two intermediate zeros, $u_1, u_2, u_1 < u_2$, and $F'(u_+) > 0$, $F'(u_1) < 0$, $F'(u_2) > 0$, $F'(u_-) < 0$. The monostable $[u_+, u_1]$-waves, i.e., the waves with the limits $u(\pm\infty) = u_\pm$ at infinity, exist for all values of the speed greater than or equal to some minimal speed c_0. The bistable $[u_1, u_-]$-wave exists for a single value of speed c_1. If $c_1 > c_0$, then there exist $[c_+, c_-]$-waves for all speeds $c \in [c_0, c_1)$. If $c_1 \leq c_0$, then such waves do not exist, and there is a system of two waves propagating one after another with different speeds and a growing distance between them. All these properties including convergence of solutions of the Cauchy problem to waves and systems of waves can be found in [40].

Delay reaction–diffusion equation

$$\frac{\partial u}{\partial t} = D\frac{\partial^2 u}{\partial x^2} + ru(1-qu) - uf(u_\tau) \tag{3}$$

is a particular case of Equation (1) where the integral $J(u)$ is replaced by u. In numerical simulations this equation is considered in a sufficiently long interval $0 < x < L$ with the homogeneous Neumann boundary conditions and with some initial condition $u(x,t) = u_0(x), -\tau \le t \le 0$, where $u_0(x) = u_0$ for $0 \le x \le x_0$ and $u_0(x) = 0$ otherwise. Here u_0 and x_0 are some positive constants. Equation (3) was introduced in [27] to study spatial models of infection development in tissues. The influence of time delay on the wave propagation manifests itself in the most spectacular way in the monostable–bistable case where $c_0 > c_1$, and there is a system of two waves propagating with different speeds. The presence of time delay can lead to the emergence of complex spatiotemporal structures between the two waves. Some examples of numerical simulations are shown in Figure 3.

Existence of waves described by this equation was proved in [41,42]. Since conventional monotonicity conditions and the maximum principle are not applicable in this case, the proof of the wave existence requires sophisticated mathematical techniques. There are only a few works where the wave existence is proved for the delay reaction–diffusion equation in the bistable case without the monotonicity condition (see also [33]).

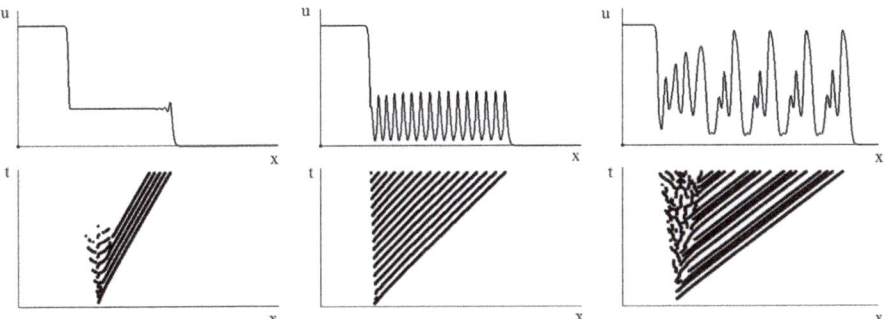

Figure 3. Snapshots of different regimes of wave propagation in numerical simulations of Equation (3) in the monostable–bistable case. The speed of the monostable wave is greater than the speed of the bistable wave, and the distance between them grows (**upper row, left**). The intermediate equilibrium between the wave becomes unstable, and the monostable wave is space periodic (**upper row, middle**). This periodic wave can be followed by complex spatiotemporal oscillations (**upper row, right**). The lower row shows the position of local maxima of the same solutions on the (x,t)-plane. Reprinted from [37] with permission.

3.1.1. Nonlocal Equation

The presence of the nonlocal term in Equation (1) can influence the regimes of wave propagation presented above for the local equation. Let us recall that the homogeneous in-space stationary solution u_- can be stable or unstable depending on the values of parameters. In particular, for a sufficiently small diffusion coefficient or for a sufficiently large N in the definition of the kernel $\phi(x)$:

$$\phi(x) = \frac{1}{2N}\begin{cases} 1, & |x| \le N \\ 0, & |x| > N \end{cases} \tag{4}$$

this solution loses its stability resulting in the bifurcation of a periodic in-space stationary solution. If this solution is stable, then the regimes of wave propagation are the same as before (monostable, bistable, monostable–bistable). Suppose now that it is unstable and consider, first, the monostable case. Then there are two transition: the first one is provided by the $[u_+, u_-]$-wave, the second one is the transition from the constant solution u_- to the periodic solution $u_p(x)$. If the speed c_0 of the former is greater than the speed c_p of the latter, then they propagate one after another one with a growing distance between them (Figure 4, left). If $c_p > c_0$, then they merge and form a single periodic wave (Figure 4, right). These different regimes are also observed in the bistable and monostable–bistable cases.

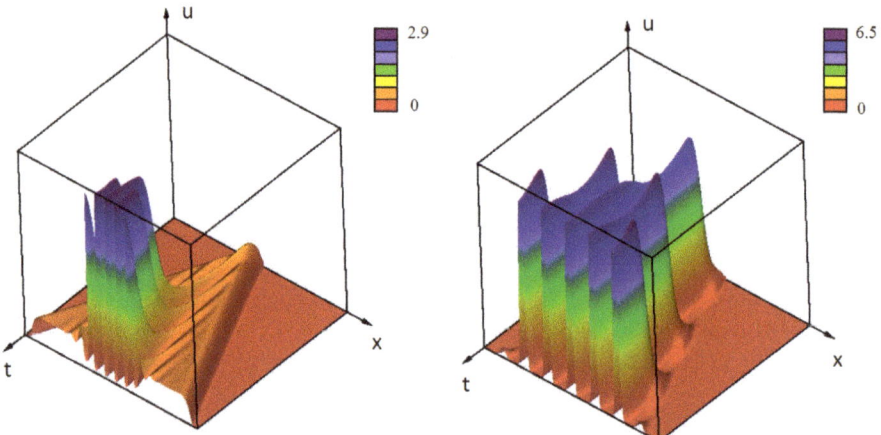

Figure 4. Numerical simulations of Equation (1) show the waves propagating from the center of the interval towards its boundaries in the monostable case. In the first monostable case (**left**) the periodic perturbation propagates slower than the $[u_+, u_-]$-wave, and the distance between them grows. In the second monostable case (**right**), the periodic perturbation propagates faster, it merges with the wave, and they form a single periodic wave. The values of parameters: $D = 10^{-5}, r = 1, q = 1, k_1 = 5, k_3 = 3$, $N = 0.035$ (**left**), $N = 0.1$ (**right**); $f(u) = k_1 u e^{-k_3 u}$, $\tau = 0$.

Next, consider Equation (1) with time delay and, for simplicity of presentation, with a linear function $f(u)$. In this case, we have a monostable equation with two stationary points $u_+ = 0$ and $u_- > 0$. Stability analysis of the homogeneous in-space stationary solution u_- with respect to spatial and temporal perturbations was carried out in Section 2. If this point is stable, then we observe propagation of a usual $[u_+, u_-]$-wave with a constant speed and a constant profile. However, this wave is not necessarily monotonic with respect to x, as it is the case for the local equation. Damped oscillation behind the wave occur for N sufficiently large (Figure 5, left).

Behavior of solutions becomes more complex if the solution u_- is unstable. If the spatiotemporal perturbation of this solution propagates in space with the speed less than the speed of the $[u_+, u_-]$-wave, then this wave propagates with a constant speed and a constant profile, possibly with decaying spatial oscillations behind the wave front. This wave is followed by the region of spatiotemporal oscillations (Figure 5, right). If the perturbations of the solution u_- propagate faster than the $[u_+, u_-]$-wave, then they merge and form an oscillating wave propagating with a variable speed (not shown).

Let us recall that in the monostable case the waves exist for all values of the speed greater than or equal to the minimal speed c_0. The value of the minimal speed is determined by the linearized problem at 0, and it does not depend on N and τ if the wave remains stable.

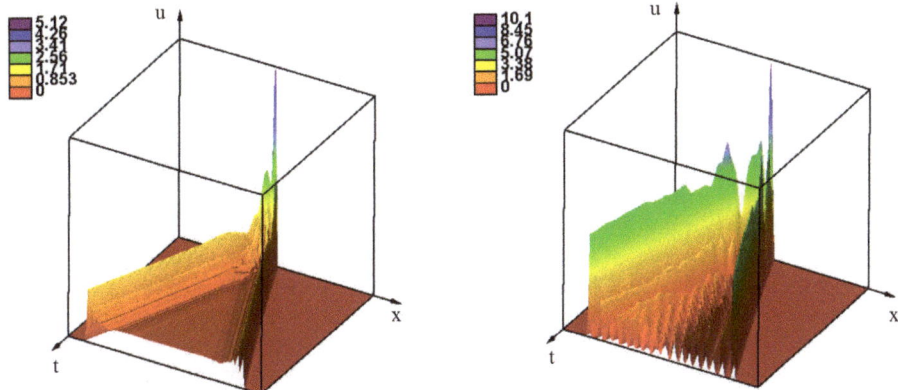

Figure 5. Numerical simulations of Equation (1) with $f(u) = k_1 u$. If the solution u_- is stable, then there is a $[w_+, w_-]$-wave propagating with a constant speed and profile with possible spatial oscillations independent of time (**left**). If this solution is unstable, then this wave is followed by spatiotemporal oscillations (**right**). The values of parameters: $r = 1, q = 1, L = 2, D = 10^{-5}, k_1 = 0.9, \tau = 3, N = 0.25$, $\tau = 2$ (**left**), $\tau = 4$ (**right**), $t = 150$.

3.1.2. Bifurcations of Waves and Pulses

Existence and stability of pulses and waves for Equation (1) depends on the width N of the support of the kernel $\phi(x)$.

Let us recall that the local reaction–diffusion equation with a bistable function $F(u)$ has a positive stationary solution decaying at infinity (pulse solution) if and only if $I_F = \int_{u_+}^{u_-} F(u) du > 0$. This stationary solution is unstable. It also has a stable $[u_+, u_-]$-wave whose speed is positive under the same condition on the integral I_F. Similar properties hold for the nonlocal equation with N sufficiently small. With an increase of N, the wave becomes non-monotone as a function of x but it still has a constant speed and profile.

Periodic structures and waves appear as the width N of the support of the kernel $\phi(x)$ exceeds a critical value N_c^1 for which the corresponding eigenvalue of the linearized problem crosses the origin. If we further increase the value of N, then instead of periodic waves we observe stable pulses. Thus, there are two bifurcations with a transition from simple waves to periodic waves (through an intermediate regime with two waves) and from periodic waves to pulses. The first bifurcation occurs due to the essential spectrum crossing the origin. The second one is a nonlocal bifurcation where the speed of the periodic wave decreases as N approaches a critical value N_c^2, and it becomes zero for N exceeding the critical value. At the same time, the spikes of the periodic wave become pulses. Let us note that multiple pulses are not stationary solutions of Equation (1), they slowly move from each other with a decaying speed.

3.2. Emergence of Strains

Virus density distribution $u(x,t)$ as a function of its genotype x and time t characterizes the existence of virus strains and their evolution. In this context, a strain is a positive localized solution of Equation (1), i.e., a solution with maximum at some x_0 (most frequent genotype) and rapidly decaying as the distance $|x - x_0|$ increases. Existence and stability of stationary localized solutions (pulses) of Equation (1) was studied in [23]. They correspond to persisting virus strains. In this section, we will study the emergence of new strains due to propagating of periodic waves. As it was discussed in the previous section, stable pulses and waves are mutually exclusive.

3.2.1. Initiation of Periodic Waves

Suppose that the initial virus distribution $u_0(x)$ represents a non-negative function with a narrow support at the center of the interval. For a properly chosen values of parameters the solution of Equation (1) with this initial condition develops to a periodic travelling wave. At the first stage of this dynamics, the solution rapidly growth remaining localized at the center of the interval (Figure 6a). It reaches a maximal level, and then the peak gradually decreases and becomes wider (Figure 6b,c). At some moment of time, two other peaks appear from each side of the first one. After a short transient period, they converge to approximately the same height. If the interval is sufficiently large, then other peaks will appear after some time gradually filling the whole interval (Figure 4, right). This is a typical dynamics of the initiation of periodic waves which occurs under the conditions presented in Section 2.

Let us also note that new strains (peaks) appear at some distance from the first one by a genetic jump and not as a gradual evolution of the original strain. The value of the virus density between them is close to zero. This is not the case for a large diffusion coefficient (mutation rate) where new strains appear continuously (Appendix A, Figure A2). If the diffusion coefficient is large enough, then the strains do not form, and viruses with any genotype exist. This is determined by the stability of the stationary points (Section 2).

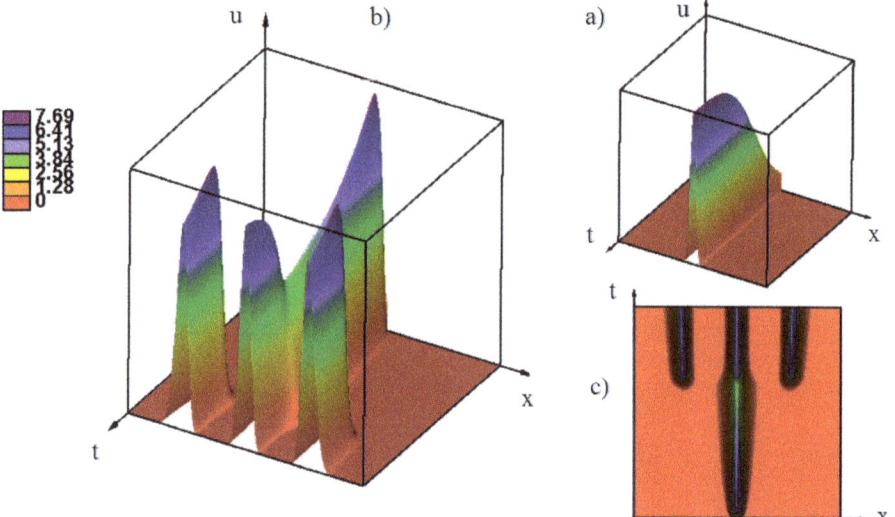

Figure 6. Emergence of a periodic wave in numerical simulations of Equation (1). (**a**) At the first stage, solution growth remaining localized at the center of the interval. (**b**) Then it decreases and widens, and after some time, other peaks of solution appear. (**c**) Another representation of the same solution as in (**b**). Values of parameters: $D = 10^{-5}, r = 1, q = 1, N = 0.2, \tau = 0, f(u) = 0$, the maximum of the initial condition $0.9, t = 75$.

3.2.2. the Influence of Immune Response

We consider the function of immune response in the form $f(u) = (k_1 u + k_2) e^{-k_3 u}$. In order to explain the influence of immune response on the emergence of strains, consider the function $F(u) = ru(1 - qu - f(u))$. If $k_2 = k_3 = 0$, then $F(u)$ has a single positive zero u_-, and this is a monostable case. For the values of k_1 sufficiently small, behavior of solution of Equation (1) is similar to the case where $f(u) \equiv 0$ with the propagation of a periodic wave and the emergence of new strains (Figure 6). If k_1 is large enough, then the equilibrium u_- becomes stable, and there is a stationary $[u_+, u_-]$-wave without emerging peaks, as it is the case of a periodic wave.

Let us recall that the growing branch of the function $f(u)$ corresponds to the antigen stimulated immune response while the decreasing characterizes death or exhaustion of immune cells due to high virus concentration. Thus, if we consider only the growing branch, then a strong immune response (large k_1) does not eliminate infection but prevents the formation of virus strains. Instead of the localized solutions with separated peaks, the virus density distribution converges to a constant positive solution.

In the case where the decreasing branch of the immune response function is present ($k_2 = 0$, $k_3 \neq 0$), the function $F(u)$ can have up to three positive zeros $u_1 < u_2 < u_-$. As we discussed above, this is a monostable–bistable case where the behavior of solutions depends on the values of parameters and on the choice of initial condition. Set $u_0(x) = u_0$ for $x_1 \leq x \leq x_2$ and $u_0(x) = 0$ otherwise. If u_0 is small enough, then the solution represents a monostable wave (Figure 7, left) without strain formation. If u_0 is sufficiently large, then for the same values of parameters as before, the central peak is formed followed after some time by the appearance of two monostable waves of low amplitude (Figure 7, right).

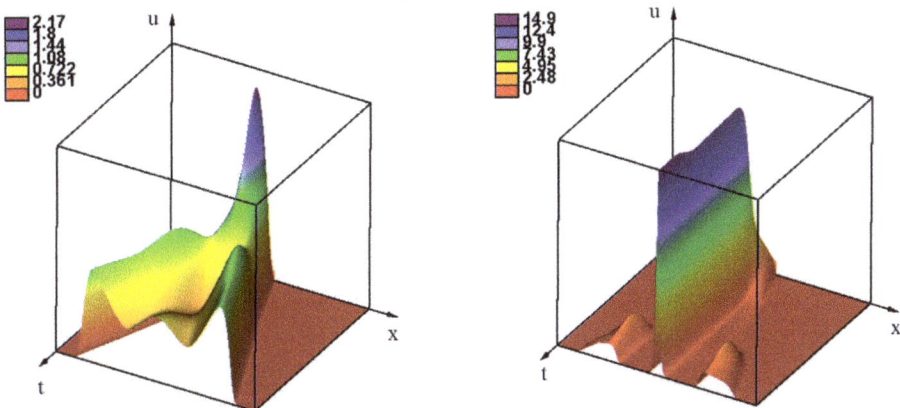

Figure 7. Numerical simulations of Equation (1) with two different initial conditions and the same values of parameters: $D = 10^{-5}, r = 1, q = 1, N = 0.2, \tau = 0, f(u) = k_1 e^{-k_3 u}, k_1 = 1, k_3 = 0.6$, the maximum of the initial condition equals 0.1 (**left**) and 0.9 (**right**), $x_1 = 0.48, x_2 = 0.52, t = 75$.

Furthermore, the width $x_2 - x_1$ of the support of the initial condition can also influence this behavior. A counterintuitive result is that increasing the support of the initial condition leads to the disappearance of the high amplitude peak and to the convergence of solution to the low amplitude monostable wave. The explanation of this effect is that two pulses (peaks) form if the support is sufficiently wide. They compete with each other, their amplitude becomes less than for a single pulse, it is not sufficient to overcome the threshold and to form a stable central pulse.

Under further increase of k_3, propagation of a periodic wave, as it is described above, is observed. If $k_2 \neq 0$, then $f(0) > 0$, i.e., immune response is nonzero even without antigen due to memory cells. Depending on the values of parameters, solution can form a stable pulse, vanish, or initiate simple or periodic waves described above.

3.2.3. Effect of the Delay of the Antiviral Immune Response

In the case of a nonlocal equation with time delay in the immune response term, spatial structures presented in the previous paragraph can become oscillating. Some examples of virus strain evolution are shown in Figure 8. The left and middle figures are obtained for the same values of parameters with different initial conditions. If the initial virus load is large enough, then there is a dominating virus strain and some other strains with low virus density and a complex spatiotemporal behavior.

If the initial virus load is sufficiently small, then the dominating central strain does not exist, and there is only a variety of different genotypes with low densities. Changing the properties of the immune response (function $f(u)$), we observe stable stationary strains similar to those in Figure 4 (right image) after the initial front propagation.

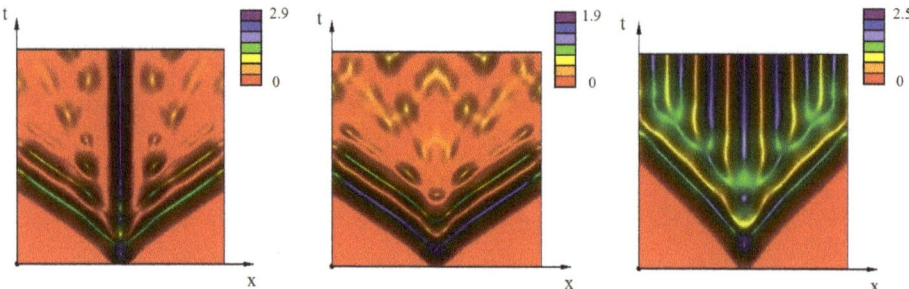

Figure 8. Numerical simulations of Equation (1). Virus evolution with time delay in the term describing the immune response represented as level lines of the solution $u(x,t)$ on the (x,t)-plane. Different regimes coexist for the same values of parameters depending on the initial conditions, with high initial viral load (**left**) and low initial viral load (**middle**). Values of parameters: $D = 10^{-4}, r = 1, q = 1$, $N = 0.1, f(u) = k_1 e^{-k_3 u}, k_1 = 8, k_3 = 3, t = 80$ (**left** and **middle**), $k_3 = 6, t = 50$ (**right**); the maximum of the initial condition 0.9 (**left**), 0.1 (**middle** and **right**).

3.2.4. the Influence of Genotype-Dependent Mortality

We will finish this section with the analysis of the genotype-dependent mortality on the emergence and evolution of virus quasi-species. In order to show this influence more precisely, we consider it in the case without immune response, $f(u) \equiv 0$. Set $\sigma(x) = 0$ for $x_1^* \leq x \leq x_2^*$ and $\sigma(x) = 1$ otherwise. Figure 9 (left) shows the emergence of virus strains for $x_1^* = 0.3, x_2^* = 0.7$ and $N = 0.09$. The initial condition has a support at the center of the interval. Similar to the case of initiation of a periodic wave, there is one strain in the beginning of the simulation, and two other strain appear sometime later. These three strains fill the whole admissible interval where $\sigma(x) = 0$, and new strains do not appear outside of this interval because virus mortality rate is greater there that its reproduction rate.

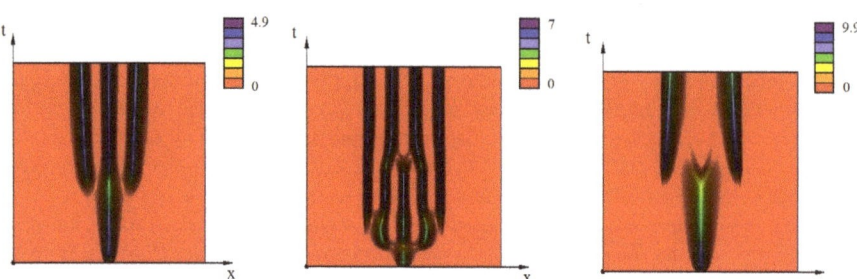

Figure 9. Numerical simulations of Equation (1). Virus evolution without immune response and with the genotype-dependent mortality $\sigma(x)$ represented as level lines of the solution $u(x,t)$ on the (x,t)-plane. Values of parameters: $D = 10^{-5}, r = 1, q = 1, N = 0.09$ (**left**), $N = 0.08$ and 0.09 (**middle**), $N = 0.2$ (**right**).

In the middle figure, we begin the simulation with $N = 0.08$ with five emerging strains. When they become steady and do not evolve any more, we change the value of N to $N = 0.09$, as in the previous simulation. However, this time we observe the regime with four strains instead of three strains observed previously. Hence, different stationary regimes can exist for the same values of parameters, and the initial condition determines the convergence to each of them.

Finally, consider the case where $N = 0.2$ (Figure 9, right). As before, there is single peak of solution at the center of the interval in the beginning of the simulation. However, after some time, it disappears giving rise to two other peaks. Such behavior is determined by the size of the admissible interval: it cannot support three wide peaks, and the two of them from the sides suppress the one at the center of the interval. Thus, virus tends to fill the admissible interval in the most efficient way, that is, to maximize its total density.

4. Discussion

4.1. Virus Quasi-Species

Speciation is considered to be a general property of the living matter [43]. It manifests itself in the emergence of biological species and in a variety of other systems [44]. In the framework of mathematical modelling, speciation appears due to a non-homogeneous density distributions $u(x,t)$. In biological populations, x can be a morphological characteristic or some characterization of the genotype.

Describing virus quasi-species dynamics, we observe some similarities with the general speciation theory due to the competition for host cells but also some specific features because of the presence of the immune response and of the genotype-dependent mortality. If we consider the virus density distribution $u(x,t)$ as a function of its genotype x and of time t, then a virus quasi-species (strain or variant genome) corresponds to a localized solution with a maximal value at some genotype x_0 and rapidly decaying as the distance $|x - x_0|$ increases. In the case of persistent strains, the most frequent genotype x_0 is fixed but it can be also time-dependent for the evolving strains.

Existence of virus strains considered to be a positive stationary density distribution decaying at infinity (in the approximation of an infinite genotype space) is not a priori given. In the previous work [23] we revealed two mechanisms providing the existence and stability of such solutions. In the first case, the existence of virus strains is determined by the genotype-dependent mortality where the virus can survive only inside some viability interval. The maximum of the density distribution is achieved in the middle of the corresponding viability interval. The second mechanism is determined by the immune response under the assumption that the immune response function $f(u)$ decreases for large u. In this case, the virus can survive and form a persistent strain if its concentration is sufficiently high, and if the competition with other strains for host cells occurs in a sufficiently wide range of genotypes.

In mathematical terms, virus strains correspond to stable pulse solutions of the corresponding nonlocal reaction–diffusion equations. Existence of stable pulses does not occur for the conventional local equations.

4.2. Emergence of New Quasi-Species: Summary of the Results

It is important to note that stable pulses and periodic travelling waves are mutually exclusive, they are not observed for the same values of parameters. In this work we study periodic travelling waves. Emergence of new peaks in the virus density distribution during the wave propagation corresponds to the emergence of new virus strains.

From the mathematical point of view, the conditions of the emergence of periodic travelling waves can be determined by the linear stability analysis of the homogeneous in-space stationary solutions. In order to show the influence of different factors on the stability conditions, this analysis is carried out in Section 2 for the single nonlocal term, for both nonlocal terms and for the nonlocal delay equation. In the presence of a single nonlocal term, periodic spatial structures bifurcate from the constant solution if the diffusion coefficient D is sufficiently small and if the kernel $\phi(x)$ of the integral satisfies certain conditions. In particular, for a piece-wise constant kernel, its support should be sufficiently large.

In the case of two nonlocal terms, the qualitative behavior of solutions is similar. The interaction of the nonlocal terms can have stabilizing or destabilizing effect depending on the values of parameters. The influence of time delay (and a single nonlocal term) on the stability conditions depends on the immune response function resulting in temporal or spatiotemporal oscillations.

The transition between a virus-free equilibrium and an infected equilibrium is provided by travelling waves (Section 3). In the case without nonlocal terms or time delay, this is a conventional wave with a constant speed and profile or two waves propagating one after another with different speeds. The nonlocal terms can result in the emergence of periodic waves, while time delay can lead to a complex spatiotemporal pattern formation.

Let us note that the qualitative behavior of solutions is quite robust, and it is not very sensitive to the particular choice of the immune response function and of the integral kernels. The range of genotypes is supposed to be sufficiently large to neglect the influence of the boundaries. Mathematically, we can consider the whole real axis. In numerical simulations, we consider a bounded sufficiently large interval. In most cases, we stop the simulations before the wave approaches the boundary of the interval. In some cases (Figure 8), we continue the simulations to reveal spatiotemporal pattern formation inside the interval. In this case, periodic boundary conditions are more convenient because they do not influence the behavior of the solution inside the interval. Dirichlet or Neumann boundary conditions can influence the behavior of solutions. Thus, periodic boundary conditions do not have here biological significance, but they are more appropriate for mathematical modelling.

4.3. Biological Interpretations

Suppose that the initial viral load is localized in a narrow interval of genotypes (e.g., a founder virus in the HIV case). Due to its multiplication and mutations, the density distribution grows and widens. Viruses with different genotype begin to compete for the host cells leading to the appearance of new strains at some distance to avoid the competition with the existent strains. This description of speciation is generic [22], it is not specific for virus quasi-species. Specific features of virus diversification and strain emergence are related to the immune response.

If we take into account the cross-reactivity in the immune response, i.e., different antigens can stimulate the same immune cells, then the immune response interferes with virus competition for the host cells. This interaction is quite complex and can act in different ways on different strains. However, if the mutation rate (diffusion in the genotype space) is sufficiently small, then the speciation of virus quasi-species will necessarily occur. The critical conditions leading to the emergence of new strains depend on the parameters of the problem, and the immune response can have both stabilizing and destabilizing effect.

The influence of immune response becomes easier to determine if we neglect cross-reactivity. Assuming that the immune response function $f(u)$ is increasing due to the stimulation of immune response by the virus antigens, the model predicts that immune response acts to suppress the formation of new strains. If the growth rate of the function $f(u)$ is sufficiently high, then the speciation of the virus density distribution completely disappears. In this case, instead of a discrete set of virus strains our study predicts that a uniform density distribution as a function of virus genotype will take place.

The situation becomes again more complex if we consider also a decreasing branch of the immune response function, which appears due virus-induced death of immune cells, and time delay in the immune response required for the clonal expansion of immune cells. In this case, along with periodic travelling waves described above, complex nonlinear dynamics of solutions can take place with various patterns of emerging and disappearing strains.

Genotype-dependent virus mortality restrains the evolution of virus species to the admissible interval. The emergence and the evolution of virus strains within the viability interval depend on the values of parameters and on the initial viral load. Let us note that different virus density distributions can be observed for the same values of parameters.

Concluding this discussion, we point out the limitations of the model considered in this work due to various simplifying assumptions. We do not take into account the presence of different immune cells and of cytokines participating in the immune response, complex intracellular regulation of cell fate and of virus multiplication. On the other hand, these and other simplifications allow us to reveal some generic properties of the evolution of virus quasi-species, which would be more difficult to identify in a more complex model. This modelling framework provides a starting basis for further investigations and for the introduction of more detailed models.

Author Contributions: Conceptualization, G.B. and V.V.; methodology, G.B. and A.M.; software, N.B. and V.V.; formal analysis, V.P. and V.V.; writing—original draft preparation, G.B. and V.V. All authors have read and agreed to the published version of the manuscript.

Funding: This research received no external funding.

Acknowledgments: The research was funded by the Russian Science Foundation (Grant no. 18-11-00171) to N.B., G.B. A.M. and V.V. V.P. and V.V. were partially supported by the "RUDN University Program 5-100". A.M. was also supported by a grant from the Spanish Ministry of Economy, Industry and Competitiveness and FEDER grant no. SAF2016-75505-R (AEI/MINEICO/FEDER, UE) and the "Maria de Maeztu" Programme for Units of Excellence in R&D (MDM-2014-0370).

Conflicts of Interest: The authors declare no conflict of interest.

Appendix A. Additional Simulations

Patterns bifurcating due to the instability of the homogeneous in-space solutions.

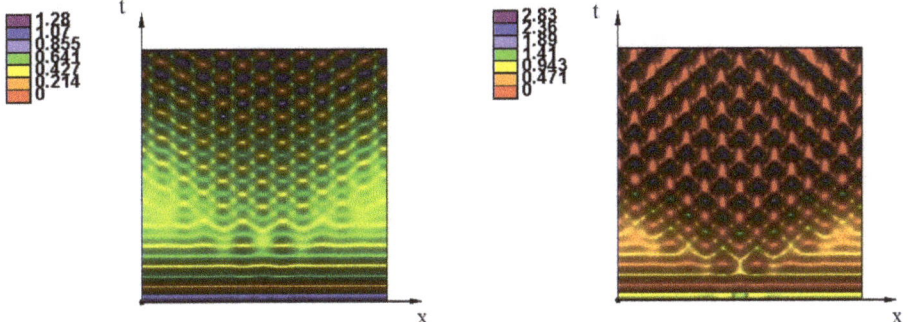

Figure A1. Numerical simulations of Equation (13) for the linear function $f(u) = k_1 u$. Level lines of the solution $u(x,t)$ on the plane (x,t) (**left**). Two snapshots of solution (**right**). The value of parameters: $r = 1, q = 1, k_1 = 0.95, N = 0.1, \tau = 4, D = 0.0001$ (**left**), $D = 0.00001$ (**right**), $t = 150$.

Initiation and propagation of periodic waves.

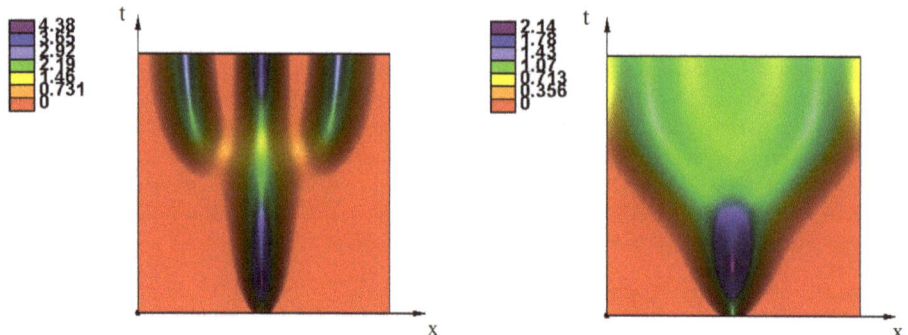

Figure A2. Level lines of the solution $u(x,t)$ of Equation (1) on the (x,t)-plane. Values of parameters: $r = 1, q = 1, N = 0.2, \tau = 0, f(u) = 0$ (left and middle), the maximum of the initial condition 0.9, $D = 0.0001, t = 35$ (**left**) and $D = 0.0005, t = 20$ (**right**).

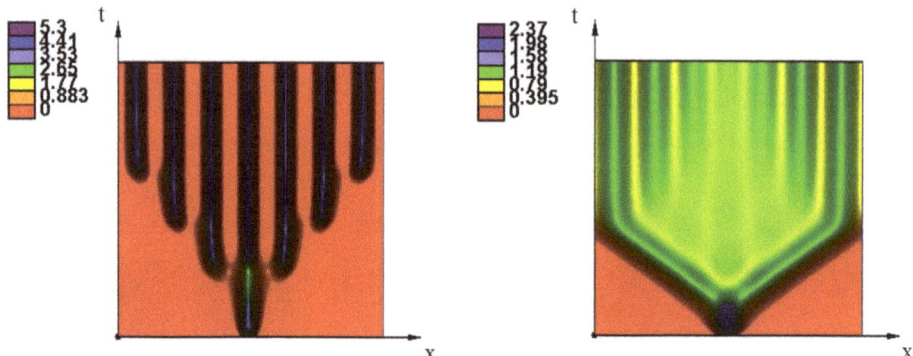

Figure A3. Level lines of the solution $u(x,t)$ of Equation (1) on the (x,t)-plane. Values of parameters: $r = 1, q = 1, N = 0.1, \tau = 0, f(u) = 0$ (left and middle), the maximum of the initial condition 0.9, $D = 0.00001, t = 130$ (**left**) and $D = 0.0001, t = 75$ (**right**).

Propagation of waves in the case of time delay in the immune response term.

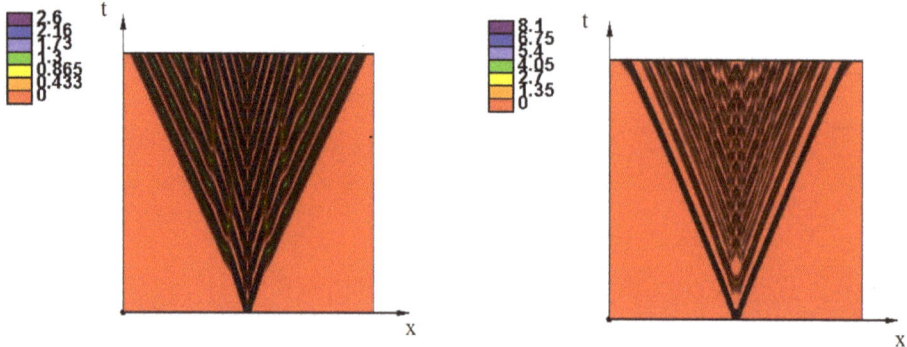

Figure A4. Level lines of the solution $u(x,t)$ of Equation (1). The values of parameters: $r = 1, q = 1$, $L = 2, D = 10^{-5}, k_1 = 0.9, \tau = 4, N = 0.05$ (**left**), $N = 0.1$ (**right**), $t = 150$.

References

1. Keele, B.H.; Giorgi, E.E.;Salazar-Gonzalez, J.F.; Decker, J.M.; Pham, K.T.; Salazar, M.G.; Sun, C.; Grayson, T.; Wang, S.; Li, H.; et al. Identication and characterization of transmitted and early founder virus envelopes in primary HIV-1 infection. *Proc. Natl. Acad. Sci. USA* **2008**, *105*, 7552–7557. [CrossRef] [PubMed]
2. Plikat, U.; Nieselt-Struwe, K.; Meyerhans, A. Genetic drift can dominate short-term human immunodeficiency virus type 1 nef quasispecies evolution in vivo. *J. Virol.* **1997**, *71*, 4233–4240. [CrossRef] [PubMed]
3. Biebricher, C.K.; Eigen, M. What is a quasispecies? *Curr. Top. Microbiol. Immunol.* **2006**, *299*, 1–31. [PubMed]
4. Domingo, E.; Perales, C. Viral quasispecies. *PLoS Genet.* **2019**, *15*, e1008271. [CrossRef] [PubMed]
5. Goodenow, M.; Huet, T.; Saurin, W.; Kwok, S.; Sninsky, J.; Wain-Hobson, S. HIV-1 isolates are rapidly evolving quasispecies: Evidence for viral mixtures and preferred nucleotide substitutions. *J. Acquir. Immune Defic. Syndr.* **1989**, *2*, 344–352. [PubMed]
6. Holland, J.J.; De La Torre, J.C.; Steinhauer, D.A. RNA virus populations as quasispecies. *Curr. Top. Microbiol. Immunol.* **1992**, *176*, 1–20.
7. Meyerhans, A.; Cheynier, R.; Albert, J.; Seth, M.; Kwok, S.; Sninsky, J.; Morfeldt-Manson, L.; Asjo, B.; Wain-Hobson, S. Temporal fluctuations in HIV quasispecies in vivo are not reflected by sequential HIV isolations. *Cell* **1989**, *58*, 901–910. [CrossRef]
8. Phillips, R.E.; Rowland-Jones, S.; Nixon, D.F.; Gotch, F.M.; Edwards, J.P.; Ogunlesi, A.O.; Elvin, J.G.; Rothbard, J.A.; Bangham, C.R.; Rizza. C.R.; et al. Human immunodeficiency virus genetic variation that can escape cytotoxic T cell recognition. *Nature* **1991**, *354*, 453–459. [CrossRef]
9. Larder, B.A.; Kemp, S.D. Multiple mutations in HIV-1 reverse transcriptase confer high-level resistance to zidovudine (AZT). *Science* **1989**, *246*, 1155–1158. [CrossRef]
10. Vignuzzi. M.; Stone, J.K.; Arnold, J.J.; Cameron, C.E.; Andino, R. Quasispecies diversity determines pathogenesis through cooperative interactions in a viral population. *Nature* **2006**, *439*, 344–348. [CrossRef]
11. Collier, D.A.; Monit, C.; Gupta, R.K. The Impact of HIV-1 drug escape on the global treatment landscape. *Cell Host Microbe* **2019**, *26*, 48–60. [CrossRef] [PubMed]
12. Esposito, I.; Trinks, J.; Soriano, V. Hepatitis C virus resistance to the new direct-acting antivirals. *Expert Opin. Drug Metab. Toxicol.* **2016**, *12*, 1197–1209. [CrossRef] [PubMed]
13. Kiepiela, P.; Leslie, A.J.;Honeyborne, I.; Ramduth, D.; Thobakgale, C.; Chetty, S.; Rathnavalu, P.; Moore, C.; Pfaerott, K.J.; Hilton, L. Dominant inuence of HLA-B in mediating the potential co-evolution of HIV and HLA. *Nature* **2004**, *432*, 769–775. [CrossRef] [PubMed]
14. Haas, G.; Plikat, U,; Debré, P.; Lucchiari, M,; Katlama, C.; Dudoit, Y.; Bonduelle, O.; Bauer, M.; Ihlenfeldt, H.G.; Jung, G.; et al. Dynamics of viral variants in HIV-1 Nef and specific cytotoxic T lymphocytes in vivo. *J. Immunol.* **1996**, *157*, 4212–4221. [PubMed]
15. Ganusov, V.V.; Goonetilleke, N.; Liu, M.K.; Ferrari, G.; Shaw, G.M.; McMichael, A.J.; Borrow, P.; Korber, B.T.; Perelson, A.S. Fitness costs and diversity of the cytotoxic T lymphocyte (CTL) response determine the rate of CTL escape during acute and chronic phases of HIV infection. *J. Virol.* **2011**, *85*, 10518–10528. [CrossRef] [PubMed]
16. Turnbull, E.L.; Wong, M.; Wang, S.; Wei, X.; Jones, N.A.; Conrod, K.E.; Aldam, D.; Turner, J.; Pellegrino, P.; Keele, B.F.; et al. Kinetics of expansion of epitope-specific T cell responses during primary HIV-1 infection. *J. Immunol.* **2009**, *182*, 7131–7145. [CrossRef]
17. Ganusov, V.V.; Neher, R.A.; Perelson, A.S. Mathematical modeling of escape of HIV from cytotoxic T lymphocyte responses. *J. Stat. Mech.* **2013**, *2013*, P01010. [CrossRef]
18. Bocharov, G.A.; Ford, N.J.; Edwards, J.; Breinig, T.; Wain-Hobson, S.; Meyerhans, A. A genetic algorithm approach to simulating human immunodeficiency virus evolution reveals the strong impact of multiply infected cells and recombination. *J. Gen. Virol.* **2005**, *86*, 3109–3118. [CrossRef]
19. Swanstrom, R.; Coffin, J. HIV-1 pathogenesis: The virus. *Cold Spring Harb. Perspect. Med.* **2012**, *2*, a007443. [CrossRef]
20. Bessonov, N.; Reinberg, N.; Volpert, V. Mathematics of Darwin's diagram. *Math. Model. Nat. Phenom.* **2014**, *9*, 5–25. [CrossRef]
21. Bessonov, N.; Reinberg, N.; Banerjee, M.; Volpert, V. The origin of species by means of mathematical modelling. *Acta Bioteoretica* **2018**, *66*, 333–344.

22. Genieys, S.; Volpert, V.; Auger, P. Adaptive dynamics: Modelling Darwin's divergence principle. *Comptes Rendus Biol.* **2006**, *329*, 876–879. [CrossRef] [PubMed]
23. Bessonov, N.; Bocharov, G.; Meyerhans, A.; Popov, V.; Volpert, V. Nonlocal reaction-diffusion model of viral evolution. Existence and dynamics of strains. *Preprints* **2019**. [CrossRef]
24. Volpert, V. *Elliptic Partial Differential Equations. Volume 2. Reaction-Diffusion Equations*; Birkhäuser: Basel, Switzerland, 2014.
25. Bessonov, N.; Bocharov, G.; Meyerhans, A.; Volpert, V. Interplay between reaction and diffusion processes in governing the dynamics of virus infections. *J. Theor. Biol.* **2018**, *457*, 221–236.
26. Bocharov, G,; Volpert, V.; Ludewig, B.; Meyerhans, A. *Mathematical Immunology of Virus Infections*; Springer: Cham, Switzerland, 2018.
27. Bocharov, G.; Meyerhans, A.; Bessonov, N,; Trofimchuk, S.; Volpert, V. Spatiotemporal dynamics of virus infection spreading in tissues. *PLoS ONE* **2006**. [CrossRef]
28. Bocharov, G.; Meyerhans, A.; Bessonov, N,; Trofimchuk, S.; Volpert, V. Modelling the dynamics of virus infection and immune response in space and time. *Int. J. Parallel Emerg. Distrib. Syst.* **2019**, *34*, 341–355. [CrossRef]
29. Perthame, B.; Genieys, S. Concentration in the nonlocal Fisher equation: The Hamilton-Jacobi limit. *Math. Model. Nat. Phenom.* **2007**, *4*, 135–151. [CrossRef]
30. Banerjee, M.; Volpert, V. Spatio-temporal pattern formation in Rosenzweig–Macarthur model: Effect of nonlocal interactions. *Ecol. Complex.* **2017**, *30*, 2–10. [CrossRef]
31. Lorz A.; Lorenzi, T.; Clairambault, J.; Escargueil, A.; Perthame, B. Modeling the effects of space structure and combination therapies on phenotypic heterogeneity and drug resistance in solid tumors. *Bull. Math. Biol.* **2015**, *77*, 1–22. [CrossRef]
32. Gourley, S.A.; Chaplain, M.A.J.; Davidson, F.A. Spatio-temporal pattern formation in a nonlocal reaction-diffusion equation. *Dyn. Syst.* **2001**, *16*, 173–192. [CrossRef]
33. Alfaro, M.; Ducrot, A.; Giletti, T. Travelling waves for a non-monotone bistable equation with delay: Existence and oscillations. *Proc. Lond. Math. Soc.* **2018**, *116*, 729–759. [CrossRef]
34. Alfaro, M.; Coville, J.; Raoul, G. Bistable travelling waves for nonlocal reaction diffusion equations. *Discret. Contin. Dyn. Syst.* **2014**, *34*, 1775–1791. [CrossRef]
35. Apreutesei, N.; Bessonov, N.; Volpert, V.; Vougalter, V. Spatial structures and generalized travelling waves for an integro-differential equation. *Discret. Contin. Dyn. Syst. Ser. B* **2010**, *13*, 537–557. [CrossRef]
36. Nadin G.; Rossi, L.; Ryzhik, L.; Perthame, B. Wave-like solutions for nonlocal reaction-diffusion equations: A toy model. *Math. Model. Nat. Phenom.* **2013**, *8*, 33–41. [CrossRef]
37. Bessonov, N.; Bocharov, G.; Touaoula, T.M.; Trofimchuk, S.; Volpert, V. Delay reaction-diffusion equation for infection dynamics. *Discret. Contin. Dyn. Syst. B* **2019**, *24*, 2073–2091. [CrossRef]
38. Volpert, V. Pulses and waves for a bistable nonlocal reaction-diffusion equation. *Appl. Math. Lett.* **2015**, *44*, 21–25. [CrossRef]
39. Britton, N.F. Spatial structures and periodic travelling waves in an integro-differential reaction-diffusion population model. *SIAM J. Appl. Math.* **1990**, *6*, 1663–1688. [CrossRef]
40. Volpert, V. Asymptotic behavior of solutions of a nonlinear diffusion equation with a source term of general form. *Sib. Math. J.* **1989**, *30*, 25–36. [CrossRef]
41. Trofimchuk, S.; Volpert, V. Traveling waves for a bistable reaction-diffusion equation with delay. *SIAM J. Math. Anal.* **2018**, *50*, 1175–1199. [CrossRef]
42. Trofimchuk, S.; Volpert, V. Global continuation of monotone waves for a unimodal bistable reaction-diffusion equation with delay. *Nonlinearity* **2019**, *32*, 2593. [CrossRef]
43. Coyne J.A.; Orr, H.A. *Speciation*; Sinauer Associates: Sunderland, MA, USA, 2004.
44. Volpert, V. Branching and aggregation in self-reproducing systems. *ESAIM Proc. Surv.* **2014**, *47*, 116–129. [CrossRef]

© 2020 by the authors. Licensee MDPI, Basel, Switzerland. This article is an open access article distributed under the terms and conditions of the Creative Commons Attribution (CC BY) license (http://creativecommons.org/licenses/by/4.0/).

www.ingramcontent.com/pod-product-compliance
Lightning Source LLC
LaVergne TN
LVHW070408100526
838202LV00014B/1409

MDPI
St. Alban-Anlage 66
4052 Basel
Switzerland
Tel. +41 61 683 77 34
Fax +41 61 302 89 18
www.mdpi.com

Mathematics Editorial Office
E-mail: mathematics@mdpi.com
www.mdpi.com/journal/mathematics